CURRENT TOPICS IN

DEVELOPMENTAL BIOLOGY

VOLUME 20

COMMITMENT AND INSTABILITY IN CELL DIFFERENTIATION

CURRENT TOPICS IN

DEVELOPMENTAL BIOLOGY

EDITED BY

A. A. MOSCONA

CUMMINGS LIFE SCIENCE CENTER
THE UNIVERSITY OF CHICAGO
CHICAGO, ILLINOIS

ALBERTO MONROY

STAZIONE ZOOLOGICA
NAPLES, ITALY

VOLUME 20

COMMITMENT AND INSTABILITY IN
CELL DIFFERENTIATION

VOLUME EDITOR

T. S. OKADA

INSTITUTE FOR BIOPHYSICS
FACULTY OF SCIENCE
KYOTO UNIVERSITY
KYOTO, JAPAN

ASSISTANT EDITOR

H. KONDOH

INSTITUTE FOR BIOPHYSICS
FACULTY OF SCIENCE
KYOTO UNIVERSITY
KYOTO, JAPAN

1986

ACADEMIC PRESS, INC.

Harcourt Brace Jovanovich, Publishers

Orlando   San Diego   New York   Austin
London   Montreal   Sydney   Tokyo   Toronto

ACADEMIC PRESS JAPAN, INC.
Hokoku Bldg. 3-11-13, Iidabashi, Chiyoda-ku, Tokyo 102

United States Edition published by ACADEMIC PRESS, INC.
Orlando, Florida 32887

United Kingdom Edition published by ACADEMIC PRESS, INC. (LONDON) LTD.
24/28 Oval Road, London NW1   7DX

LIBRARY OF CONGRESS CATALOG CARD NUMBER: 66−28604

Conference on Transdifferentiation and Instability
in Cell Commitment

Held by Yamada Science Foundation

ISBN 0−12−153120−1

PRINTED IN THE UNITED STATES OF AMERICA

86 87 88 89   9 8 7 6 5 4 3 2 1

CONTENTS

CHAPTER 1. Conversion of Retina Glia Cells into Lenslike
Phenotype Following Disruption of
Normal Cell Contacts
A. A. MOSCONA

CHAPTER 2. Instability in Cell Commitment of Vertebrate
Pigmented Epithelial Cells and Their
Transdifferentiation into Lens Cells
GORO EGUCHI

CHAPTER 16.   Transitory Differentiation of Matrix Cells and Its
              Functional Role in the Morphogenesis of the
              Developing Vertebrate CNS
              SETSUYA FUJITA

CHAPTER 17.   Prestalk and Prespore Differentiation during
              Development of Dictyostelium discoideum
              IKUO TAKEUCHI, TOSHIAKI NOCE,
              AND MASAO TASAKA

CHAPTER 18.   Transdifferentiation Occurs Continuously
              in Adult Hydra
              HANS BODE, JOHN DUNNE, SHELLY HEIMFELD,
              LYDIA HUANG, LORETTE JAVOIS,

CHAPTER 22.   Probable Dedifferentiation of Mast Cells in Mouse
               Connective Tissues
               YUKIHIKO KITAMURA, TAKASHI SONODA,
               TORU NAKANO, AND YOSHIO KANAYAMA

CHAPTER 23.   Instability and Stabilization in Melanoma
               Cell Differentiation
               DOROTHY C. BENNETT

CHAPTER 24.   Differentiation of Embryonal Carcinoma Cells:
               Commitment, Reversibility, and Refractoriness
               MICHAEL I. SHERMAN

CHAPTER 29.   Instability of Chromosomes and Alkaloid Content
               in Cell Lines Derived from Single Protoplasts of
               Cultured *Coptis japonica* Cells
               YASUYUKI YAMADA AND MASANOBU MINO

CONTRIBUTORS

Numbers in parentheses indicate the pages on which the authors' contributions begin.

HANSJÜRG ALDER,* *Institute of Zoology, University of Basel, CH-4051 Basel, Switzerland* (117)

CHRISTIANE AYER-LE LIÈVRE,† *Institut d'Embryologie, Annexe du Collège de France, Centre National de la Recherche Scientifique, 94736 Nogent-sur-Marne, France* (111)

DOROTHY C. BENNETT, *Department of Anatomy, St. George's Hospital Medical School, London SW17 0RE, England* (333)

IRA B. BLACK, *Division of Developmental Neurology, Cornell University Medical College, New York, New York 10021* (165)

HANS BODE, *Developmental Biology Center and Department of Developmental and Cell Biology, University of California, Irvine, Irvine, California 92717* (257)

D. J. BOWER,‡ *Institute of Animal Genetics, University of Edinburgh, Edinburgh EH9 3JN, Scotland* (137)

ALLAN BRADLEY, *Department of Genetics, University of Cambridge, Cambridge CB2 3EH, England* (357)

R. M. CLAYTON, *Institute of Animal Genetics, University of Edinburgh, Edinburgh EH9 3JN, Scotland* (137)

KURT DROMS,§ *Department of Molecular, Cellular and Developmental Biology, University of Colorado, Boulder, Colorado 80309* (211)

JOHN DUNNE, *Developmental Biology Center and Department of Devel-*

*Present address: Friday Harbor Laboratories, Friday Harbor, Washington 98250.

†Present address: Department of Histology, Karolinska Institute, S-10401 Stockholm, Sweden.

‡Present address: M. R. C. Clinical and Population Cytogenetics Unit, Edinburgh, Scotland.

§Present address: School of Pharmacy, University of Colorado, Boulder, Colorado 80309.

opmental and Cell Biology, University of California, Irvine, Irvine, California 92717 (257)

GORO EGUCHI, *Division of Morphogenesis, Department of Developmental Biology, National Institute for Basic Biology, Myodaijicho, Okazaki 444, Japan* (21)

L. H. ERRINGTON, *Institute of Animal Genetics, University of Edinburgh, Edinburgh EH9 3JN, Scotland* (137)

TOSHITAKA FUJISAWA, *Department of Developmental Genetics, National Institute of Genetics, Mishima, Shizuoka-ken 411, Japan* (281)

SETSUYA FUJITA, *Department of Pathology, Kyoto Prefectural University of Medicine, Kawaramachi, Kyoto 602, Japan* (223)

HIROSHI HARADA, *Institute of Biological Sciences, The University of Tsukuba, Sakura-mura, Ibaraki 305, Japan* (397)

SHELLY HEIMFELD, *Developmental Biology Center and Department of Developmental and Cell Biology, University of California, Irvine, Irvine, California 92717* (257)

LYDIA HUANG, *Developmental Biology Center and Department of Developmental and Cell Biology, University of California, Irvine, Irvine, California 92717* (257)

HIROYUKI IDE, *Biological Institute, Tôhoku University, Sendai 980, Japan* (79)

JUN IMAMURA, *Institute of Biological Sciences, The University of Tsukuba, Sakura-mura, Ibaraki 305, Japan* (397)

TOMOICHI ISHIKAWA, *Department of Anatomy, Kochi Medical School, Kochi 781-51, Japan* (99)

LORETTE JAVOIS, *Developmental Biology Center and Department of Developmental and Cell Biology, University of California, Irvine, Irvine, California 92717* (257)

J.-C. JEANNY,* *Institute of Animal Genetics, University of Edinburgh, Edinburgh EH9 3JN, Scotland* (137)

G. MILLER JONAKAIT,† *Division of Developmental Neurology, Cornell University Medical College, New York, New York 10021* (165)

YOSHIO KANAYAMA, *Institute for Cancer Research and Department of Medicine, Osaka University Medical School, Kita-ku, Osaka 530, Japan* (325)

CHIAKI KATAGIRI, *Zoological Institute, Faculty of Science, Hokkaido University, Sapporo 060, Japan* (315)

*Present address: Unité de Recherches Gérontologiques, Paris, France.

†Present address: Department of Zoology and Physiology, Rutgers University, Newark, New Jersey 07102.

YUKIHIKO KITAMURA, *Institute for Cancer Research and Department of Medicine, Osaka University Medical School, Kita-ku, Osaka 530, Japan* (325)

OSAMU KOIZUMI, *Developmental Biology Center and Department of Developmental and Cell Biology, University of California, Irvine, Irvine, California 92717* (257)

HISATO KONDOH, *Institute for Biophysics, Faculty of Science, Kyoto University, Kyoto 606, Japan* (153)

MASAHARU KYO, *Institute of Biological Sciences, The University of Tsukuba, Sakura-mura, Ibaraki 305, Japan* (397)

N. LE DOUARIN, *Institut d'Embryologie, Annexe du Collège de France, Centre National de la Recherche Scientifique, Nogent-sur-Marne 94736, France* (291)

MITSUGU MAÉNO, *Zoological Institute, Faculty of Science, Hokkaido University, Sapporo 060, Japan* (315)

MICHAEL T. MCMANUS, *Department of Biochemistry, University of Oxford, Oxford OX1 3QU, England* (383)

FREDERICK MEINS, JR., *Friedrich Miescher-Institut, CH-4002 Basel, Switzerland* (373)

MASANOBU MINO,* *Research Center for Cell and Tissue Culture, Faculty of Agriculture, Kyoto University, Sakyo-ku, Kyoto 606, Japan* (409)

A. A. MOSCONA, *Laboratory for Developmental Biology, Cummings Life Science Center, The University of Chicago, Chicago, Illinois 60637* (1)

HARUKAZU NAKAMURA, *Department of Anatomy, Hiroshima University School of Medicine, Minami-ku, Hiroshima 734, Japan* (111)

TORU NAKANO, *Institute for Cancer Research and Department of Medicine, Osaka University Medical School, Kita-ku, Osaka 530, Japan* (325)

MARK A. NATHANSON, *Department of Anatomy, New Jersey Medical School, Newark, New Jersey 07103* (39)

CHIEMI NISHIMIYA, *Department of Developmental Genetics, National Institute of Genetics, Mishima, Shizuoka-ken 411, Japan* (281)

TOSHIAKI NOCE, *Department of Botany, Faculty of Science, Kyoto University, Kyoto 606, Japan* (243)

MASAHARU OGAWA, *Department of Physiology, Kochi Medical School, Kochi 781-51, Japan* (99)

HITOSHI OHTA, *Department of Neuropsychiatry, Kochi Medical School, Kochi 781-51, Japan* (99)

*Present address: ZEN-NOH Agricultural Technical Center, Kanagawa 254, Japan.

T. S. Okada,* *Institute for Biophysics, Faculty of Science, Kyoto University, Kyoto 606, Japan* (153)

Daphne J. Osborne, *Department of Plant Sciences, The School of Botany, University of Oxford, Oxford OX1 3RA, England* (383)

M. Sambasiva Rao, *Department of Pathology, Northwestern University Medical School, Chicago, Illinois 60611* (63)

Janardan K. Reddy, *Department of Pathology, Northwestern University Medical School, Chicago, Illinois 60611* (63)

Elizabeth Robertson, *Department of Genetics, University of Cambridge, Cambridge CB2 3EH, England* (357)

Dante G. Scarpelli, *Department of Pathology, Northwestern University Medical School, Chicago, Illinois 60611* (63)

Volker Schmid, *Institute of Zoology, University of Basel, CH-4051 Basel, Switzerland* (117)

Michael I. Sherman, *Roche Institute of Molecular Biology, Roche Research Center, Nutley, New Jersey 07110* (345)

Takashi Sonoda, *Institute for Cancer Research and Department of Medicine, Osaka University Medical School, Kita-ku, Osaka 530, Japan* (325)

Noboru Sueoka, *Department of Molecular, Cellular and Developmental Biology, University of Colorado, Boulder, Colorado 80309* (211)

Tsutomu Sugiyama, *Department of Developmental Genetics, National Institute of Genetics, Mishima, Shizuoka-ken 411, Japan* (281)

Ikuo Takeuchi, *Department of Botany, Faculty of Science, Kyoto University, Kyoto 606, Japan* (243)

Masao Tasaka,† *Department of Botany, Faculty of Science, Kyoto University, Kyoto 606, Japan* (243)

Shin Tochinai, *Zoological Institute, Faculty of Science, Hokkaido University, Sapporo 060, Japan* (315)

Kenji Watanabe, *Department of Anatomy, Fukui Medical School, Matsuoka-cho, Yoshida-gun, Fukui 910–11, Japan* (89)

John Westerfield, *Developmental Biology Center and Department of Developmental and Cell Biology, University of California, Irvine, Irvine, California 92717* (257)

James A. Weston, *Department of Biology, University of Oregon, Eugene, Oregon 97403* (195)

Yasuyuki Yamada, *Research Center for Cell and Tissue Culture, Fac-*

*Present address: The National Institute for Basic Biology, Myodaiji, Okazaki 444, Japan.

†Present address: Division of Developmental Biology, The National Institute for Basic Biology, Okazaki 444, Japan.

ulty of Agriculture, Kyoto University, Sakyo-ku, Kyoto 606, Japan (409)

MARCIA YAROSS, *Developmental Biology Center and Department of Developmental and Cell Biology, University of California, Irvine, Irvine, California 92717* (257)

CATHERINE ZILLER, *Institut d'Embryologie, Annexe du Collège de France, Centre National de la Recherche Scientifique, 94736 Nogent-sur-Marne, France* (177)

FOREWORD

The Yamada Science Foundation was established about 8 years ago in Osaka by the good offices of the late Kiro Yamada, the founder and the former president of Rohto Pharmaceutical Company Limited, in order to promote research in the basic natural sciences. Since then, the foundation has provided generous support to promote innovative research projects in a variety of natural science disciplines, to provide travel funds for scientists, and to organize international conferences.

The past seven conferences sponsored by the Yamada Science Foundation have covered a range of subjects in chemistry, physics, and biology. This year, "Transdifferentiation and Instability in Cell Commitment" was selected simply because this is one of the most exciting and rapidly developing subjects in cell and developmental biology. The presentations from the conference form the basis for this volume of *Current Topics in Developmental Biology*. It has been a pleasure to acknowledge the efforts of Professor Okada and the other members of the organizing committee for arranging such a stimulating and exciting program. I am sure that the fruits of this meeting and the volume will be a great source of inspiration and stimulation of scientists working in the field of developmental biology.

Osamu Hayaishi

PREFACE

Diverse cell types can arise without alteration of genomic constitution. Animal cells are characterized by the stable nature of their differentiated state, in spite of preservation of the genome throughout development. Convertibility of differentiated cells is considered to be a rare and exceptional occurrence. But, looking back on the history of developmental biology, we note that the discovery of instability of differentiated cells was one of the earliest findings to stimulate the interest of many investigators of embryonic development. In 1891, Colucci discovered that adult *Triturus* can regenerate lens from the iris epithelium. This definitive, historically first example of transdifferentiation highlights the potential instability of differentiated cells.

The main aim of the present volume is to give an overview of recent studies on instability of the differentiated state of cells, with emphasis on description of representative systems (experimental and normal) for investigating cell type modification. The publication of this volume is timely; it is now clear that events such as DNA rearrangements and modification occur during differentiation in certain cell types; at the same time, it is becoming evident that in other cells differentiation does not completely foreclose their potential to change. Thus, a comprehensive, general theory of cell differentiation requires a critical evaluation of evidence for the instability of the differentiated state.

The special volume editor is most grateful to Drs. Moscona and Monroy, the editors of *Current Topics in Developmental Biology,* for the opportunity to collaborate with them on this special volume. He also thanks the contributors and the staff of Academic Press and of Academic Press Japan for expert cooperation in producing this volume. Finally, he is indebted to the Yamada Science Foundation, Osaka, Japan, for sponsoring the conference at which the contributions to this volume were presented.

T. S. Okada

INTRODUCTION: CAN SPECIALIZED CELLS CHANGE THEIR PHENOTYPE?

T. S. Okada

I. Examination of Convertibility of Differentiated Cells

It has been established that in the immune system gene reorganization accompanies cell differentiation (cf. Joho *et al.*, 1983). However, the view that differentiated cells retain most of the genomic DNA present in the fertilized egg without any structural changes is still valid for most cases of embryonic cell differentiation (cf. DiBerardino *et al.*, 1984). Since differentiated cells are, as a rule, stable, most of their genomic information must be inactive or in a latent condition. To understand the nature of the differentiated state in terms of gene expression, two approaches can be considered. First, we can ask about the mechanisms of selective inactivation of genomic information in the course of cell differentiation (cf. Caplan and Ordahl, 1978). Second, we can inquire whether it is possible to reactivate latent information in differentiated cells and convert them into alternative cell types (Moscona and Linser, 1983).

In appropriate test systems, the second approach has certain advantages over studies starting with uncommitted cells. Here, we deal with cells that already are differentiated and investigate if they can change into another phenotype. In favorable situations, we might expect to be able to identify sets of genes which become newly reactivated and/or inactivated. Fortunately, as described in chapters in this volume, experimental and other systems are now available that permit the appearance of new cell types from populations of other well-identified cells (e.g., chapters by Eguchi, Moscona, Nathanson, Schmid and Alder, Rao *et al.*). There is excellent probability that such systems might bring closer the ultimate goal of understanding mechanisms of cell differentiation and modification.

Early embryonic cells are generally pluripotent, with several options as to their future commitment and differentiation. Even in adult organisms pluripotent cells are present as so-called stem cells. In this respect, the situation is especially remarkable in plants in which some types of differentiated cells retain a potentiality for further changes of phenotypic modification. Differentiation of stem cells (or development of unexpected cell types from minor cell populations) can obscure occurrence of cell type conversion, as pointed out in several chapters in this volume (e.g., by Weston, Nakamura and Ayer-Le Lièvre).

In the present volume, several topics related to commitment of pluripotential cells are also included for more comprehensive documentation of the unstable nature of cell differentiation. It seemed particularly important and worthwhile to compare the plasticity of the differentiated state of plant cells with that of animal cells. In fact, readers may find that recent studies on invertebrates (e.g., chapter by Schmid and Alder) are beginning to link these two situations which previously appeared irreconcilable.

II. Systems of Cell Type Conversion or Transdifferentiation

Instability of phenotypic cell characteristics has been observed at different levels, not only with respect to the prospective fate of undifferentiated cells, but also with respect to conversion of differentiated cells into another type distinct in morphology, function, and molecular constitution from the original type. In the present volume, the plasticity of phenotypic expression in differentiated neurons is demonstrated by examining the synthesis of various neurotransmitters as molecular markers (chapters by Jonakait and Black, Ziller, and Weston). These studies provide evidence for potential instability of neurotransmitter phenotypes that are normally expressed late in differentiation, when the cells had already become committed to a definite type. Conversion of pigment cells provides an example at still another level of cell changes. Different kinds of pigment cells differ functionally and morphologically, but they share many common properties both structural and metabolic (Bagnara et al., 1979). As reviewed in the chapter by Ide, the convertibility between different types in the pigment cell class is most evident in cell culture systems.

There are several well-established examples of cell conversion into other types which belong to the same cell lineages in normal development. The transdifferentiation of chromaffin cells into neurons (as described by Ogawa et al.) is a representative case in that it is an example of conversion of a neural crest-derived cell into another type that normally originates from the neural crest. Melanogenesis in cultures of peripheral nervous tissues is another example of conversion representing cell types normally of neural crest origin (Nicholas *et al.*, 1977). In fact, it has been suggested that marker molecules which characterize different cell types of neural crest origin share common initial metabolic steps in their biosynthesis (Bagnara *et al.*, 1979). The conversion of muscle into cartilage, reviewed by Nathanson, is also an example of a switch between two distinct cell types with a common mesodermal origin. Although these two cell types are very different in morphology and function, some similarity exists in their molecular constitution. Reddy *et al.* describe the presence of hepatocytes in rat pancreas, a conversion affecting different cell types both of which normally originate from embryonic endoderm.

Cell type conversion (or cellular metaplasia, in classical terminology) is known to occur in ocular tissues during regeneration of eye parts in certain amphibians. Such cell type conversion has been repeatedly demonstrated also in cell cultures from ocular tissues of other vertebrates (Moscona, 1957; Eguchi, 1979; Okada, 1980, 1983). In these situations, conversion can occur not only between developmentally related cells, as in the transformation of the pigment epithelium cells into neural retina cells, but also between cell types which early in development segregate into different cell lineages. An example of the latter is modification of neural retina cells and pigment epithelium cells (both of neuroepithelium origin) into lens-type cells (normally of ectodermal origin). Recent *in vitro* studies on this kind of cell conversion are reviewed in the chapters by Eguchi and by Moscona. In these studies, the phenotypes of the original and of the modified cells have been convincingly identified by several markers. The changes described occur in cell culture systems started with homogeneous cell populations. Thus, these results provide unequivocal demonstration that definitive cell types can change, under appropriate conditions, into other phenotypes.

Except in the case of regeneration of amphibian eye tissues, there are few convincingly documented examples of cell transdifferentiation in the organism (cf. Slack, 1980; Stocum, 1984). However, utilizing refined techniques of cell marking, cell type conversion was reported to occur in the course of amphibian limb regeneration, including transdifferentiation of muscle into cartilage (cf. Steen, 1973; Namenwirth, 1974).

The source of cells for regeneration in invertebrates has long been controversial (Slack, 1980). As in vertebrates, also in invertebrates, there are not many critical studies demonstrating occurrence of cell type conversion during regeneration, mainly due to a lack of proper techniques to identify the origin of particular cells. In recent studies on planaria, Gregmini and Miceli (1980) utilized naturally occurring cell mosaics and demonstrated that, during regeneration in this organism, germ cells can give rise to muscle cells. As summarized by Slack (1980), in several of these systems, some cells can dedifferentiate and their progeny appears in the regenerates as different histological types.

Instability of differentiated cells in invertebrates is highlighted by regeneration studies in the medusa *Podocoryne*. In this organism, a nearly complete manubrium (including germ cells) can be restored from a cultured piece of striated muscle (Schmid and Alder, this volume). This is, undoubtedly, an extreme example of capacity of differentiated animal cells to express a wide repertoire of diversification; conceptually, it provides a link between the generally limited plasticity of differentiated vertebrate cells and the greater plasticity of plant cells. In this context, it should be stressed that irreversible commitment occurs also in some plant cells as discussed in the chapters by Meins and by Osborne and McManus. All of this speaks against assuming that there are fundamental differences between plant and animal cells in the nature and stability of the differentiated state.

Phenotype instability has been observed not only under experimental conditions of cell culture or regeneration, but also in normal development. In the hepatopancreas of the crayfish *Procambarus,* cells originating from the same precursors and belonging to the same cell series regularly change phenotypes. The term "cellular ontogeny" was used to denote such serial conversion in the life history of cells (Davis and Burnett, 1964). In the silkworm, *Antheraea polyphemus,* cuticle-producing cells in the larval silk gland transdifferentiate into salt-secreting cells during metamorphosis into the adult moth (Selman and Kafatos, 1974). In the present volume, regular changes in cell phenotypes during normal development are described in cellular slime molds, *Dictyostelium* (Takeuchi *et al.*), and hydra (Bode *et al.*). In summary, there is growing evidence from diverse systems (experimental and also normal) that the outcome of differentiation in animal cells may be less stable than conventionally assumed.

Reversible interconversion between two different cell types is uncommon. Conversion occurs as a one-way progressive change from A to B, but rarely from B to A. In the case of ocular cells, a two-step sequence of transdifferentiation was proposed, starting with neural retina cells and terminating in a lens phenotype (Okada *et al.*, 1975). In this respect, the situation appears to resemble transdetermination of imaginal discs of *Drosophila;* it is well-known that changes in the state of determination in this system occur sequentially (Hadorn, 1978). This is obviously another example of instability of differentiation, but it could not be included in the limited space of the present volume.

We can assume that cells, which had attained definitive phenotypes of particular kind, nevertheless retain "options" for potential conversion into another cell type. However, it should be emphasized that the range of options appears to be generally limited in animal cells. It is likely that most cells can express only a certain set of specific phenotypes, although they arrest at a stage of nonterminal commitment (cf. discussion by Osborne and McManus, this volume).

III. Gene Expression in the Process of Cell Type Conversion

Studies on expression of genes coding for so-called lens-specific proteins during transdifferentiation of ocular tissue cells revealed that the appearance of lens cells in nonlens cell cultures is preceded by a change at the transcriptional level (cf. Okada, 1983, for review). Recent studies suggested that the change is not qualitative but quantitative. As reported by Clayton *et al.* in this volume (see also Agata *et al.*, 1983), a low-level transcription of genes coding for δ-crystallin, one of the representative marker proteins of lens differentiation, is detected in several nonlens tissues *in situ* in chick embryos, many of which have been known to transdifferentiate into lens-like cells. Further study will be required to determine whether such low-level transcription is actually prerequisite for options for further changes which differentiated cells retain.

Results from a different approach, i.e., transfer of cloned genes coding for δ-crystallin, are also relevant to discussion of molecular basis of phenotype instability. The chapter by Kondoh and Okada describes that expression of δ-crystallin genes xenoplastically introduced into a variety of mammalian cells differs quantitatively depending on the recipient cell type; it is particularly high in lens cells, but also in retinal glial cells and epidermis.

Finally, we have to ask what are the cues that elicit cell conversion. Significantly, there is general agreement among authors in this volume studying systems that range from transdifferentiation of medusa muscle cells (Schmid and Alder) to conversion of retinal glia into lens-like cells (Moscona). The concensus view is that disruption of original tissue relationships of cells triggers a cascade of reactions that terminates in the appearance of new cell types. There seems little doubt that subtle changes at the cell-surface level may initiate the activation of a set of latent, if not completely inactive, genes, thus leading to a change of phenotype (see chapter by Moscona). The future challenge is to identify the nature of these cell-surface changes, in addition to characterizing the various steps in the subsequent cascade and their interrelationships.

In spite of the fundamental importance of this problem, the instability of the differentiated state of cells has so far not been intensively investigated, particularly in animal cells. This is largely due to difficulties in establishing suitable systems to be probed for interpretable results. These difficulties are now being rapidly overcome and this problem will soon be ready for new approaches utilizing contemporary techniques of molecular biology.

ACKNOWLEDGMENT

I thank Dr. R. Kelley for reading the original manuscript and providing constructive advice. Original investigations from the author's laboratory on the instability of differentiation have been supported by a Special Research Grant, Multicellular Organization, and a Cancer Research Grant from the Japan Ministry of Education, Science and Culture.

REFERENCES

Agata, K., Yasuda, K., and Okada, T. S. (1983). *Dev. Biol.* **100,** 222–226.

Bagnara, J. T., Matsumoto, J., Ferris, W., Frost, S. K., Turner, W. A., Jr., Tchen, T. T., and Taylor, J. D. (1979). *Science* **203,** 410–415.

Caplan, A. I., and Ordahl, C. P. (1978). *Science* **201,** 120–130.

Davis, L. E., and Burnett, A. L. (1964). *Dev. Biol.* **10,** 122–153.

DiBerardino, M. A., Hoffner, N. J., and Etkin, L. D. (1984). *Science* **224,** 946–954.

Eguchi, G. (1979). *In* "Mechanisms of Cell Change" (J. D. Ebert and T. S. Okada, eds.), pp. 273–291. Wiley, New York.

Gregmini, V., and Miceli, C. (1980). *Wilhelm Roux's Arch. Dev. Biol.* **188,** 107–113.

Hadorn, E. (1978). *In* "The Genetics and Biology of *Drosophila*" (M. Ashburner and T. R. F. Wright, eds.), Vol. 2, pp. 556–617. Academic Press, New York.

Joho, R., Nottenburg, C., Coffman, R. L., and Weisman, I. L. (1983). *Curr. Top. Dev. Biol.* **18,** 15–58.

Moscona, A. A. (1957). *Science* **125,** 598–599.

Moscona, A. A., and Linser, P. (1983). *Curr. Top. Dev. Biol.* **18,** 155–188.

Namenwirth, M. (1974). *Dev. Biol.* **41,** 42–56.

Nicholas, D. H., Kaplan, R. A., and Weston, J. A. (1977). *Dev. Biol.* **60,** 226–237.

Okada, T. S. (1980). *Curr. Top. Dev. Biol.* **16,** 349–380.

Okada, T. S. (1983). *Cell Differ.* **13,** 177–183.

Okada, T. S., Itoh, Y., Watanabe, K., and Eguchi, G. (1975). *Dev. Biol.* **45,** 318–329.

Selman, K., and Kafatos, F. C. (1974). *Cell Differ.* **3,** 81–94.

Slack, M. W. (1980). *Nature (London)* **286,** 760.

Steen, P. (1973). *Am. Zool.* **13,** 1349.

Stocum, D. L. (1984). *Differentiation* **27,** 13–28.

ACKNOWLEDGMENTS

Yamada Science Foundation was established in February 1977 in Osaka through the generosity of Mr. Kiro Yamada. Mr. Yamada was president of Rohto Pharmaceutical Company Limited, a well-known manufacturer of medicines in Japan. He recognized that creative, unconstrained, basic research is indispensable for the future welfare and prosperity of mankind and he has been deeply concerned with its promotion. Therefore, funds for this Foundation were donated from his private holdings.

The principal activity of the Yamada Science Foundation is to offer financial assistance to creative research in the basic natural sciences, particularly in interdisciplinary domains that bridge established fields. Projects which promote international cooperation are also favored. By assisting in the exchange of visiting scientists and encouraging international meetings, this Foundation intends to greatly further the progress of science in the global environment.

In this context, Yamada Science Foundation sponsors international Yamada Conferences once or twice a year in Japan. Subjects to be selected by the Foundation should be most timely and stimulating. These conferences are expected to be of the highest international standard so as to significantly foster advances in their respective fields.

The editor wishes to acknowledge the executive members of Yamada Science Foundation, including Leo Esaki, Kenichi Fukui, Osamu Hayaishi, Noburo Kamiya, Jiro Kondo, Takeo Nagamiya, Shuntaro Ogawa, Syûzô Seki, Tomoji Suzuki, Jin-ichi Takamura, and Yasusada Yamada.

CHAPTER 1

CONVERSION OF RETINA GLIA CELLS INTO LENSLIKE PHENOTYPE FOLLOWING DISRUPTION OF NORMAL CELL CONTACTS

A. A. Moscona

LABORATORY FOR DEVELOPMENTAL BIOLOGY
CUMMINGS LIFE SCIENCE CENTER
THE UNIVERSITY OF CHICAGO
CHICAGO, ILLINOIS

I. Introduction

One of the dominant principles in contemporary developmental biology is that communication and interactions among cells play a crucial role in embryonic differentiation and morphogenesis. There is growing evidence that *contact-mediated* interactions between cells are involved in regulation of gene expression, and that, in turn, the genome controls processes which determine cell contact, cell associations, and interactions of cells (Moscona, 1974; Moscona and Linser, 1983; Moscona *et al.*, 1981). Sequences of such outside–inside regulatory reciprocities are considered to be of principal importance in driving cell differentiation and morphogenesis during embryonic development. Here I propose to take this concept a step further and suggest that cell contacts and contact-dependent cell interactions may function also in stablizing the phenotype of differentiated cells. This suggestion is based on our evidence that, if differentiated glia cells of the retina are removed from their normal contact relationships in the tissue, their phenotype characteristics become altered (Moscona and Linser, 1983). The specific conditions and results discussed here are not ex-

1

pected to apply rigidly to all other types of cells, but the proposed concept may turn out to be quite generally valid, at least in its key aspects, and deserves consideration.

Except in cases of DNA loss, differentiated cells retain much, perhaps most, of the genomic information present in the original zygote, albeit in a latent form (DiBernardino *et al.*, 1984). Nevertheless, cells in the organism ordinarily persist in their specialized states and do not change their established phenotypes (significant exceptions are found in certain pathological conditions and neoplasias). This indicates existence of mechanisms which keep the phenotype of differentiated cells stable and safeguard it from modification. This is an important problem, with broad biomedical significance. However, its experimental investigation has been hindered by paucity of systems in which it is possible to alter phenotypic characteristics of differentiated cells under conditions amenable to analysis. Information gained from such systems would advance knowledge not only of phenotype controls in specialized cells, but also of cell differentiation, and of cell modification in pathological conditions.

There are only a few well-documented experimental examples of phenotype conversion. Among the best-known classical cases are transdetermination of imaginal discs in *Drosophila* (Hadorn, 1966) and lens regeneration from the iris and from the pigment epithelium in some species of newts (Yamada, 1977; Reyer, 1977). The discovery of these systems drew attention to the problem of phenotype control, but their inherent complexity limited their detailed analysis. Our work has been concerned with two other cases of phenotype modification. The first will be described briefly; the second represents the main subject of this chapter.

II. Keratogenesis in the Chorioallantoic Membrane

The chorioallantoic membrane (CAM) is the respiratory organ of the chick embryo. The outer surface of the CAM consists of a single layer of cells which are in contact with the eggshell and are specialized for CO_2 and O_2 transfer, and for transport of calcium from the shell into the embryo (Terepka *et al.*, 1976; Tuan, 1984). However, if this epithelium is detached from the shell and is exposed to a higher than normal oxygen level, the cells undergo a striking transformation; they multiply rapidly, make keratin, and in 4–8 days form a stratified and keratinized skinlike structure (Fig. 1). Thus, cells which are normally specialized for respiratory functions can become converted into a keratogenic phenotype (Moscona, 1959). The transformation is elicited by two sets of changes, and both are essential: detachment of the CAM

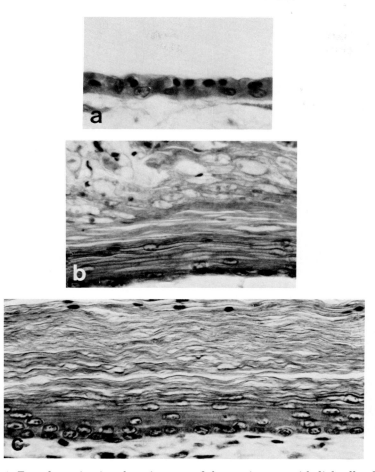

FIG. 1. Transformation into keratinocytes of the respiratory epithelial cells of the chorioallantoic membrane (CAM), 8-day chick embryo. For description of experimental conditions see text. (a) CAM epithelium (top) detached from the eggshell. (b) CAM epithelium exposed for 5 days to higher than normal O_2 level. Extensive proliferation, stratification, and keratinization of the epithelial cells. On top, parakeratotic cells; lower, keratinizing cells. (c) After 8 days: layers of keratin and keratinizing cells; CAM epithelium transformed into keratinized structure. ×360. (Modified from Moscona, 1959; Moscona and Carneckas, 1959.)

from the shell without simultaneous exposure to an altered $CO_2:O_2$ ratio (and vice versa) does not cause keratinization. Apparently, a synergistic action of both stimuli triggers changes in these respiratory cells conducive to their conversion into keratinocytes (Moscona and Carneckas, 1959). It is possible that still other stimuli may cause this end effect in these cells. But the principal point here is that, under appropriate conditions, specialized cells can drastically alter their growth characteristics and macromolecular synthesis and convert into a different phenotype.

III. Modification of Retina Cells into Lenslike Phenotype

A. BACKGROUND

It has long been known that cells dissociated from the neural retina of chick embryo can form *in vitro* lenslike structures or *lentoids* (Moscona, 1957, 1960). Since retina and lens pursue very different programs of differentiation in the embryo and have very different structural and biochemical characteristics and functions, such a striking change in cell type raised a puzzling problem. Indeed, when this finding was first reported in 1957, it met with considerable skepticism, since it did not conform to the then-prevailing view which was that cell differentiation represented a stable and irreversible commitment. It had been known that some species of newts can regenerate lens from the iris, a tissue which is contiguous with the retina; however, such regeneration does not occur in higher vertebrates and, moreover, the neural retina is a much more complex structure than the iris. At that time, there were no satisfactory methods for identifying in the retina the exact cellular precursors of these lentoids, or for investigating this system in biochemical detail.

This problem attracted renewed interest some years later, when T. S. Okada, G. Eguchi, and their associates confirmed by immunohistochemical and other methods that cultured retina cells of chick embryo can, in fact, give rise to a cell type which accumulates lens proteins (Okada *et al.*, 1975). These workers used retina cells from very early chick embryos and cultured them in monolayer. After several weeks they noted spontaneously arising clusters, referred to as "lentoid bodies," that consisted of cells which contained δ-crystallin, a protein characteristic of chicken lens. Their appearance was designated as "transdifferentiation," meaning conversion of a differentiated cell into another cell type (Okada, 1980).

This work raised further questions. The cultures were started with cells from early (3- to 4-day-old) embryonic retina, which is only in the beginning stages of development; the cells are undergoing growth, programming, and diversification, and most have not yet attained definitive-type characteristics. This, and the very long culture time it took before lentoid bodies began to appear, prevented resolution of two prime issues: (1) whether the lentoid bodies in these cultures arose from differentiated or from undifferentiated retina cells and (2) which of the several cell types in the retina was the progenitor of the lentoid bodies. Without answers to these questions, a convincing case for "transdifferentiation" was difficult to make, and it could not be decided whether this system was suitable for investigating the broader problem of phenotype stability and modifiability.

B. MÜLLER GLIA CELLS

In order to examine these issues, my associates and I chose to work with differentiated cells from postmitotic retina of 13- and 16-day chick embryos. By the thirteenth day of development the retina had completed its growth and overall histological organization (Moscona and Moscona, 1979); neurons and glia are phenotypically defined and are already in the process of functional maturation. Our first objective was to find out if such cells would give rise to lentoids.

The avian retina contains several kinds of neurons, but only one type of glia, known as Müller cells. As will become apparent, these glia cells are of special interest here. Definitive Müller glia cells in 13- and 16-day retina are distinguished by several characteristics: they are postmitotic, they contain *carbonic anhydrase-C* (CA-C) (Linser and Moscona, 1981b), they are inducible for *glutamine synthetase* (GS), and they are susceptible to the gliatoxic effect of α-aminoadipic acid (Linser and Moscona, 1979, 1981a). Both glia and neurons express a retina-specific, cell-surface antigen in the retina, *R-cognin* (Moscona, 1980), that determines mutual recognition among retina cells and mediates their adhesive affinity for each other (Hausman and Moscona, 1979; Ben-Shaul *et al.,* 1979; Ophir *et al.,* 1983).

The induction of GS in Müller glia cells requires a comment (for a review, see Moscona, 1983). GS is induced by 11β-hydroxycorticosteroids (we use cortisol); the hormone elicits accumulation of GS mRNA, and this results in *de novo* synthesis of GS. However, the important point here is that, in addition to the hormonal inducer, induction and expression of GS in Müller cells also require contact (contact-mediated interactions) between the glia cells and neurons (Linser and Moscona, 1983). Thus, if the retina is dissociated and the cells are cultured monodispersed in monolayer or in suspension, GS is not inducible. However, if the dissociated cells reassemble into multicellular clusters or are reaggregated, GS can be induced in the glia cells, but only if the glia and neurons restore close contacts or reconstruct retinotypic tissue architecture (Linser and Moscona, 1979, 1983).

The lesson from GS induction is that specific cell–cell contacts (contact-mediated cell interations) function in sustaining phenotypic characteristics in differentiated cells, in this case in maintaining Müller glia cells inducible for GS. We suggested (Moscona and Linser, 1983) that contact with neurons maintains the glia cell surface in a particular condition which is prerequisite for GS inducibility. According to this suggestion, "signals" relayed from the cell surface into the glia cell capacitate it for GS induction, and contact with neurons is essential for generating these signals. Disruption of normal cell contacts abrogates the signals and renders the glia cells not inducible. We further suggested that continued cell separation might lead to further changes in the cell surface and within the glia cells that could result in phenotype modification. There is no evidence for gap junctions between neurons and glia in chicken retina at the ages used here, nor is there evidence that a diffusable neuronal substance renders the glia cells inducible for GS (Linser and Moscona, 1983). Apparently, the glia–neuron contact interactions required for GS induction depend largely on membrane molecules which mediate the mutual adhesion and association of these cells, molecules such as R-cognin (Linser and Moscona, 1983).

C. MODIFICATION OF RETINA GLIOCYTES INTO LENTOIDAL CELLS

It has long been known that, if freshly dissociated embryonic retina cells are immediately reaggregated (by rotation in flasks; Moscona, 1961), glia and neurons restore a retinotypic architecture in the multicellular aggregates and reestablish histological relationships that closely resemble those in the retina (Fig. 3a). To accomplish this, the cells "regenerate" recognition–adhesion mechanisms on the cell surface that normally hold them together in the tissue and which are degraded during cell dissociation (Ben-Shaul et al., 1980). It has been demonstrated that R-cognin is depleted from the cell surface during cell dissociation and that its reexpression on the cell surface is essential for restoration of retinotypic cell associates and cell organization in the aggregates, i.e., for restitution of neuron–glia contact relationships (Hausman and Moscona, 1979; Ben-Shaul et al., 1980).

Very different results are obtained if dissociated retina cells are maintained dispersed in monolayer culture for several days and then are reaggregated. Before describing the results, some of the conditions should be explained. In these experiments (Moscona and Degenstein, 1981), we used cells from postmitotic retina of 13-day chick embryos for the reasons detailed above. In monolayer culture, the dispersed glia cells flatten out, assume the shape of large epithelioid gliocytes, reinitiate DNA synthesis, and multiply; we refer to them as large epithelioid retinal *(LER) gliocytes.* The neuronal cells remain nondividing, form small clusters, and extend elongated processes. After 5–7 days, the culture consists of a carpet of LER gliocytes and neuronal clusters (Fig. 2). CA-C, an enzyme marker of Müller glia cells, continues to be expressed in the LER gliocytes (Linser and Moscona, 1983), which confirms their derivation from Müller cells. We also showed that, if Müller cells are selectively destroyed in the retina, monolayer cultures prepared from such glia-depleted retina contain no LER cells (Moscona and Degenstein, 1982). It was important to identify the progenitors of LER gliocytes because, as explained below, LER gliocytes give rise to lentoids.

After 5–7 days, the monolayer culture was dispersed, and the cells were aggregated by rotation in flasks. In the numerous small aggregates that formed, there was no restoration of retinotypic tissue architecture. Instead, the cells segregated in two distinct regions: in each aggregate LER gliocytes (identified by immunostaining for CA-C) assembled in the center into a compact spherical core, and the neuronal cells were localized externally as a loosely adhering outer zone (Fig. 3b–d) (Moscona and Degenstein, 1981). Therefore, these two kinds of cells no longer were mutually adhesive, i.e., the contact affinities of one or both had become altered during separation in monolayer culture. In the retina, glia cells are associated with neurons; in these cell aggregates, the glia-derived LER cells segregated from the neurons and displayed preferential adhesiveness to each other. The overall structure of these aggregates is shown in scanning electron micrographs (Fig. 4), which clearly demonstrate the segregation of the large core cells that closely adhere to each other from the much smaller neuronal cells which surround the core almost completely (Fig. 4a) or partly (Fig. 4b). The neuronal cells were so weakly attached to the inner core that they could be dislodged with ease. The "naked" cores were translucent (Fig. 3d).

From earlier studies it had been known that such a distinct cell segregation occurs typically and consistently in composite aggregates made up of cells obtained from different tissues (Moscona, 1956, 1962, 1974; Steinberg, 1964); cells of each tissue segregate from the rest and assemble in a distinct region. Tissue-specific cell assortment demonstrates that cells belonging to the same tissue express self-recognition and mutual contact affinity, whereas cells from different tissues recognize their disparities and display a "disaffinity" (Moscona, 1980). Accordingly, LER gliocytes and neuronal cells reacted in the aggregates as if they had originated in different tissues. This suggested that one or both had acquired a new phenotypic (histotypic) identity and no longer

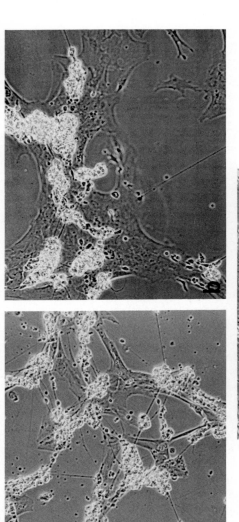
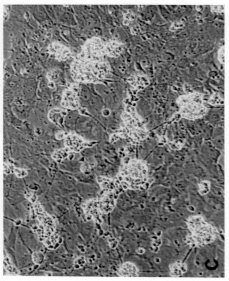

recognized the other as belonging to the same tissue (retina) (Moscona et al., 1983).

That the LER gliocytes had indeed become modified was further suggested by finding that the inner core cells in these aggregates strongly immunostained with antiserum to chicken lens proteins; the neuronal cells did not immunostain with this antiserum (Fig. 5). For this reason (and others to follow), the cores were designated as "lentoids" and their constituent cells referred to as "lentoidal" cells. These lentoidal cells are identical with the LER gliocytes which, in turn, are directly derived from definitive postmitotic Müller glia cells of 13-day retina. Therefore, under these experimental conditions, the separated, cultured Müller cells became modified into a lenslike lentoidal cell type. In fact, lens antigens can be immunocytochemically detected in LER gliocytes while these cells are still in monolayer culture (Moscona et al., 1983). This indicates that gliocyte modification begins shortly after the cells are separated from the tissue and plated in culture.

Gliocytes from 16-day retina also give rise to lentoidal cells when cultured in monolayer for 5–7 days and form lentoids in aggregates (Moscona and Degenstein, 1981). Therefore, Müller cells at an advanced stage of specialization can become lentoidal as a result of disruption of cell contacts and continued separation in culture.

As mentioned above, the derivation of LER and lentoidal cells from Müller glia was initially determined by showing that they contained the enzyme CA-C which, in 13-day chicken embryo retina, is a typical Müller glia marker. Expression of CA-C in lentoidal cells is not surprising, since this enzyme is also present in normal embryonic lens cells (Linser and Moscona, 1981b). However, there is additional evidence for glial origin of lentoidal cells. Definitive Müller glia in chicken retina can be selectively destroyed by the gliatoxic agent α-aminoadipic acid (Linser and Moscona, 1981a). Monolayer cultures prepared from 13-day or 16-day retina tissue that had been pretreated with this agent consisted mostly of neuronal cells and contained only a few LER gliocytes. When cells from these cultures were aggregated, they formed no lentoids, or only a few minute ones (Moscona and Degenstein, 1982). Thus, selective destruction of Müller glia in the retina

---

FIG. 2. Monolayer cultures of dissociated neural retina cells from 13-day chick embryo [in Medium 199 with 10% fetal bovine serum (FBS)]. (a) At 2 days: clusters of neuronal cells. Müller glia cells begin to spread out and assume the shape of large epithelioid retinal gliocytes (LER gliocytes). (b) At 4 days: area of expanding and multiplying LER gliocytes and clusters of neuronal cells. (c) At 7 days: carpet of LER gliocytes with neuronal cell clusters. ×260.

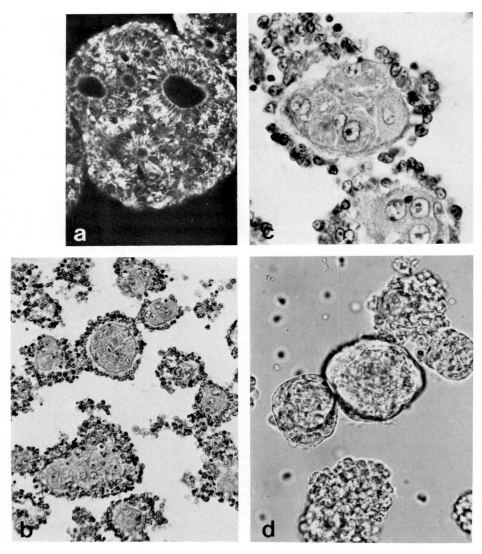

FIG. 3. (a) Tissue reconstruction in aggregate of freshly dissociated embryonic retina cells. Restoration of retinotypic tissue architecture characterized by formation of retinal rosettes that consist of neuronal and glia cells. The culture medium contained cortisol to induce GS. This histological section was immunostained with anti-GS antiserum and FITC-GAR. GS immunofluorescence localized in Müller glia cells (the white cells in this figure), which are seen to have assumed in the aggregate characteristic position. (For details, see Linser and Moscona, 1979.) (b–d) Aggregates of monolayer-precultured retina cells; compare with Fig. 3a. Dissociated retina cells of 13-day chick embryo were

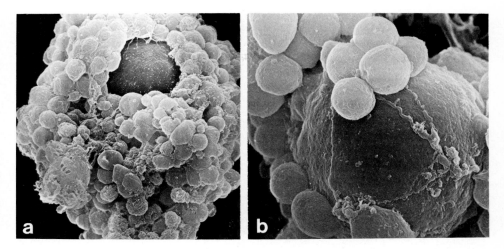

FIG. 4. Scanning electron micrographs of lentoid-containing cell aggregates, obtained as in Fig. 3b and c. (a) The core (lentoid) surrounded almost completely by loosely attached neuronal cells. ×172. (b) Lentoid with a few neuronal cells. ×387. (Details in Ophir *et al.*, 1985.)

depleted it of cells that can give rise to LER gliocytes and, hence, to lentoids.

In other experiments, neurons in monolayer cultures of 13-day reti-na cells were selectively destroyed by short treatment with the quinol-ine compounds chinoform (Shinde and Eguchi, 1982) or broxyquinoline (Moscona *et al.*, 1983); such cultures contained only LER gliocytes. When these cells were aggregated, they formed lentoids that incorpo-rated virtually all the gliocytes; because of absence of neurons, such lentoids were devoid of the outer zone cells (Moscona *et al.*, 1983).

Taken together, the above results provide conclusive evidence that, in this experimental system, lentoids arise from Müller glia-derived LER gliocytes. Therefore, definitive, postmitotic Müller glia cells from 13-day or 16-day retina can undergo modification into a lentoidal phe-notype. It is not yet clear if reinitiation of DNA replication and cell

---

cultured in monolayer for 7 days and then were dispersed and aggregated by rotation. (b) Histological section of cell aggregates (48 hours). Note segregation of LER gliocytes from neuronal cells. Each aggregate contains a compact core (lentoid) composed of LER glio-cytes, surrounded by loosely attached neuronal cells. ×160. (c) Higher magnification of aggregates obtained as in Fig. 3b, showing core (lentoidal) cells and loosely attached neuronal cells. ×1060. (d) "Naked" cores (center) obtained by removing the neuronal cells; live culture. ×460.

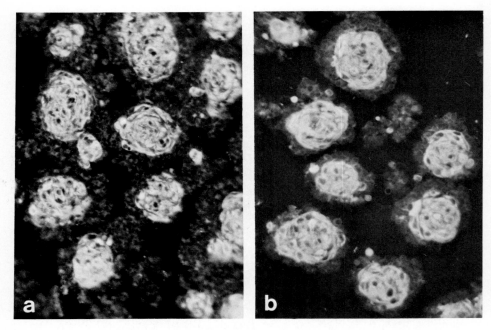

Fig. 5. Immunostaining with antiserum to lens proteins of cell aggregates similar to those shown in Fig. 3b. FITC-fluorescence. (a) Immunostaining with antiserum to chicken whole lens. Immunofluorescence localized in core cells (lentoid); no reaction in surrounding neuronal cells. ×376. (b) Immunostaining with anti-α-crystallin antiserum localized in lentoidal (core) cells; in this specimen a few straggling LER gliocytes remained outside the lentoids. ×376.

divisions are obligatory for this modification. Since virtually all the LER cells become lentoidal, it is highly unlikely that the lentoids arise by selection or clonal growth from some small subpopulation of Müller glia cells.

D. CHANGE IN CELL ADHESIVENESS: LOSS OF R-COGNIN

An important and striking feature of gliocyte modification into lentoidal cell type is the change in contact affinity: loss of adhesiveness to neurons and acquisition of mutual adhesivity. An explanation for this change is suggested by recent evidence that LER gliocytes no longer express R-cognin, a cell-surface antigen implicated in the mechanisms of mutual recognition and adhesion of retina cells (Hausman and Moscona, 1979). Loss of R-cognin was discovered by treating monolayer cultures of retina cells with antiserum to R-cognin and complement; this treatment caused rapid lysis of neuronal cells, while LER gliocytes remained intact (Linser and Moscona, 1983). The same treatment applied to noncultured retina cells resulted in lysis of virtually all the glia and neuronal cells (Hausman and Moscona, 1979). Cell immunolysis (in the presence of complement) by antibodies specific for a surface antigen is acceptable evidence for the presence of this antigen on these cells. Accordingly, lysis of neurons in the monolayer cultures demonstrated that these cells continued to express R-cognin, whereas nonlysis of LER gliocytes indicated that these cells no longer displayed this antigen on the cell surface. This conclusion was further strengthened by a more recent observation (Ophir *et al.*, 1985): by means of immunolabeling and scanning electron microscopy of 7-day monolayer cultures of 13-day retina cells, R-cognin was detected on neurons, while on LER gliocytes this antigen was absent or greatly reduced (Fig. 6). It is not yet known whether the absence of R-cognin is due to its actual loss, or to its "masking," or to modification of its antigenic sites.

In view of the role of R-cognin in mediating the mutual affinity and

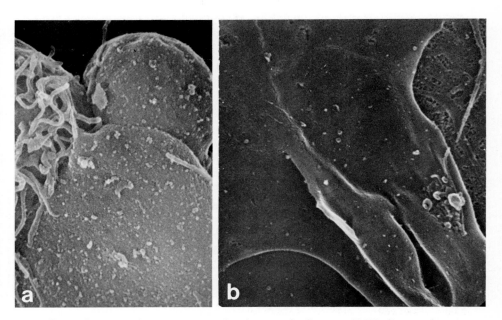

FIG. 6. Scanning electron micrographs of neuronal cells (a) and LER gliocytes (b) in 7-day monolayer culture of retina cells from 13-day embryo. Immunolabeled for R-cognin with R-cognin antiserum and with GAR-hemocyanin. Labeling evident on neuronal cells (a), but almost completely absent on LER gliocytes (b). ×9600. (Details in Ophir *et al.*, 1985.)

adhesivity between retina cells, its absence from the modified LER gliocytes (lentoidal cells) offers a likely explanation for their loss of contact affinity for neurons. While we cannot state that the neuronal cell surface does not become modified in monolayer culture, the difference between neurons and LER cells with respect to immunodetection of R-cognin is very striking. Since R-cognin is a retina-specific cell-surface marker, loss of this antigen is consistent with modification of the glial phenotype into a different, nonretinal cell type. The fact that these modified gliocytes (lentoidal cells) preferentially adhere to each other and form compact lentoids raises the possibility that they now express some other affinity-recognition molecule(s), perhaps characteristic for lens cells. This possibility remains to be tested. In any case, it is already known that the modified gliocytes and lentoidal cells newly express a membrane antigen normally found in differentiated lens cells (see below).

E. MP26: A LENS CELL MEMBRANE ANTIGEN

The intrinsic cell membrane protein MP26 (M_r 26K) is characteristic of differentiated lens cells (Waggoner and Maisel, 1978; Kibbelaar and Bloemendal, 1979). MP26 (also known as MIP26) is not detectable immunohistochemically in the retina of 13-day chick embryo (Moscona *et al.*, 1983). However, in monolayer cultures of retina cells this antigen begins to appear in the cell membrane of LER gliocytes and can already be detected in some of these cells on the third day by immunostaining with anti-MP26 antiserum (Moscona *et al.*, 1983). After 5 days in culture, virtually all the LER gliocytes express MP26; neurons do not (Fig. 7a). Following cell aggregation, MP26 is present in the cell membrane in the lentoids (Fig. 7b–d). The *de novo* appearance of this lens membrane antigen in lentoidal cells provides persuasive evidence of their phenotypic similarity to normal lens cells and thus reinforces the conclusion that definitive Müller glia cells can undergo phenotype modification into a lenslike cell type.

It is not yet known if the appearance of MP26 in LER gliocytes is due to gene activation or to expression of gene products that are inac-

---

FIG. 7. Immunostaining for MP26. (a) Section of stripped 5-day monolayer culture of 13-day retina cells, immunostained with anti-MP26 antiserum and FITC-RAM. Reaction with antiserum in cell membrane of LER gliocytes (upper part of figure), not in neuronal cells. ×512. (b–d) MP26 in lentoids; immunofluoresence in cell membrane of lentoidal cells. Elongated shape of lentoidal cells outlined by staining for MP26. (b, c) ×485; (d) ×1160. (e) Retina tissue of 13-day embryo immunostained for MP26, showing absence of reaction with the antiserum. ×270. (Modified from Moscona *et al.*, 1983.)

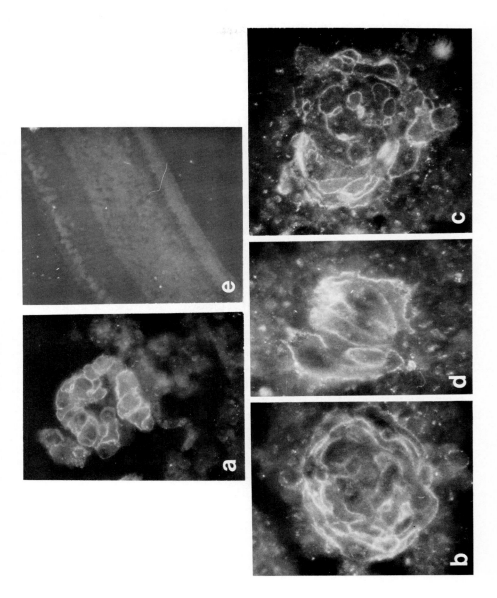

tive in normal glia cells. In any case, its appearance is part of the syndrome of modifications elicited in these glia cells by changes that follow disruption of normal cell contacts and cell separation. This underscores again the importance of cell contacts in maintaining phenotype stability of Müller cells in the retina.

F. COMMENTS AND SUMMARY

The exact nature of the changes in regulatory processes and in macromolecular synthesis that lead to the modification of Müller gliocytes into lentoidal cells remains to be elucidated. Detailed analysis may shed light on the mechanisms that stabilize the phenotype of these glia cells, and on the relations of cell-contact and cell-surface properties to these mechanisms.

Another important and semmingly puzzling question is, why do Müller glia convert into lenslike cells? Among conceivable possibilities, a likely one is that Müller glia normally express a low level of (some) lens gene products; this might predispose them to a lens-type change following destabilization of their original phenotype. In fact, it has been known that antisera to lens crystallins cross react with nonlens tissues, and there is evidence of crystallin mRNA in the retina (Clayton, 1982; Agata et al., 1983). However, the exact cellular localization of these products in retina tissue has not been reported.

Recently we began to investigate this question by immunostaining sections of retina with antiserum to a lens fraction highly enriched for α-crystallin. The antiserum reacted with Müller glia cells; neurons showed no reaction. We cannot yet conclude with certainty that the reacting antigen in Müller cells is α-crystallin and not another protein with homologous antigenic sites, or some other product which is coantigenic with a contaminant present in the original immunogen (Moscona et al., 1985).

These preliminary observations raise the possibility that Müller glia cells normally express a low level of lens gene products, possibly α-crystallin. If confirmed, this finding might suggest a basis for the tendency of these cells to assume lenslike characteristics following decontrol of their original phenotype. It now becomes of interest to determine at what embryonic and postembryonic stages such lens gene products are found in Müller cells (Moscona et al., 1985).

Finally, a comment is needed about the lentoid bodies which arise in long-term monolayer cultures of early embryonic retina cells (Okada, 1980). These structures have been identified mainly by immunostaining with antiserum to δ-crystallin, and the identity of their cellular progenitors in the retina is not clear. Nomura et al. (1980)

suggested that the progenitors may be neuronal. In light of our findings, it becomes important to determine if these lentoid bodies contain still other lens markers, especially if they express the lens cell membrane antigen MP26, or if their only characteristic is accumulation of δ-crystallin. The same questions apply also to the lentoid bodies which appear in monolayer cultures of pigmented epithelium (tapetum) cells from chick embryo eye (Eguchi, 1979). These lentoid bodies were also identified by immunostaining with antiserum to δ-crystallin. Since this antiserum cross reacts with various nonlens cells (Kodama and Eguchi, 1982), it would be of interest to find out if pigment epithelium cells normally contain low levels of δ-crystallin, and if the lentoid bodies derived from these cells express also other lens cell markers. Such information would help to comparatively characterize and classify the lentoidal structures described by various workers and would advance our understanding of the biochemical changes that accompany phenotype modification in separated retina cells.

In summary, the following working hypothesis is proposed.

1. Phenotype stability of definitive Müller glia cells in the retina is dependent on contact with neurons (contact-mediated interactions). Since Müller cells are associated in the retina with several types of neurons, these interactions may involve multiple heterotypic cell contacts.

2. Disconnection and persistent separation of these glia cells can lead to their modification into a lenslike type (lentoidal cells). We suggest that changes in the cell surface resulting from cell separation are "signaled" internally and alter regulatory processes, which results in phenotype modification.

The suggestion that cell–cell contacts and contact-mediated cell interactions play an important role in stabilizing and controlling phenotype characteristics of differentiated cells may not apply equally to every type of cell and situation; however, it is supported by evidence from other systems (e.g., Sidman, 1974; Holton and Weston, 1982; Fisher, 1984). This concept may also be heuristically useful in investigations of pathological cell transformations and of phenotype modification in metastases of neoplastic cells. The importance of cell contacts and contact-mediated cell interactions focuses increased attention to the cell surface as the site of cell affinity and cell association mechanisms, and as a source of signals that influence phenotype controls in differentiated cells (Moscona, 1974, 1980; Edelman, 1983). The obvious future task is to identify such signals.

ACKNOWLEDGMENT

The author's work was supported by these grants: PCM8209360 from the National Science Foundation; HDO1253 from NICHD, USPHS. Travel expenses to the Yamada Symposium at which this material was presented were defrayed in part by the Louis Block Fund, The University of Chicago.

REFERENCES

Agata, K., Yasuda, K., and Okada, T. S. (1983). *Dev. Biol.* **100**, 222.
Ben-Shaul, Y., Hausman, R. E., and Moscona, A. A. (1979). *Dev. Biol.* **72**, 89.
Ben-Shaul, Y., Hausman, R. E., and Moscona, A. A. (1980). *Dev. Neurosci.* **3**, 66.
Clayton, R. M. (1982). *In* "Stability and Switching in Cellular Differentiation" (R. M. Clayton and D. E. S. Truman, eds.), pp. 23–38. Plenum, New York.
DiBernardino, M. A., Hoffner, N. J., and Etkin, L. D. (1984). *Science* **224**, 946.
Edelman, G. M. (1983). *Science* **219**, 450.
Eguchi, G. (1979). *In* "Mechanisms of Cell Change" (J. D. Ebert and T. S. Okada, eds.). Wiley, New York.
Fisher, M. (1984). *Proc. Natl. Acad. Sci. U.S.A.* **81**, 4414.
Hadorn, E. (1966). *In* "Major Problems in Developmental Biology" (M. Locke, ed.), pp. 85–104. Academic Press, New York.
Hausman, R. E., and Moscona, A. A. (1979). *Exp. Cell Res.* **119**, 191.
Holton, B., and Weston, J. A. (1982). *Dev. Biol.* **89**, 72.
Kibbelaar, M. A., and Bloemendal, H. (1979). *Exp. Eye Res.* **29**, 679.
Kodama, R., and Eguchi, G. (1982). *Dev. Biol.* **91**, 221.
Linser, P., and Moscona, A. A. (1979). *Proc. Natl. Acad. Sci. U.S.A.* **76**, 6476.
Linser, P., and Moscona, A. A. (1981a). *Dev. Brain Res.* **1**, 103.
Linser, P., and Moscona, A. A. (1981b). *Proc. Natl. Acad. Sci. U.S.A.* **78**, 7190.
Linser, P., and Moscona, A. A. (1983). *Dev. Biol.* **96**, 529.
Moscona, A. A. (1956). *Proc. Soc. Exp. Med.* **92**, 410.
Moscona, A. A. (1957). *Science* **125**, 598.
Moscona, A. A. (1959). *Dev. Biol.* **1**, 1.
Moscona, A. A. (1960). *In* "Developing Cell Systems and their Control" (D. Rudnick, ed.), pp. 45–70. Ronald Press, New York.
Moscona, A. A. (1961). *Exp. Cell Res.* **22**, 455.
Moscona, A. A. (1962). *J. Cell. Comp. Physiol.* **60**, 65.
Moscona, A. A. (1974). *In* "The Cell Surface in Development" (A. A. Moscona, ed.), pp. 67–99. Wiley, New York.
Moscona, A. A. (1980). *In* "Membranes, Receptors, and the Immune Response" (E. P. Cohen and H. Köhler, eds.), Vol. 42, pp. 171–188. Liss, New York.
Moscona, A. A. (1983). *In* "Progress in Retinal Research" (N. Osborne and G. Chader, eds.), Vol. 2, pp. 111–135. Pergamon, Oxford.
Moscona, A. A., and Carneckas, Z. I. (1959). *Science* **129**, 1743.
Moscona, A. A., and Degenstein, L. (1981). *Cell Differ.* **10**, 39.
Moscona, A. A., and Degenstein, L. (1982). *In* "Stability and Switching in Cellular Differentiation" (R. M. Clayton and D.E.S. Truman, eds.), pp. 187–197. Plenum, New York.
Moscona, A. A., and Linser, P. (1983). *Curr. Top. Dev. Biol.* **18**, 155.
Moscona, A. A., Brown, M., Degenstein, L., Fox, L., and Soh, B. M. (1983). *Proc. Natl. Acad. Sci. U.S.A.* **80**, 7239.

Moscona, A. A., Fox, L., Smith, J., and Degenstein, L. (1985). *Proc. Natl. Acad. Sci. U.S.A.*, in press.

Moscona, M., and Moscona, A. A. (1979). *Differentiation* **13**, 165.

Moscona, M., Degenstein, L., Byun, K. Y., and Moscona, A. A. (1981). *Cell Differ.* **10**, 317.

Nomura, K., Tagaki, S., and Okada, T. S. (1980). *Differentiation* **16**, 141.

Okada, T. S. (1980). *Curr. Top. Dev. Biol.* **16**, 349.

Okada, T. S., Itoh, Y., Watanabe, K., and Eguchi, G. (1975). *Dev. Biol.* **45**, 318.

Ophir, I., Moscona, A. A., and Ben-Shaul, Y. (1983). *Cell Differ.* **13**, 133.

Ophir, I., Moscona, A. A., Loya, N., and Ben-Shaul, Y. (1985). *Cell Differ.*, in press.

Reyer, R. W. (1977). *In* "Handbook of Sensory Physiology VII/5" (F. Crescitelli, ed.), pp. 338–373. Springer-Verlag, Berlin and New York.

Shinde, S. L., and Eguchi, G. (1982). *In* "Problems of Normal and Genetically Abnormal Retinas" (R. M. Clayton, J. Haywood, H. W. Reading, and A. Wright, eds.). Academic Press, New York.

Sidman, R. L. (1974). *In* "The Cell Surface in Development" (A. A. Moscona, ed.), pp. 221–253. Wiley, New York.

Steinberg, M. (1964). *In* "Cellular Membranes in Development" (M. Locke, ed.), pp. 321–366. Academic Press, New York.

Terepka, A., Coleman, J., Ambrecht, H., and Gunther, T. (1976). *Symp. Soc. Exp. Biol.* **30**, 117.

Tuan, R. S. (1984). *Ann. N. Y. Acad. Sci.* **429**, 459.

Waggoner, P. R., and Maisel, H. (1978). *Exp. Eye Res.* **27**, 151.

Yamada, T. (1977). "Control Mechanisms in Cell-Type Conversion in Newt Lens Regeneration." Karger, Basel.

CHAPTER 2

INSTABILITY IN CELL COMMITMENT OF VERTEBRATE PIGMENTED EPITHELIAL CELLS AND THEIR TRANSDIFFERENTIATION INTO LENS CELLS

Goro Eguchi

DIVISION OF MORPHOGENESIS
DEPARTMENT OF DEVELOPMENTAL BIOLOGY
NATIONAL INSTITUTE FOR BASIC BIOLOGY
MYODAIJICHO, OKAZAKI, JAPAN

I. Introduction

Considering potential for growth and differentiation, animal tissues can be classified into three groups: renewing, stable, and static (Hay, 1974). Fully differentiated functional cells in renewing tissue are periodically replaced by differentiating progenitor cells which are produced by continual growth of their stem cells. This phenomenon is observed in the epidermis, as one of clear examples. In static tissues, such as nerve and muscle, differentiated functional cells lose their growth potential and maintain their differentiated functions through-

21

out life. Stable tissues, represented by various glandular and connective tissues, maintain certain functions but retain proliferative abilities as well. Many functional parenchymal cells, e.g., hepatocytes, exhibit dormant growth potential. Such tissues are classified as stable tissues (Goss, 1964).

The pigmented epithelium of vertebrate eyes can be classified as a stable tissue. Pigmented epithelial cells (PECs) readily synthesize DNA and express high growth potential when exposed to selected pathological or culture conditions. However, this growth potential is completely repressed by regulatory mechanisms which are in effect when the same cells form tissues *in situ*. PECs of urodelean amphibia have been widely studied as a particularly interesting example of this phenomenon. The PECs of dorsal iris and the pigmented epithelium of the retina of newts can regenerate both lens and neural retina, respectively, by cellular metaplasia (transdifferentiation) (reviewed in Scheib, 1965; Reyer, 1977; Yamada, 1977, 1982). The dorsal retinal pigmented epithelia of newts and some frogs also have the ability to transdifferentiate into lens (Sato, 1951, 1953). The ability to regenerate lens is also observed in fish and in avian embryos (Sato, 1961; reviewed in Scheib, 1965).

Since 1970, studies of lens regeneration and transdifferentiation of PECs *in vitro* have been extensive (see reviews in Eguchi, 1976, 1979, 1983; Eguchi and Ito, 1981, 1982; Clayton, 1978, 1979, 1982; Okada, 1980, 1983; Yamada, 1982). In 1973, Eguchi and Okada first demonstrated, using clonal cell culture, that the progeny of retina PECs isolated from older chick embryos could switch their specificity to lens cells (Eguchi and Okada, 1973). Similar *in vitro* switching of cell type in PECs has also been observed in newts (Eguchi *et al.*, 1974) and in human fetuses (Yasuda *et al.*, 1978). Thus, evidence so far accumulated substantiates the assumption that a dormant ability of PECs to switch their differentiated state is widely maintained in vertebrates.

More recently, we have succeeded in establishing a new culture system to study the mechanism of transdifferentiation of chick embryonic PECs to lens cells (Eguchi, 1983; Eguchi and Itoh, 1981, 1982; Eguchi *et al.*, 1982; Itoh and Eguchi, 1981, 1982, 1985a,b). In this system, we can obtain a large number of dedifferentiated, multipotent PECs which can readily and synchronously redifferentiate into either lens or pigment cells under given culture conditions.

In this chapter, our recent studies of cellular and molecular events during stabilization and transdifferentiation of chick embryo PECs *in vivo* and *in vitro* will be reviewed.

II. Stable Differentiated State of Pigmented Epithelial Cells (PECs)

A. THE STABLE STATE OF PIGMENTED EPITHELIUM AS A MONOLAYER
EPITHELIUM OF PECs IN THE EYE *in Situ*

During development of vertebrate eyes, when mesenchymal cells cover the surface of the optic cup, cells in the posterior pole of the outer layer of the optic cup begin to synthesize melanin pigment and differentiate into pigmented epithelial cells (PECs). These cells constitute the pigmented epithelium. Differentiation of the pigmented epithelium, which starts at the posterior pole of the optic cup and is directed by the ectomesoderm of the choroid, continues to propagate centrifugally toward the equatorial region.

PECs of the well-differentiated eye *in situ* adhere firmly to the collagen layer at their basal surface and keep close contact with each other at their apex to form a simple epithelium with hexagonal cellular arrangement. The shape of PECs is maintained both by cell–cell contact and cell–substrate adhesion, and, in addition, by circumferential actin bundles which are organized just underneath the opposed cell membrane at the apical border of each PEC. Under normal conditions in the eye *in situ*, PECs never express their growth potential, but maintain a stable, functional differentiated state.

B. STABILIZATION OF THE DIFFERENTIATED STATE OF PECs

By the 8- to 9-day stage of development, PECs in the posterior pole of the eye of chick embryos have differentiated and their growth potential is repressed. However, these cells readily proliferate and dedifferentiate when dissociated and maintained *in vitro*. When cultures of PECs reach confluence, dedifferentiated cells form a simple epithelial sheet with a polygonal cellular pattern and redifferentiate to pigment cells (Eguchi and Okada, 1973). When such cultures of PECs are maintained further, they stabilize as a monolayer epithelium with typical hexagonal arrangement of pigmented cells (Fig. 1) (Honda and Eguchi, 1980; Honda *et al.*, 1984).

Electron microscopy of ultrathin sections cut perpendicularly through a monolayer revealed that differentiated PECs *in vitro* develop and adhere to settle on a thick layer of collagen at their basal surface. In addition, cells form close association with each other at the apical boundary. Many cytoplasmic processes project from the free surface of each PEC, and bundles of actin microfilaments form circumferential rings at the apical boundary of each cell (Figs. 1 and 2) (Owaribe *et al.*, 1981; Owaribe and Masuda, 1982). These ultrastruc-

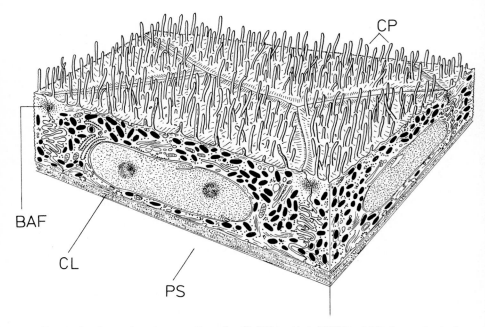

FIG. 1. A schematic representation of well-differentiated PECs which form a typical monolayer of simple epithelium *in vitro*. Each PEC differentiated *in vitro* retains structural characteristics identical to those of the same cell type differentiated in the eye *in situ*. BAF, Bundle of actin filaments; CL, collagen layer synthesized by PECs monolayer; CP, cytoplasmic process; PS, plastic substrate of culture dish.

tural characteristics are all identical to those observed in PECs differentiated in the normal eye *in situ*. Experimental and theoretical analyses suggest that these circumferential actin bundles participate both in the formation of a monolayer epithelial pattern by PECs and in the structural and functional stabilization of each cell in the monolayer (Honda and Eguchi, 1980; Honda *et al.*, 1984).

III. Transdifferentiation of PECs *in Vitro*

A. ADULT NEWT PECs

We have clearly demonstrated by clonal culture analysis that PECs dissociated from the pigmented epithelia of both the iris and the retina of fully grown newts transdifferentiate into lens cells (Abe and Eguchi, 1977; Eguchi *et al.*, 1974, 1982; Eguchi, 1976, 1979, 1983; Eguchi and Itoh, 1982). In spite of the fact that the lens-regenerating potential *in situ* is restricted to the dorsal half of the pigmented epithelium, the

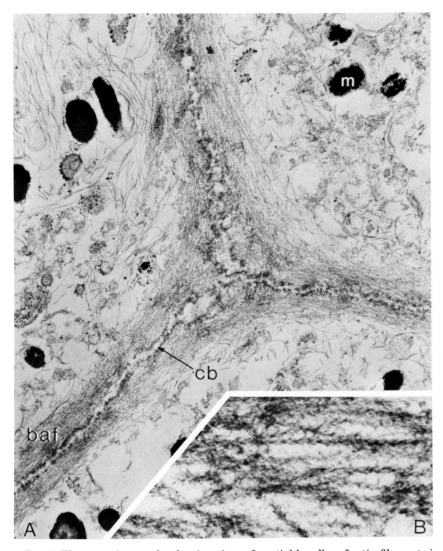

FIG. 2. Electron micrographs showing circumferential bundles of actin filaments in differentiated PECs *in vitro*. (A) Thick bundles of actin filaments (baf) along the cell boundary (cb) are clearly revealed in an ultrathin section through the apical border of glycerinated PECs. m, Melanosome. ×27,000. (B) Filaments decorated with heavy meromyosin develop an arrowhead-like appearance, confirming their actin composition. ×75,000.

ventral PECs can transdifferentiate into lens cells in a manner similar to that of dorsal PECs when dissociated and cultured *in vitro*. This suggests that the differentiated state of the ventral PECs is more stable than that of the dorsal PECs *in situ* and that they must be unstabilized when dissociated *in vitro*.

B. Avian Embryo PECs

Retinal PECs of 8- to 9-day-old chick embryos grow vigorously when dissociated and cultured with medium supplemented with fetal bovine serum (FBS). Although PECs in cultures can maintain their original specificity as pigment cells through at least two generations in culture, completely dedifferentiated PECs (which no longer express melanogenic properties) appear usually at the end of the third culture generation (the second subcultivation or passage). These dedifferentiated cells still continue to grow and eventually develop into lentoid, spherical aggregates of fully differentiated lens cells with a full complement of crystallin proteins, 60–90 days after the primary inoculation. Evidence for the transdifferentiation of lens cells from retinal PECs in chick embryos has been obtained in clonal culture (Fig. 3)

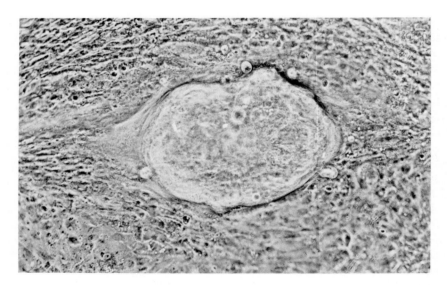

Fig. 3. A typical, well-differentiated lentoid exhibiting a full complement of lens-specific characteristics. This structure developed in the culture of clonal progeny derived from a single PEC of an 8.5-day-old chick embryo. ×100. (From Eguchi and Okada, 1973.)

(Eguchi, 1979; Eguchi and Okada, 1973). A similar study showed that retinal PECs of 8- to 9-day-old quail embryos can transdifferentiate into lens cells faster than chick embyro PECs (Eguchi, 1976).

C. Human Fetus PECs

After demonstrating transdifferentiation of PECs to lens cells *in vitro* in 8- to 9-day-old chick and quail embryos (which can no longer regenerate lens *in vivo*), we have extended our studies to human fetal PECs.

Clean pieces of pigmented epithelium were isolated separately from the iris and the retina of human fetal eyes at approximately 12 weeks after conception and were dissociated into single cells for culture.

When PECs were isolated from iris and cultured, tiny lentoid bodies consisting of several elongated cells with crystallins were formed in some colonies about 30 days after inoculation. When PECs were isolated from retina and cultured, the center of each colony of once-depigmented cells started repigmentation about 30 days after inoculation. No lentoid formation was observed by this stage. At 20 days following the secondary inoculation at clonal cell density, two types of colonies were distinguished; the first type consisted of only fibroblastic-shaped cells and the second consisted of central epithelial cells with peripheral fibroblastic-like cells. In many of the latter colonies, lentoids were formed at the boundary between the epithelial center and peripheral fibroblastic zone from about 35 days onward (Yasuda *et al.*, 1978). Thus, PECs of both iris and retina in human fetuses are able to transdifferentiate into lens cells *in vitro* when cultures are started from dissociated cells.

IV. Dedifferentiated State of Chick Embryo PECs

A. Microenvironments Regulating the Transdifferentiation of PECs

As described previously, PECs in the eye *in situ* settle on a supportive collagenous substrate and are in close contact with each other, forming a cohesive simple epithelium with typical polygonal cellular pattern. Under *in vivo* and *in vitro* conditions, the differentiated state of PECs is stably maintained and their growth potential is repressed. In fact, transdifferentiation of PECs to lens cells was effectively suppressed by a collagen extracted from rat tails and applied to plastic culture dishes prior to seeding PECs into the dish (Eguchi, 1976; Yasuda, 1979). From these results, we speculated that cell-surface properties of PECs might be modified during the process of transdifferentiation by factors both intrinsic and extrinsic to the cell which affect both adhesion and motility.

An effective inhibitor of melanogenesis, phenylthiourea (PTU), significantly enhances the lentoid differentiation of chick embryo PECs cultured *in vitro* (Eguchi, 1976). This substance was also found to promote cell membrane permeability for divalent cations by chelating Cu^{2+} ions (Masuda and Eguchi, 1984). When PECs are cultured in standard culture medium containing PTU (PTU medium), depigmented PECs constituting the confluent cell sheet no longer develop melanosomes and do not redifferentiate as do cultures in standard culture medium without PTU. It is characteristic for dedifferentiated cells (dePECs) in PTU medium that they continue to grow and eventually form multicellular layers. Lentoid differentiation always takes place in such multicellular layers, even in primary cultures which are maintained for only 25 days (Fig. 4) (Eguchi and Itoh, 1981, 1982; Itoh and Eguchi, 1981). In this context, PTU can actually enhance lens transdifferentiation of PECs without suppressing cell growth.

We assume that PTU affects the stability of the differentiated state of PECs by altering cell-surface properties. Based on this assumption, PECs were cultured for 20 days in PTU medium. The completely depig-

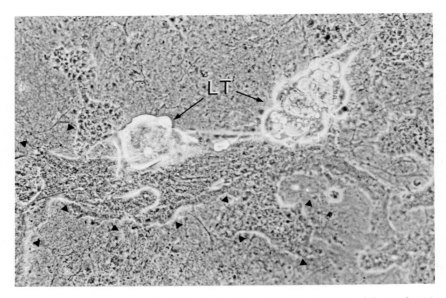

FIG. 4. A phase micrograph of primary cultures of PECs maintained for 25 days in the presence of PTU (1.0 m*M*). Lentoids (LT) are developing from multilayered areas in the culture (closed triangles) ×50.

mented area which formed in the monolayer was cut out with a stain-less-steel tube of 1.0- to 3.0-mm diameter. These disks dissected from the monolayer, after the treatment with Dispase II, detached readily from the plastic substratum without losing cell-to-cell adhesion and turned into a tiny vesicle. These vesicles were cultured on soft agar for 5 days with the PTU medium. All vesicles developed into lentoids with well-differentiated lens fibers surrounded by epithelium. This interest-ing result strongly suggests that the effect of PTU in enhancing trans-differentiation of PECs can be amplified by controlling the microen-vironment of dePECs (Eguchi and Itoh, 1982; Eguchi *et al.*, 1982).

B. TESTICULAR HYALURONIDASE AMPLIFIES THE EFFECT OF PTU ON TRANSDIFFERENTIATION OF PECs

To modify the microenvironment of PECs cultured *in vitro,* we have introduced testicular hyaluronidase (HUase) into our culture system to selectively decompose hyaluronate and proteoglycans. The applica-tion of HUase to our culture system of PECs was based on the findings by Itoh, who demonstrated that this enzyme remarkably promoted lens transdifferentiation in cultures of neural retina cells of chick embryos (Itoh, 1978).

Freshly inoculated PECs were maintained in standard medium and then transferred to PTU medium containing HUase (PTU-HU medi-um). Differentiation of lentoids was remarkably amplified by HUase in the presence of PTU, depending upon the concentration of the en-zyme. In contrast, HUase alone induced the loss of differentiated prop-erties of PECs, but inhibited their transdifferentiation into lens. Semi-quantitative analyses estimating the number of lentoids developed in a unit area of the culture dish have revealed that the efficiency of lens transdifferentiation in PTU-HU medium in optimum concentrations was amplified more than 50-fold when compared with control cultures maintained in PTU medium (Eguchi and Itoh, 1982; Eguchi *et al.,* 1982; Itoh and Eguchi, 1981, 1982, 1985a).

C. ESTABLISHMENT OF THE BIPOTENT DEDIFFERENTIATED STATE OF PECs

It is likely that the serum supplementing our culture media con-tains a number of unknown factors which may influence the growth and differentiation of tissue cells and directly or indirectly affect the range of transdifferentiation. Itoh (1976) clearly demonstrated that transdifferentiation of neural retina cells from chick embryos into lens and pigment cells can be remarkably enhanced by supplementing MEM with dialyzed fetal bovine serum (dFBS) in place of fetal bovine serum and with ascorbic acid (AsA). On the basis of his discovery, we have tested the effects of dFBS and AsA on transdifferentiation from PECs to lens cells.

When PECs were cultured in medium (PTU-HU-dFBS medium) which was prepared by supplementing Eagle's minimum essential medium (MEM) with dFBS (6%), PTU (0.5 mM), and HUase (150–250 U/ml medium), the cells dedifferentiated rapidly, grew vigorously, and eventually formed dense, multilayered cell sheets within about 2 weeks (Eguchi, 1983; Eguchi and Itoh, 1982; Eguchi et al., 1982; Itoh and Eguchi, 1981, 1982, 1985). These dePECs with epithelial morphology could be maintained in the dedifferentiated state as long as they were cultured in this medium. However, dePECs readily expressed lens cell specificities when dissociated, inoculated at especially high cell density, and maintained with PTU-HU-dFBS medium containing AsA. Almost all the dePECs began to differentiate very synchronously to lens cells, forming numerous transparent lentoids within about 10 days. In contrast, sister populations of dePECs redifferentiated into pigment cells when cultured in medium without PTU and HUase (Fig. 5). Quantitative estimation of lens specificity by measuring δ-crystallin content revealed that at the final stage of lens cell differentiation from dePECs cultured in PTU-HU-dFBS medium, more than 50% of the total protein of cultures is δ-crystallin (the major specific protein for the developing chick lens) (Eguchi and Itoh, 1982; Eguchi et al., 1982; Eguchi, 1983). These results show that the dedifferentiated state of PECs is bipotent and the expression of either of two different phenotypes can be readily manipulated by changing environmental conditions (Fig. 6).

V. Gene Expression in the Process of Transdifferentiation

To approach the mechanisms which control transdifferentiation, molecular studies are needed for each cell change during lens transdifferentiation of PECs in vitro. In the first step toward this goal, we have attempted to describe the transcriptional patterns of genes which should be specifically expressed by both lens cell and the pigmented cell, using our established culture system of chick embryo PECs. Recent results on the transcription of the δ-crystallin gene during transdifferentiation of the PEC in vitro are relevant not only to characterize the bipotent dedifferentiated state but also to understand modulation in cell phenotype in terms of gene expression.

A large number of primary cultures of PECs isolated from 8- to 9-day-old chick embryos were prepared using standard culture medium supplemented with 6% FBS. After 1 week of culture, cells were dissoci-

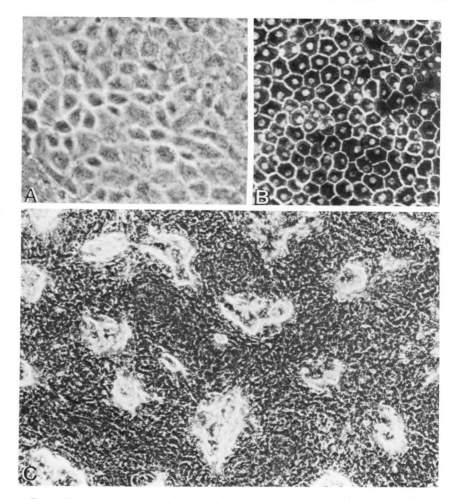

FIG. 5. Expression of either pigment or lens cell specificities by bipotent dedifferentiated pigmented epithelial cells (dePECs) of chick embryos *in vitro*. (A) Dedifferentiated
PECs in culture. ×200. (B) PECs in a monolayer epithelium, redifferentiated from
dePECs shown in (A). ×200. (C) Lentoids developing in high-density cultures of dePECs
a week after culturing. ×35.

ated and transferred to the PTU-HU-dFBS medium. After at least two
additional subcultivations (passages) a large number of dePECs were
collected. These cells were dissociated and cultured under two different conditions: one permissive for transdifferentiation into lens and
the other permissive for redifferentiation into pigment cells as pre-

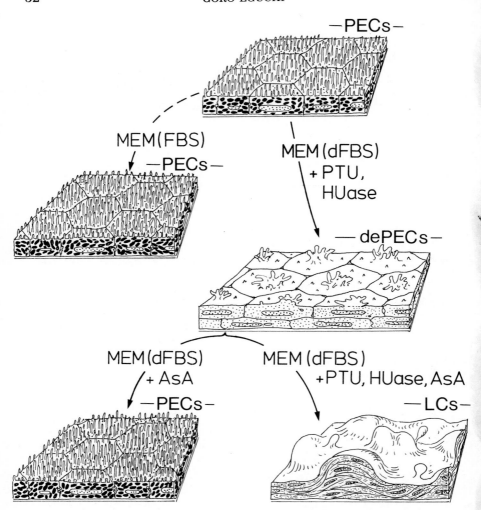

Fɪɢ. 6. Transdifferentiation *in vitro* of retinal PECs from chick embryos. Pigmented epithelial cells (PECs) maintain their differentiated properties when cultured with Eagle's minimum essential medium (MEM) supplemented with fetal bovine serum MEM(FBS). They dedifferentiate readily when cultured with MEM supplemented with dialyzed fetal bovine serum MEM(dFBS), phenylthiourea (PTU), and testicular hyaluronidase (HUase). Dedifferentiated cells (dePECs) transdifferentiate into lens cells (LCs) when cultured at high cell density under permissive conditions. In addition, the same population of dePECs readily redifferentiates to PECs under conditions of culture which lack both PTU and HUase. AsA, Ascorbic acid.

viously described. Lens transdifferentiation in cultures in PTU-HU-dFBS medium with AsA was observed within 1 week after initiation of cultured dePECs in this medium (Fig. 5), whereas redifferentiation of pigment cells was achieved much faster. Cells were harvested at each state in the process of dedifferentiation, transdifferentiation, and re-differentiation for analyses of transcripts of δ-crystallin genes.

Poly(A)$^+$ RNAs were extracted from cells harvested from cultures and isolated by binding to oligo(dT)-cellulose. They were separated by electrophoresis on 1.0% agarose gels and then blotted onto nylon filters (Agata *et al.*, 1983). The RNAs thus blotted were hybridized with a ^{32}P-labeled *Pst*I fragment of cloned δ-crystallin cDNA (pBδ11) (Yasuda *et al.*, 1984).

In our Northern blot analysis, transcripts hybridizing with δ-crystallin cDNA were detected in RNAs extracted from dePECs, al-though no mature mRNA in lens cells was detected (Fig. 7, lane 2). A clear band at the level of 9.5 kb and a diffuse smear were observed in dePECs RNAs. The 9.5-kb transcript was also found in lens RNAs (Fig. 7, lane 4). We interpret this to be a precursor RNA of δ-crystallin mRNA. In contrast to dePECs, RNAs extracted from cultures main-

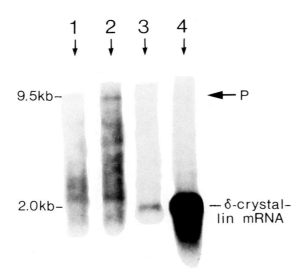

FIG. 7. Northern blot analysis of poly(A)$^+$ RNAs purified from PECs during *in vitro* transdifferentiation into lens cells. *Pst*I fragments of δ-crystallin cDNA (pBδ11) were used as a probe. (1) 5 μg of RNA from PECs freshly dissected from 8.5-day-old chick embryos. (2) 5 μg of RNA from dePECs in culture. (3) 0.5 μg of RNA from transdifferenti-ated lens cells from dePECs. (4) 0.5 μg of RNA from lens cells freshly dissected from lenses of 8.5-day-old chick embryos (control). P, Precursor of δ-crystallin mRNA.

tained in conditions permissive for transdifferentiation into lens cells were found to contain only small amounts of the 9.5-kb transcript, in addition to mature mRNA for δ-crystallin as the major component (Fig. 7, lane 3). In addition, we have observed the rapid disintegration of the 9.5-kb transcript in cells placed in the conditions permissive for the redifferentiation. In contrast, we have also demonstrated the gradual accumulation of δ-crystallin mRNA in dePECs which were supposed to be committed to transdifferentiation into lens cells (K. Agata and G. Eguchi, unpublished). Neither δ-crystallin mRNA nor the 9.5-kb transcript was detected in both freshly dissociated PECs of 8- to 9-day-old chick embryos and PECs redifferentiated from dePECs.

These results indicate that (1) the δ-crystallin gene is transcribed, but not processed to the mature δ-crystallin mRNA in dePECs; (2) δ-crystallin transcripts detected in the dePECs are rapidly processed to the mature δ-crystallin mRNA soon after exposure of the cells to conditions permissive for transdifferentiation into lens cells; and (3) δ-crystallin transcripts in dePECs are rapidly disintegrated during redifferentiation of dePECs into pigment cells.

VI. Concluding Remarks

Although the cell type conversion of PECs can be observed in the process of Wolffian lens regeneration and of neural retina regeneration in a limited number of species, results of cell culture works since 1973 (Eguchi, 1976, 1979; Okada, 1980, 1983; Clayton, 1979, 1982) permit us to conclude that the population of PECs capable of transdifferentiation is widely conserved in vertebrates including humans, at least in the embryonic period. However, this multipotential property is well stabilized in physiological conditions *in situ*. That close cell–cell contact may be one of the requisites for the stabilization of PECs is revealed by observations of a cohesive epithelial cell sheet supported by a continuous collagen substrate (Eguchi, 1979; Yasuda, 1979).

We have now established a culture system in which a pure population of dePECs can be maintained, by introducing PTU, HUase, and dialyzed fetal bovine serum in an appropriate way. With this system, we are able to investigate cellular properties of dePECs in detail. These cells exhibit properties similar to neoplastic cells. Dedifferentiated PECs do not suppress growth at confluence and continue to proliferate to form multilayers. Under such conditions, significant reduction of cell–cell and cell–substrate adhesion is observed (Okada *et al.*, 1973). In addition, it has been shown that dePECs consume glucose in place of oxygen and release lactate as a metabolic product (Y. Itoh and G. Eguchi, unpublished). Furthermore, dePECs are bipotent cells with the ability to become either lens cells (transdifferentiation) or PECs (redifferentiation). These two directions can now be reversibly manipulated by regulating culture conditions. The conditions which permit differentiation into given directions have been clearly identified.

Preliminary efforts to elucidate mechanisms of transdifferentiation in terms of the regulation of gene expression have been started. Our culture system fulfills conditions necessary for such studies, since it is possible to collect large numbers of both bipotent cells and cells in the process of commitment to either lens or pigment cells. Northern blot analysis showed that the δ-crystallin gene, specifically expressed in lens cells *in situ,* is transcribed in bipotent dePECs, although the transcripts are not processed to the mature mRNA in the dePECs. These results clearly show that dePECs, which are bipotent and do not express any other differentiated trait of lens cells, have been partially committed to becoming lens cells by initiation of the specific gene expression.

It has been demonstrated that δ-crystallin mature mRNA can be detected in some nonlenticular tissues such as pigment epithelium, neural retina, and others in early chick embryos *in situ* (Bower *et al.,* 1983; Agata *et al.,* 1983). The ectopic translation of δ-crystallin at a low level is observed also in tissues at early stages in their development (Baravanov, 1977; Clayton, 1979, 1982; Kodama and Eguchi, 1982). On the basis of these observations Clayton has advanced the hypothesis that such a preexistence of the δ-crystallin mRNA should be one of the necessary conditions for cells to transdifferentiate into lens cells *in vitro* (Clayton, 1982; and Clayton, this volume). Experimental evaluation of this hypothesis is in progress, since the behavior of dePECs may be explained if such a suggestion is valid.

Finally, to further understand the multipotent dedifferentiated state of PECs at the level of gene expression, it is necessary to extend investigation to the transcription of genes which are specifically expressed in the melanogenic cell and to determine whether the transcripts are also present in dePECs. Further analysis of the molecular and cellular mechanisms of transdifferentiation using the culture system described in this chapter will contribute further to our understanding the molecular basis of phenotypic instability and cell commitment in differentiation.

ACKNOWLEDGMENTS

I greatly acknowledge the following people for their collaboration to the recent works reviewed in this article: R. Kodama, K. Agata, Y. Karasawa (from my laboratory), and Y. Itoh (Aichi Medical University). My sincere thanks are also due to Prof. T. S. Okada and his colleagues, Drs. K. Yasuda and H. Kondoh (Kyoto University) who very kindly provided δ-crystallin cDNA for our experiments and gave us invaluable advice, and also to Dr. R. M. Clayton, Institute of Animal Genetics, University of Edinburgh, for constructive discussion, and to Prof. R. O. Kelley, University of New Mexico for his kind critical reading of the manuscript.

The work in my laboratory reviewed here was supported by Grants-in-Aid for Basic Cancer Research (Project Nos. 560110033, 57010031, 580110034), for Special Project Research, "Multicellular Organization" (Project Nos. 58119005, 59113007), and also for General Research (Project No. 59480023) from the Ministry of Education, Science and Culture to G. E.

REFERENCES

Abe, S., and Eguchi, G. (1977). *Dev. Growth Differ.* **19**, 309.

Agata, K., Yasuda, K., and Okada, T. S. (1983). *Dev. Biol.* **100**, 222.

Barabanov, V. M. (1977). *Dokl. Akad. Nauk SSSR* **234**, 195.

Bower, D. J., Errington, L. H., Pollock, B. J., Morris, S., and Clayton, R. M. (1983). *EMBO J.* **2**, 333.

Clayton, R. M. (1978). *In* "Stem Cell and Tissue Homeostasis" (B. I. Lord, C. S. Potten, and R. J. Cole, eds.), pp. 115–138. Cambridge Univ. Press, London and New York.

Clayton, R. M. (1979). *Ophthalmic Res.* **11**, 324.

Clayton, R. M. (1982). *In* "Differentiation *in Vitro*" (M. M. Yeoman and D. E. S. Truman, eds.), pp. 83–120. Cambridge Univ. Press, London and New York.

Eguchi, G. (1976). *Ciba Found. Symp.* **40**, 242–258.

Eguchi, G. (1979). *In* "Mechanisms of Cell Change" (J. D. Ebert and T. S. Okada, eds.), pp. 273–291. Wiley, New York.

Eguchi, G. (1983). *In* "Developmental Biology. Afro-Asian Perspective" (C. S. Goel and R. Bellair, eds.), pp. 97–108. Indian Society of Developmental Biologists, Poona.

Eguchi, G., and Itoh, Y. (1981). *In* "Pigment Cell 1981: Phenotypic Expression in Pigment Cells" (M. Seiji, ed.), pp. 271–278. Univ. of Tokyo Press, Tokyo.

Eguchi, G., and Itoh, Y. (1982). *Trans. Ophthalmol. Soc. U.K.* **102**, 308.

Eguchi, G., and Okada, T. S. (1973). *Proc. Natl. Acad. Sci. U.S.A.* **71**, 1495.

Eguchi, G., Abe, S., and Watanabe, K. (1974). *Proc. Natl. Acad. Sci. U.S.A.* **72**, 5052.

Eguchi, G., Masuda, A., Karasawa, Y., Kodama, R., and Itoh, Y. (1982). *In* "Stability and Switching in Cellular Differentiation" (R. M. Clayton and D. E. S. Truman, eds.), pp. 209–221. Plenum, New York.

Goss, R. J. (1964). "Adaptive Growth." Academic Press, New York.

Hay, E. D. (1974). *In* "Concept of Development" (J. Lash and J. R. Whittaker, eds.), pp. 404–428. Sinauer, Stamford, Connecticut.

Honda, H., and Eguchi, G. (1980). *J. Theor. Biol.* **84**, 575.

Honda, H., Kodama, R., Takeuchi, T., Yamanaka, H., Watanabe, K., and Eguchi, G. (1984). *J. Embryol. Exp. Morphol.* **83** (Suppl.), in press.

Itoh, Y. (1976). *Dev. Biol.* **54**, 157.

Itoh, Y. (1978). *Zool. Mag. Tokyo.* **87**, 370.

Itoh, Y., and Eguchi, G. (1981). *Dev. Growth Differ.* **23**, 449.

Itoh, Y., and Eguchi, G. (1982). *Dev. Growth Differ.* **24**, 396.

Itoh, Y., and Eguchi, G. (1985a). *Cell Differ.*, in press.

Itoh, Y., and Eguchi, G. (1985b). *Dev. Biol.*, in press.

Kodama, R., and Eguchi, G. (1982). *Dev. Biol.* **91**, 221.

Masuda, A., and Eguchi, G. (1984). *Cell Struct. Funct.* **9**, 25.

Okada, T. S. (1980). *Curr. Top. Dev. Biol.* **16,** 349–390.

Okada, T. S. (1983). *Cell Differ.* **13,** 177.

Okada, T. S., Eguchi, G., and Takeichi, M. (1973). *Dev. Biol.* **45,** 318.

Owaribe, K., and Masuda, H. (1982). *J. Cell Biol.* **95,** 310.

Owaribe, K., Kodama, R., and Eguchi, G. (1981). *J. Cell Biol.* **90,** 507.

Reyer, R. W. (1977). *In* "Handbook of Sensory Physiology VII/5: The Visual System in Vertebrates" (F. Crescitelli, ed.), pp. 309–390. Springer-Verlag, Berlin and New York.

Sato, T. (1951). *Embryologia* **1,** 21.

Sato, T. (1953). *Wilhelm Roux' Arch. Entwicklungsmech. Org.* **146,** 487.

Sato, T. (1961). *Embryologia* **6,** 251.

Scheib, D. (1965). *Ergeb. Anat. Entwicklungsgesch.* **38,** 45.

Yamada, T. (1977). *Monogr. Dev. Biol.* **13.**

Yamada, T. (1982). *In* "Cell Biology of the Eye" (D. S. McDevitt, ed.), pp. 193–242. Academic Press, New York.

Yasuda, K. (1979). *Dev Biol.* **68,** 618.

Yasuda, K., Okada, T. S., Hayashi, M., and Eguchi, G. (1978). *Exp. Eye Res.* **26,** 591.

Yasuda, K., Nakajima, N., Isobe, T., Okada, T. S., and Shimura, Y. (1984). *EMBO J.* **3,** 1397.

CHAPTER 3

TRANSDIFFERENTIATION OF SKELETAL MUSCLE INTO CARTILAGE: TRANSFORMATION OR DIFFERENTIATION?

Mark A. Nathanson

DEPARTMENT OF ANATOMY
NEW JERSEY MEDICAL SCHOOL
NEWARK, NEW JERSEY

I. Introduction

The great variety of experimental systems in which differentiated cells modulate their phenotype, patterns of biosyntheses, or both (this volume) suggests that "stability" may not characterize the process we call differentiation. Rather, the "acquisition of stability" appears to be a more accurate definition of this process. One may envision a given cell, in a defined microenvironment, exhibiting a *major* biosynthetic pattern, with alternative syntheses held in abeyance for lack of an appropriate stimulus (i.e., a permissive microenvironment). The work discussed in this chapter utilizes the well-characterized bone-inductive system to study the mechanism whereby skeletal muscle apparently "transforms" into cartilage. A brief review of the state of differentiation of the source skeletal muscle precedes this discussion. It is not possible to cite all of the outstanding reports in each subject, although it is acknowledged that a large group of scientists have made contributions on which this work is based.

CURRENT TOPICS IN
DEVELOPMENTAL BIOLOGY, VOL. 20

Demineralized bone-mediated formation of cartilage from skeletal muscle and the regeneration of amphibian lenses and neural retina from iris and retinal pigmented epithelium, respectively (see Urist, 1965; Okada, 1980, for review), are two major examples of phenotypic alteration (metaplasia). The occurrence of metaplasia is rare but highly instructive in that it cautions developmental biologists that conditions external to a differentiated cell are capable of imposing a bias toward one phenotype or another. To understand development, then, one must understand (1) the morphological and biochemical differentiation of a tissue *in situ,* (2) the metabolic needs of a tissue *in situ,* (3) the development of a tissue under standard and nonstandard conditions, and (4) the embryology of a tissue in relation to its neighboring tissues. In most cases we understand 1 and 2 above and have only a limited understanding of 3 and 4. This is due, in part, to a lack of knowledge regarding the means to elicit a particular phenotype and, in part, from a bias toward regarding a particular tissue as the descendant of a committed precursor cell type. The problems inherent in this view are twofold. First, this view of development requires that an organism have a wide variety of precursor cell types, each requiring its own "inductive" agent. It additionally requires a means of "inducing" the inducers, and gives rise to a level of complexity which this investigator finds cumbersome and uncharacteristic of development. Second, the inducer must be of short range, be inactivated at the outer limits of this range, or it must interact with specific receptors. The development of such receptors gives rise to a level of complexity similar to that discussed above. This investigator's studies suggest that a more realistic view of development is that cells are capable of displaying a range of phenotypes, depending upon limited criteria, such as its class as ectodermal, endodermal, or mesodermal, and that environmental influences select from the range of phenotypes available to each class.

II. Skeletal Muscle Differentiation

Whether or not embryonic limb mesenchymal cells are predetermined remains a viable question in developmental biology, because investigators have yet to agree on an adequate definition of differentiation. The results of investigations in separate laboratories may vary depending upon the conditions under which the experiment is conducted. For example, it has been shown that myogenic cells will delay fusing into syncytial myotubes and, in some cases, enter an additional cell cycle if fresh culture medium is added or if initial inoculum densities are low (Konigsberg, 1971; O'Neill and Stockdale, 1972a). The quality of the resulting skeletal muscle also depends upon its substratum (Hauschka and Konigsberg, 1966; Konigsberg, 1970) and nutritional environment (Coon and Cahn, 1966; White and Hauschka, 1971; Yaffe, 1971; Konigsberg, 1971; Ramirez and Aleman, 1972; Hauschka, 1974; de la Haba *et al.*, 1975).

Investigations of developing avian and mammalian limbs have led to the formulation of several hypothetical mechanisms regarding the differentiation of skeletal muscle and cartilage. One hypothesis suggests that early limb mesenchyme contains precursor cells, which are already committed to a myogenic or chondrogenic fate (Holtzer and Sanger, 1972; Holtzer *et al.*, 1974). The myogenic stem cell, or presumptive myoblast, is viewed as a cell with only two developmental options: (1) to replicate, giving rise to other presumptive myoblasts, or (2) to undergo a critical or "quantal" mitosis, which prepares the daughter cells (now termed myoblasts) to become postmitotic and competent to fuse with other myoblasts into syncytial myotubes. The existence of a presumptive myoblast is inferred from experiments in which blocks of chick limb mesenchyme, from stages of chick development prior to overt myogenesis, are cultured *in vitro* and analyzed for the appearance of differentiated progeny. In these experiments, stage 17–18 (2–3 days of embryonic development) was the earliest stage at which skeletal muscle could be detected (Dienstman *et al.*, 1974). Similar mesenchyme in monolayer culture (Dienstman *et al.*, 1974) and in clonal culture (Bonner and Hauschka, 1974) failed to yield differentiated progeny until stages 21–22 (3.5–4 days *in vivo*). Insofar as these stages occur prior to the appearance of skeletal muscle in the embryonic chick limb *in vivo* (stage 25, 4.5–5 days), these data are interpreted to mean that an element of predetermination exists with respect to the differentiation of limb mesenchyme (Dienstman *et al.*, 1974). However, the precursor cell type (presumptive myoblast) does not display a definitive morphology, nor does it synthesize specific products, and cannot be precisely identified. Its identity can only be inferred as part of a hypothetical scheme. It must also be remembered that in each case the tissues are cultured *in vitro* for varying periods of time, such that stage 17–18 chick limb mesenchyme is no longer 2–3 days old, but actually 5–6 days old in the experiments reported by Dienstman *et al.*, and thus follow the *in vivo* schedule quite closely. Investigators rarely document the initial appearance of differentiated cells in these types of experiments and data are clearly lacking with regard to aging *in vitro* prior to differentiation. In the single study dealing with this point, of which this investigator is aware, older myogenic cells preferentially fused with myogenic cells of a similar age, rather than younger myogenic cells (cells cultured for a lesser period of time) (Yaffe,

1971; see below). In light of these data it may be argued that limb mesenchymal cells are not predetermined, but acquire differentiated traits over a period of time, in response to presently unknown factors, and only create the impression that undifferentiated limb mesenchyme contains "determined" cells.

The "quantal" mitosis is viewed as a critical event which, via presently unknown means, prepares the presumptive myoblast to become a true myoblast, postmitotic and fusion competent. This hypothesis is based upon the finding that [^{3}H]thymidine-labeled cells, from 11-day chick breast muscle, delay fusing for up to 5–8 hours past the end of mitosis (Okazaki and Holtzer, 1966). Insofar as myogenic cells are not found to fuse in either the G_2, S, or M phases of the cell cycle, these cells would be in the G_1 phase. G_1 typically lasts 2–3 hours *in vitro* (Bischoff and Holtzer, 1969) or approximately 4 hours *in vivo* (see Herrmann *et al.*, 1970), suggesting that the cells are in an extended G_1 phase (or G_0, since they are thought to have withdrawn from the cell cycle). In contrast, other investigators detected fusion as early as 3 hours after pulsing myogenic cultures with [^{3}H]thymidine (O'Neill and Stockdale, 1972b) and failed to detect myogenic cells which withdraw from the proliferative pool (Buckley and Konigsberg, 1977). It is well known that the G_1 phase lasts for longer periods in older embryos (Herrmann *et al.*, 1970), and there remains some doubt whether myoblasts actually withdraw from the cell cycle prior to fusion. The validity of the quantal mitosis concept remains unproven.

An alternative hypothesis contends that each mesenchymal cell is endowed with the potential for a limited number of cellular programs and that its final position in the limb determines how this potential is realized (stabilized) (Wolpert, 1969). For example, a program for "limb mesenchyme" may be expressed as skeletal muscle, fibrous connective tissue, or cartilage, depending upon the location of a cell within the developing limb. The factors which evoke these phenotypes are unknown. *In vivo* experiments suggest that limb mesenchymal cells are initially phenotypically unstable and become stabilized shortly after definitive myogenic and chondrogenic regions appear (Zwilling, 1966; Searls and Janners, 1969). A considerable body of data suggests that a cell's immediate environment plays a large part in determining its ultimate fate (for review see Hall, 1970). The subsequent acquisition of phenotypic stability has also been noted *in vitro* (Coon, 1965; Konigsberg, 1963; Richler and Yaffe, 1970) and is agreed upon by all investigators, irrespective of the early mode of differentiation.

An elegant series of experiments, using transplants of quail somite tissue into a chick host, has shown that limb musculature derives primarily from somitic tissue (Christ *et al.*, 1977; Chevallier *et al.*, 1977) and indicates separate lineages for muscle and cartilage. Limb musculature could form in the absence of a somitic contribution, albeit to a limited extent (McLachlan and Hornbruch, 1979), and some of the somite-derived cells were able to form cartilage (Kiney *et al.*, 1981). It is clear that, in the absence of experimental intervention, development proceeds according to a clearly defined pattern and provides the impression of distinct lineages for muscle and cartilage. However, experimental data suggest that at least all mesenchymal cells may have equivalent differentiative potential.

Several lines of evidence suggest that myogenic cells do not remain completely unchanged prior to fusion: (1) fusion is blocked by inhibition of protein synthesis prior to its onset (Shainberg *et al.*, 1969); (2) older myogenic cultures exhibit fusion very shortly after subculture, whereas younger cultures delay fusion until they have been cultured for approximately as long as the older cultures (Yaffe, 1971); and (3) when fusion is blocked by culturing myogenic cells in media containing lowered calcium (Shainberg *et al.*, 1969), subculturing at intervals, and returning to media containing adequate calcium, the myogenic cells still fuse at approximately the same time as parallel cultures not exposed to lowered calcium (Yaffe, 1971). These results suggest that the subcultured cells have memory of previous time in culture. Most likely they underwent some changes, as yet undefined, which precede fusion and these changes need only be elaborated upon rather than. begun anew.

It is evident from the above discussion that there is little likelihood of detecting a myogenic precursor cell population, since by definition, and in the absence of discrete populations within limb mesenchyme (Hilfer *et al.*, 1973), it cannot be identified by criteria which define differentiated skeletal muscle. The overwhelming weight of evidence suggests that limb mesenchyme is composed of similar cells, which respond to external cues and begin a sequence of events which lead to fusion.

III. Formation of Cartilage by Skeletal Muscle

The experimental system used in this investigator's studies was first discovered by Dr. Marshall Urist of the University of California. Dr. Urist found that embryonic rat skeletal muscle forms hyaline cartilage *in vitro* and bone *in vivo,* when exposed to demineralized bone (bone matrix; Urist, 1965; Nogami and Urist, 1974). These data were subsequently verified by Reddi and Huggins (1973). In this investigator's laboratory, bone matrix reproducibly elicits cartilage from skeletal muscle *in vitro*. The occurrence of phenotypic alterations *in vitro* demonstrates that hyaline cartilage arises from the explanted tissue and not from a migratory, precursor cell type. Insofar as skeletal muscle contains myoblasts, fibroblasts within fibrous connective tissue, and syncytial myotubes, the hyaline cartilage may arise from either of these sources. Either source will never form cartilage *in vivo*.

This investigator's initial studies asked two questions: (1) Does hyaline cartilage arise from skeletal muscle or from its associated fibrous connective tissue? and (2) Does the ability to form cartilage reside solely within somatic mesoderm or is it a property common to other mesodermal cell types? To answer these questions, skeletal muscle and fibroblasts from skeletal muscle were grown in clonal culture and explanted onto bone matrix substrata. For technical reasons, skeletal muscle was isolated from 11-day chick embryos and fibroblasts from 19-day rat embryo skeletal muscle. On bone matrix, both cell types formed hyaline cartilage (Nathanson *et al.*, 1978; Fig. 1). Insofar as cells were derived from animals well past the stage at which these tissues are considered to be "terminally" differentiated, the results demonstrate that differentiated cells are capable of altering their phenotype when presented with an appropriate stimulus. Additionally, it clearly demonstrates the ability of selected cell types to alter biosynthetic patterns which have previously been considered to be a fixed property of the particular cell type.

The mesodermally derived connective tissue capsules of embryonic chick thyroid and lung were also found to form hyaline cartilage, whereas their endodermally derived parenchyma did not. These results demonstrate that somatic and visceral mesenchyme are identical with respect to their developmental potency and that "cartilageness" is not a unique property of certain cells predetermined for the phenotype.

The above data show the genetic similarity of cells in an unique fashion and provide (1) a means of eliciting the formation of cartilage at will, (2) a means of characterizing the development of the tissue from its outset, (3) a means of investigating the biosynthetic potential of skeletal muscle, and (4) a means of describing cartilage differentiation in a biochemical fashion.

FIG. 1. Formation of cartilage by cloned fibroblasts and myoblasts. (A) Pooled fibroblast clones derived from 19-day embryonic rat thigh muscle. (B) Pooled skeletal muscle clones from 11-day embryonic chick thigh muscle. Preliminary experiments demonstrated that rat muscle myoblasts proliferated at a low rate and would yield insufficient tissue for these experiments. Each pool was centrifuged into pellets and transferred to substrata of bone matrix. Cultures were grown for 25–30 days *in vitro*. (Figure 1B from Nathanson *et al.*, 1978.) (A) ×80; (B) ×240.

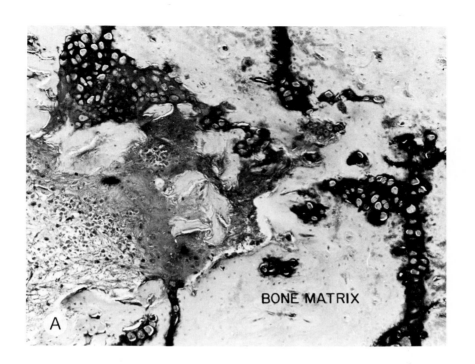

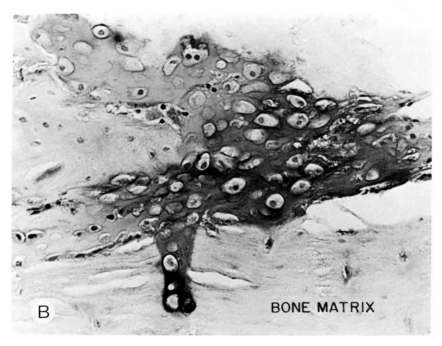

The differentiation of cartilage on bone matrix *in vitro* has been investigated at both the light and electron microscopic levels, using 19-day embryonic rat skeletal muscle as the source tissue (Nathanson and Hay, 1980a). As a control, an aliquot of skeletal muscle used for cultures onto bone matrix was explanted onto gels of type I collagen and cultured identically. The trauma of mincing skeletal muscle prior to cultivation *in vitro* causes syncytial myotubes to degenerate and nuclei of mononucleate cells to enter a heterochromatic "resting" state (Fig. 2). As the cells develop heterochromasia, it becomes increasingly difficult to distinguish myoblasts from fibroblasts. No cell death was detected among mononucleate cells. Heterochromasia was rapidly reversible under favorable culture conditions, and by 24 hours *in vitro* the explanted cells were again euchromatic. However, myoblasts and fibroblasts were still not readily distinguished. Cells having some characteristics of myoblasts appeared essentially fibroblast-like (Fig. 3A and B). Acquisition of fibroblast-like characteristics gradually encompassed the entire mononucleate population and was found to persist through 4 days *in vitro* (Fig. 3C). This is significant in that skeletal muscle cells do not transform directly into chondrocytes, but via a fibroblast-like intermediate. The intermediate is presumably not an

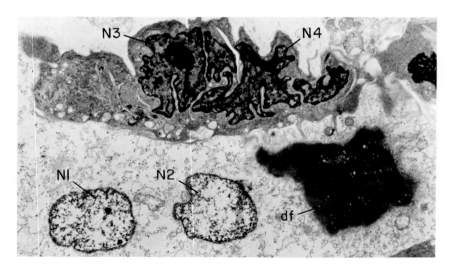

FIG. 2. Electron micrograph of skeletal muscle held *in vitro* and on ice for 4 hours. The early response of skeletal muscle to excision and mincing was found to be degeneration of myotube nuclei (N1, N2) and myofilaments (df) and acquisition of heterochromasia by mononucleate cells (N3, N4). Euchromasia is rapidly regained in organ culture. (From Nathanson and Hay, 1980a.) ×3520.

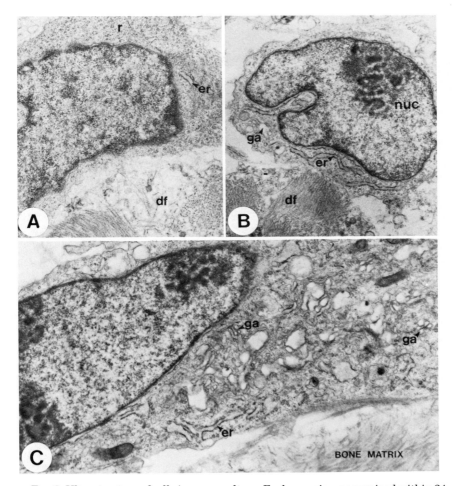

FIG. 3. Ultrastructure of cells in organ culture. Euchromasia was regained within 24 hours *in vitro,* and by 2 days surviving mononucleate cells appeared as myoblasts (A) or as cells with myoblast and fibroblast characteristics (B). Degenerate myofilaments are still present (df). In (B), the slightly heterochromatic nucleus and large ratio of nuclear-to-cytoplasmic area are reminiscent of myoblast morphology as shown in (A). Dispersed nucleoli (nuc), prominent Golgi apparatus (ga), and granular endoplasmic reticulum (er) in (B) resemble fibroblast morphology as shown in (C). Note the presence of abundant free ribosomes (r) in (A). (C) The entire mononucleate population appears fibroblast-like by 3–4 days *in vitro.* (From Nathanson and Hay, 1980a.) (A) ×10,200; (B) ×9200; (C) ×13,900.

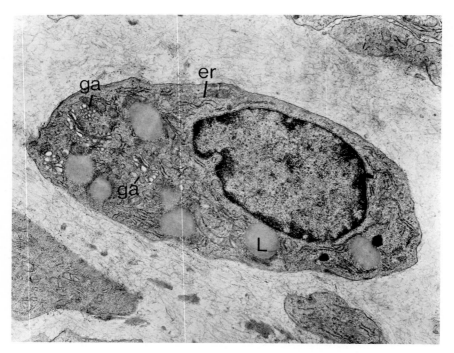

FIG. 4. Differentiation of cartilage on bone matrix. Chondroblasts appear by 6 days *in vitro* on bone matrix and increase in number thereafter. Note the presence of abundant secretory organelles, ga, er (abbreviations as in the legend to Fig. 3), and lipid (L). The extracellular matrix contains beaded, type II collagen fibrils. (From Nathanson, 1983a.) ×13,137.

additional phenotype, but one step in a series of steps during the development of a secretory morphology (i.e., as chondrocytes secrete extracellular matrix macromolecules). These morphological data add weight to the hypothesis that both skeletal muscle and cartilage derive from a common precursor cell type. Beginning on the sixth day, a few cells acquired phenotypic characteristics of chondrocytes and by 10 days masses of cartilage were found (Fig. 4). Control cultures initially regained euchromasia, but myoblasts and fibroblasts were easily discernible. These mononucleate cells began to regenerate skeletal muscle by 4 days *in vitro*. No cartilage formed.

IV. Skeletal Muscle Differentiation and Relationship to Satellite Cells

Even though cartilage has been repeatedly documented to arise from skeletal muscle grown on bone matrix in organ culture, and demonstrated in clonal culture, these studies cannot rule out the occurrence of a myoblast which has the properties of a progenitor cell type. Repeated passage of cloned chick myoblasts failed to yield evidence of undifferentiated myoblasts; the only other myogenic cell type which has putative progenitor properties is the satellite cell. Most investigators would agree that satellite cells are myoblasts, but by definition these myoblasts retain "reserve cell" properties long after the majority of myoblasts fuse to form syncytial myotubes. To resolve the question whether cartilage derives from reserve myoblasts (satellite cells), an ultrastructural study of 11-day embryonic chick skeletal muscle was undertaken. As a working hypothesis it was assumed that all myoblasts are initially identical and that some fuse and some are left behind as satellite cells. The question asked was, at what stage do satellite cells appear? The results of this study demonstrated that satellite cells appear rather late in development, at approximately 14–18 days of development in the chick (Nathanson, 1979). A survey of the literature also failed to reveal evidence of the occurrence of satellite cells until hatching or shortly thereafter (Nathanson and Hay, 1980a).

The relatively immature appearance of 11-day chick skeletal muscle, discovered during the above studies, was accompanied by a large amount of free space between myotubes and relatively few fibroblasts. Furthermore, cells were loosely arranged within the tissue, which appeared to be disorganized on both the light and electron microscopic levels. Since syncytial myotubes begin to form at approximately 5 days of development, and since fibroblasts are thought to proliferate faster than myoblasts, the tissue would be expected to be compact and contain well-developed myotubes and connective tissue. Further study revealed that young myotubes were arranged tangentially with respect to the epimysium. Few myoblasts were found in the mid-myotube regions (Nathanson, 1979). This arrangement is reminiscent of an unipennate morphology and not of the longitudinally arranged pattern associated with the thigh musculature (source tissue). It is unclear whether the embryonal nature of this tissue is a result of its manner of morphogenesis, but it appears that myotubes form by the preferential fusion of myoblasts at the ends of myotubes. A limited number of fusion sites would cause the tissue to develop slowly. This is in contrast to *in vitro* studies in which morphogenesis of the tissue occurs by a burst of fusion, and this fusion is not related to position of the fusing cells with respect to the myotubes.

Myoblasts withdraw from the cell cycle prior to fusion, but this may not be so if the myoblast was destined to become a satellite cell. Could the nonfusing myoblasts in the mid-myotube regions be the first appearance of satellite cells? An autoradiographic study was undertaken to study this question further. Preliminary data demonstrated that all of the myoblasts surrounding a myotube are capable of incorporating tritiated thymidine and therefore do not withdraw from the cell cycle (M. A. Nathanson, unpublished observation). All of the available evidence supports the hypothesis that single myogenic cells are myoblasts, a cell type which normally forms only muscle.

V. Synthesis of Cartilage Extracellular Matrix

A. SULFATED GLYCOSAMINOGLYCANS

One of the principal components of cartilage extracellular matrix are the sulfated glycosaminoglycans (GAG-S). GAG-S are a family of sulfated, high-molecular-weight, linear polysaccharides and two of them account for the majority of the polysaccharide of cartilage matrix: chondroitin 4-sulfate (Ch-4-S) and chondroitin 6-sulfate (Ch-6-S). Since chondrocytes are specialized for matrix secretion, synthesis of GAG-S would be an accurate indication of the extent of chondrogenic differentiation. Furthermore, preliminary experiments indicated that $Na_2{}^{35}SO_4$ was specifically incorporated into GAG-S (Nathanson and Hay, 1980b). Using 19-day embryonic rat skeletal muscle as the source tissue, newly synthesized polysaccharide was labeled for the last 24 hours *in vitro* on bone matrix and on gels of type I collagen as a control. Synthesis of GAG-S was also investigated in the source tissue and in authentic hyaline cartilage from embryos of the same gestational age. As expected, skeletal muscle incorporated very little sulfate into Ch-4-S, whereas cartilage incorporated quite a lot, the ratio of Ch-4-S to Ch-6-S being 0.28 for skeletal muscle and 13.09 for hyaline cartilage.

On bone matrix, sulfate incorporation into GAG-S exceeded that of control cultures as early as 24 hours *in vitro* (Fig. 5). Sulfate incorporation leveled off from days 2 to 3 and rose dramatically to 22 times that of control cultures between days 3 and 10. This pattern correlates well with the morphological data described above; the initial increase occurs as explanted cells regain euchromasia, elaborate rough endoplasmic reticulum, and begin to acquire a fibroblast-like morphology. A plateau in sulfate incorporation coincides with the acquisition of fibroblast-like characteristics, and the subsequent rise coincides with the appearance of chondrocytes.

Quite unexpectedly, it was found that as early as 24 hours after initiation of the cultures, those on both bone matrix and collagen gels synthesized increased proportions of Ch-4-S in comparison to Ch-6-S (Fig. 6). Synthesis of Ch-4-S would be expected to correlate with chondrogenesis. These data demonstrate that a specific inducer is not necessary to alter the biosynthetic potential of skeletal muscle and

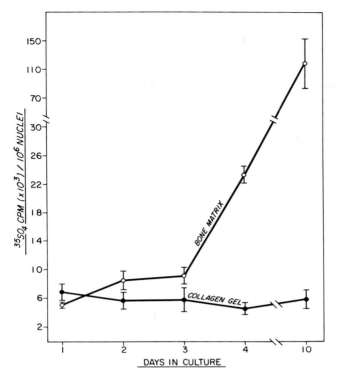

FIG. 5. Incorporation of labeled sulfate into sulfated glycosaminoglycan (GAG-S) by cultures of skeletal muscle on bone matrix and, as a control, on gels of type I collagen. The data indicate that increases in GAG-S synthesis are limited to cultures on bone matrix and occur prior to and during chondrogenesis in this experimental system. (From Nathanson and Hay, 1980b.)

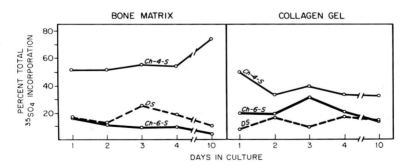

FIG. 6. Analysis of GAG-S composition. GAG-S was enzymatically depolymerized and isolated by descending paper chromatography as modified from Saito et al. (1968). Increases in Ch-4-S are thought to be elicited by trauma. However, only on bone matrix is this synthesis maintained. Increases in Ch-4-S parallel the appearance of chondrocytes on bone matrix. (From Nathanson and Hay, 1980b.)

that an environmental trauma, such as mincing of the skeletal muscle prior to placing it into culture, is sufficient to do so. This has been verified by additional experiments with traumatized skeletal muscle *in vivo* (Nathanson and Hay, 1980b). This finding is profound in that if trauma merely increases GAG-S synthesis, then levels of total sulfate incorportion should rise, but not the ratio of Ch-4-S to Ch-6-S. Cultures on bone matrix maintained high levels of Ch-4-S synthesis, whereas cultures on collagen gels did not. The effect of bone matrix is therefore not to induce Ch-4-S synthesis, but to sustain and augment it. These data should not be interpreted to mean that injured muscle undergoes similar biochemical changes in control and experimental cultures. Controls maintained a low and nearly constant sulfate incorporation, which was only approximately equivalent to that of explants onto bone matrix at 24 hours (Nathanson and Hay, 1980a). However, of this low sulfate incorporation, Ch-4-S represented the greatest initial *proportion* of the chondroitin sulfates. Regenerating skeletal muscle progressively decreased its synthesis of Ch-4-S and it would presumably return to initial levels when this process is complete. Thus, skeletal muscle normally holds Ch-4-S synthesis in abeyance, but retains a surprising readiness for its synthesis.

The experiments described above were performed in nutrient medium CMRL-1066, essentially as described by Nogami and Urist (1974). However, if an external factor, such as trauma, is sufficient to elicit the expression of cartilage in a biochemical sense, it is also likely that the composition of the nutrient medium plays a part in aiding or inhibiting this response. To test such an hypothesis, nutrient medium CMRL-1066 was tested against the commonly used media F-12 and Eagle's minimum essential medium (MEM) (Nathanson, 1983a). At 1, 2, 3, 4, 5, and 10–12 days *in vitro,* cultures were labeled with $Na_2{}^{35}SO_4$ and [^{3}H]glucosamine (to identify nonsulfated glycosaminoglycan) for the final 24 hours prior to harvesting. No differences in GAG-S synthesis were detected. With respect to nonsulfated glycosaminoglycan (GAG), MEM altered GAG synthesis such that it remained at high levels for 24 hours past the time at which its synthesis decreased in medium CMRL-1066 (Fig. 7A). Medium F-12 was found to inhibit synthesis of GAG (Fig. 7B). Histological analysis demonstrated that MEM was inefficient in eliciting chondrogenesis and that F-12 did not support the appearance of chondrocytes within the 12-day experimental period. GAG may be either chondroitin or hyaluronic acid. Following enzymatic depolymerization of GAG, it was determined that little or no chondroitin was synthesized in these cultures. Hyaluronic acid synthesis, however, was dramatically increased in MEM and decreased in

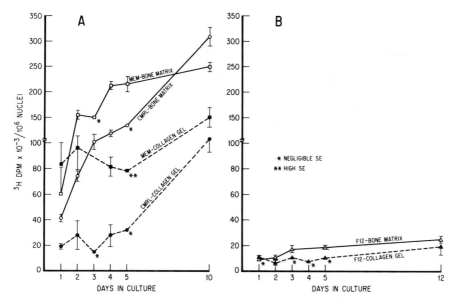

FIG. 7. Incorporation of labeled glucosamine into nonsulfated glycosaminoglycan (GAG). (A) Culture media CMRL-1066 and MEM supported incorporation of the label into GAG. Histological data demonstrated that MEM was inefficient in the production of cartilage on bone matrix. The major difference between these media lies in elevated levels of GAG synthesis between days 2 and 5. (B) Medium F-12 depressed incorporation of the label into GAG and failed to elicit chondrogenesis during the experimental period. (From Nathanson, 1983b.)

F-12 (Fig. 8). These results suggest that nutrient media are capable of altering a cell's metabolic balance, resulting in altered levels of GAG production. These data point out the care with which *in vivo* levels of GAG synthesis should be inferred from *in vitro* levels.

B. HYALURONIC ACID

It has become apparent that synthesis of hyaluronic acid (HA) may be a controlling factor in the production of cartilage extracellular matrix (Toole *et al.*, 1972). Several investigators have determined that *de novo* hyaluronidase synthesis promotes a decrease in HA prior to chondrogenesis (Toole and Gross, 1971; Toole *et al.*, 1972; Solursh *et al.*, 1974; Smith *et al.*, 1975). In the present experiments also, CMRL-1066 promoted a decrease in HA just prior to the onset of chondrogenesis (~6 days; Fig. 8), and these data are consistent with the removal of preexisting HA. Perhaps the remaining HA acts to inhibit chondrogenesis?

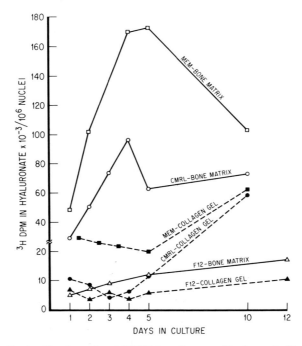

FIG. 8. Synthesis of hyaluronic acid (HA) in culture media shown in Fig. 7. Increased HA synthesis with MEM and depressed synthesis in F-12 represent divergent effects upon chondrogenesis. Data on proteoglycan synthesis suggest that increased HA synthesis and depressed chondrogenesis in MEM is correlated with the presence of molecular forms which cannot form cartilage extracellular matrix. Decreased HA synthesis in F-12 is thought to reflect altered cellular glucosamine metabolism. (From Nathanson, 1983b.)

Such an inhibition may be due to (1) synthesis of HA which cannot participate in formation of cartilage matrix or (2) feedback inhibition of high-molecular-weight HA synthesis. In an attempt to distinguish between the above alternatives, proteoglycans were isolated from a similar series of cultures, grown for 12 days in the above media, and fractionated by molecular sieve chromatography on Sepharose CL-2B.

C. CARTILAGE PROTEOGLYCAN

An extract of cartilage proteoglycan can be separated into three fractions by molecular sieve chromatography. The fraction of highest molecular size elutes at the column void volume and has been identified as an aggregate of hundreds of proteoglycan monomers with high-molecular-weight HA (Hardingham and Muir, 1972; Hascall and

Heinegard, 1974). The intermediate molecular size fraction elutes in the included volume and is generally believed to be composed of unaggregated proteoglycan monomers. The fraction of lowest molecular size is also an unaggregated monomer; however, its occurrence and chemical composition are not unique to cartilage, and it has been designated as a ubiquitous monomer (Levitt and Dorfman, 1973; Palmoski and Goetinck, 1972). The ubiquitous monomer occurs in very low concentration, typically 5% of a cartilage extract (Nathanson, 1983b); experimental analysis is difficult and little evidence is available concerning its structure. However, data from this investigator's laboratory have shown that the ubiquitous monomer may resemble proteoglycans of the cartilage type. To avoid confusion among proteoglycans, this material will be referred to as "low molecular size."

Embryonic sternal cartilage served as a control for these studies. This cartilage contained (1) proteoglycan aggregate, which eluted at the column void volume and comprised 15% of total proteoglycan, (2) proteoglycan monomer, which eluted in the included volume, and (3) material of lower molecular size, which contained greater amounts of a glucosamine label than a sulfate label (i.e., GAG-S poor). Cultures grown in CMRL-1066 produced the greatest amounts of proteoglycan aggregate, as expected. Cultures grown in MEM produced proteoglycan aggregate in amounts intermediate between those grown in CMRL-1066 and F-12. However, cultures grown in both MEM and F-12 incorporated less glucosamine than sulfate into this proteoglycan. In MEM, labeled glucosamine that would be expected to appear in the aggregate fraction appeared to comigrate with material of lower molecular size. These data establish that less high-molecular-weight HA is elicited in MEM than in other media and support the hypothesis that environmental factors impact upon the differentiation of cartilage.

The initial observation that decreases in HA synthesis correlate with chondrogenesis must now be reinterpreted. It is clear that decreases in low-molecular-weight HA favor chondrogenesis. But, how does this explain the data with medium F-12? Proteoglycan profiles from F-12 based cultures were quite similar to those of CMRL-1066, with only small amounts of proteoglycan aggregate and no apparent increase in "low-molecular-size" proteoglycan. This divergent alteration appears to be related to decreased uptake of the [^{3}H]glucosamine precursor. Thus, environmental factors may exhibit pleiotropic effects upon cellular metabolism.

Proteoglycan synthesis was investigated further at prechondrogenic (5 days *in vitro*) and postchondrogenic (6–12 days) stages using standard culture conditions in medium CMRL-1066 (Nathanson,

1983b). The data demonstrated that cultures on bone matrix contained cartilage-typical proteoglycan monomer as early as 5 days (Fig. 9). Approximately 65% of the GAG-S occurred as monomer and 5% as aggregate. However, 12-day cultures contained no greater amounts of GAG-S in either proteoglycan pool. This finding contrasts markedly with the expectation that a more extensive extracellular matrix, which histologically separates 12- from 5-day samples, would contain greater amounts of proteoglycan aggregate. Proteoglycan of authentic rat cartilage incorporates 15% of a sulfate label into aggregate. Perhaps ag-

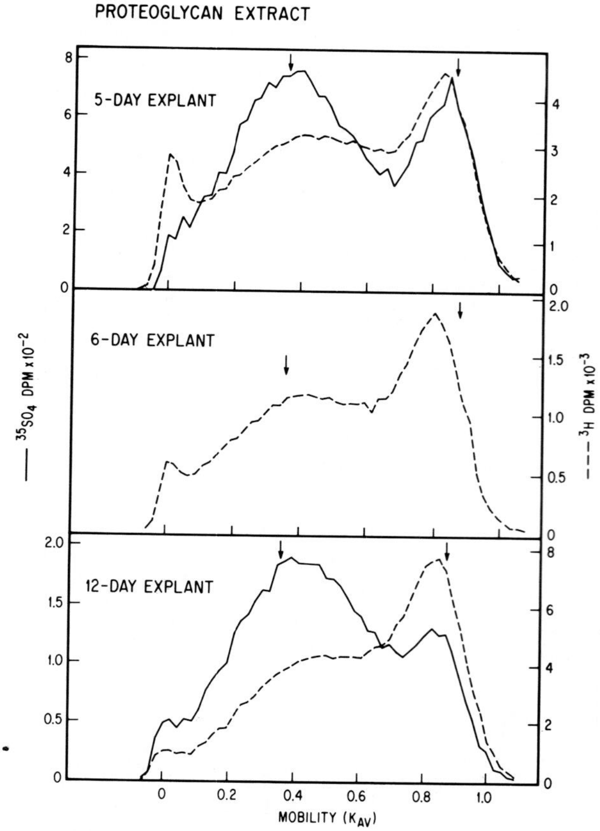

FIG. 9. Sepharose CL-2B chromatogram of proteoglycans extracted from the source skeletal muscle and cultures on bone matrix. Proteoglycan monomer (arrow at K_{av} of 0.34) was present prior to chondrogenesis. However, proteoglycan aggregate (eluting at K_{av} of 0) was present in only low amounts until later stages (note increased amounts of sulfate label at K_{av} of 0 at 12 days). (From Nathanson, 1983c.)

gregation of monomer with HA is not well advanced, even in 12-day cultures. Figure 9 shows that glucosamine-labeled material occurs at the position of monomer and in material of low molecular size. Further, the label never comprises a significant proportion of void volume material. Here again the data suggest that *in vitro* samples are limited in their ability to aggregate by the availability of high-molecular-weight HA. When HA was added to identical proteoglycan samples, 33% of the GAG-S appeared at the void volume. While the lack of HA synthesis is presently not understood, bone matrix clearly elicits the rapid and *de novo* synthesis of cartilage extracellular matrix components from a nonchondrogenic source tissue. Analyses of GAG-S chain length and chemical composition have shown that the proteoglycan is chemically indistinguishable from that of authentic cartilage.

The existence of a large pool of low-molecular-size proteoglycan deserves further comment. These fractions, isolated by guanidine extraction of freshly isolated embryonic skeletal muscle, are composed of species which label primarily with [^{3}H]glucosamine. This observation and their small size correlate with the occurrence of glycopeptides rather than proteoglycan. However, after 12 days in culture on a collagen gel, GAG-S comprises one-half to one-third of its total sulfate label and the GAG-S is composed primarily of chondroitin 6-sulfate; this environmentally induced change appears to be an enhancement of previous biosyntheses. On bone matrix, GAG-S of low-molecular-size material is chondroitin 4-sulfate, and increasing proportions bind to HA between 5 and 12 days. Thus, the stimulus to form cartilage affects all classes of GAG-S-containing material. It is likely that the aggregating low-molecular-size material is a small proteoglycan and responsible for a portion of the heterogeneity of newly differentiating proteoglycans. Heterogeneity may reflect a functional change in the extracellular matrix from one which mediated cell attachment to one of a supportive role.

VI. Origin of the Stimulus to Form Cartilage

The data presented above demonstrate that bone matrix clearly elicits formation of cartilage from a nonchondrogenic source tissue. It is appropriate to consider whether bone matrix contains a discrete factor which "induces" cartilage or whether it merely provides a microenvironment which favors secretion of extracellular matrix. A large body of data from the laboratory of Urist and colleagues suggested that bone matrix contains a diffusible substance responsible for the "activity" of bone matrix. The data were based largely upon transfilter experiments, in which bone matrix was implanted *in vivo* in a

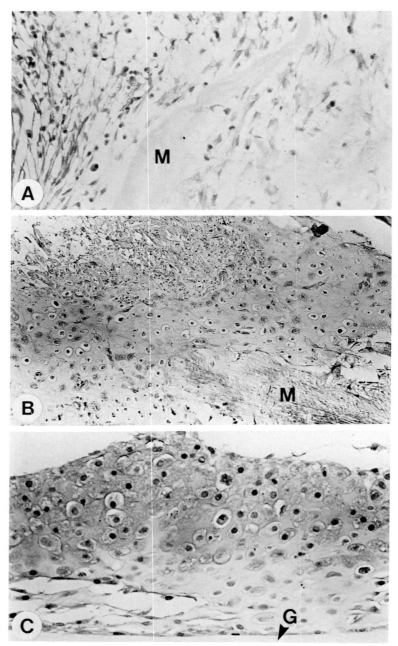

FIG. 10. Reconstitution of chondrogenic activity of bone matrix. (A) Bone matrix (M) can be inactivated by treatment with 6.0 *M* guanidine hydrochloride. (B) Alcoholic precipitation of guanidine extracts of bone matrix results in complete restoration of activity. (C) Precipitation of guanidine extracts on gels of type I collagen converts a

diffusion chamber. Under these conditions, adjacent host skeletal muscle formed regions of new bone (Nogami and Urist, 1975; Urist *et al.*, 1977). More recent data from a number of laboratories showed that guanidine extraction of bone matrix results in an inactive, residual, collagenous matrix and a complex extract, which retains the ability to elicit cartilage and bone when reconstituted with its residue or presented in an insoluble form (Takaoka *et al.*, 1981; Sampath and Reddi, 1981; Urist *et al.*, 1982; Mizutani and Urist, 1982; Yoshikawa *et al.*, 1984). A number of extracellular glycoproteins such as chondronectin, laminin, and fibronectin have been shown to mediate chondrogenesis or interactions between cells and a collagenous extracellular matrix (see Hay, 1983). Thus, it is equally likely that components of the extract mediate attachment to collagen and that this novel environment leads, via a presently unknown mechanism, to chondrogenesis.

Interactions between skeletal muscle, bone matrix, and the well-characterized glycoproteins listed above were tested by adding them to the growth medium of skeletal muscle in the standard *in vitro* system. At a concentration of 40 μg/ml, no effect of any glycoprotein was noted, irrespective of the presence of bone matrix or its substitution with a collagen gel (M. A. Nathanson, unpublished data). Further, identical concentrations of their corresponding antibodies were also without effect. If a guanidine extract acts in a similar fashion, its activity is not additive and it appears immunologically distinct from chondronectin, laminin, and fibronectin.

Perhaps, guanidine extracts do not bind to the collagen of cartilage extracellular matrix. Several laboratories have shown that extractable components of bone matrix are able to elicit sulfate incorporation into GAG-S and immunoreactive cartilage proteoglycan when the extract is simply added to a culture medium (Seyedin *et al.*, 1983; Sato and Urist, 1984; Styftestad and Caplan, 1984). In this fashion, the extract may act as a novel growth factor (see Urist *et al.*, 1983, for a discussion of growth factors). Additional experiments with collaborators T. K. Sampath and A. H. Reddi of the National Institutes of Health were designed to test the ability of soluble extracts to elicit chondrogenesis (Sampath *et al.*, 1984). In the *in vitro* assay, the chondrogenic activity of bone matrix residue could be completely restored by precipitation of the guanidine extracts upon the residue (Fig. 10A and B). Fractionation of the extract by molecular sieve chromatography resulted in four

---

previously negative control substratum (G, collagen gel) into one with chondrogenic activity. The data suggest that the "activity" of bone matrix is separable from and unrelated to the presence of bone collagen. (From Sampath *et al.*, 1984.) (A) ×320; (B) ×260; (C) ×320.

fractions, with activity residing in the last peak (peak IV; Sampath and Reddi, 1983). Precipitation of peak IV upon the residue appeared equally as effective in restoring activity. Activity in the present context was defined as the ability to elicit cartilage in sectioned and stained material. These results confirm and extend the data of previous reports.

In order for the extracted material to be a removable activity, it is also necessary to show that it is active in the absence of the residue. Thus, guanidine extracts were precipitated directly upon a control substratum of type I collagen. Under these conditions, the extract was capable of eliciting chondrogenesis (Fig. 10C), although we were unable to duplicate the data with Sepharose-fractionated material. In addition to a histological demonstration of chondrogenesis, the reconstituted collagen gels were shown to contain proteoglycan which had the chromatographic profile of cartilage extracellular matrix.

VII. Conclusions

Bone matrix has been shown to be a vehicle for the transmission of a factor or factors which cause nonchondrogenic cells to apparently transform into chondrocytes. "Transformation" is not used in the classic microbiological sense in that bone matrix appears not to impart genetic information. Rather, it appears to activate genes which are present, but not expressed. The observation that all cell types are not responsive (i.e., ectoderm and endoderm) is consistent with this view.

A more rigorous definition of the observed effect is that responsive cells "differentiate" into chondrocytes. But here again the term is not used in the classical sense, insofar as a diverse array of respondent cell types (cells derived from somatopleure, splanchnopleure, and somite) are previously differentiated. This poses a dilemma; reference to the mode of action of the signal may be of use in its resolution. Extracts of demineralized bone contain a material with the chemical properties of a glycoprotein, although final characterization must await further purification of extracted material. The extracts "stabilize" a response (synthesis of GAG-S) which appears to be nonspecific, even though the source tissue is differentiated to the extent that it will never form cartilage in the absence of extract. Stabilization, then, reflects the acquisition of histological and/or biochemical characteristics which indicate differentiation of the tissue. In the final analysis it is clear that responsive cells do not actually transform, and they may be differentiated long prior to the application of a signal to change their phenotype. The most appropriate definition of their response is that they have continued to differentiate within constraints imposed by an early and generalized primary stimulus.

ACKNOWLEDGMENT

This work was supported by United States Public Health Service, National Institutes of Health Grants AM-28240 and AM-01040.

REFERENCES

Bischoff, R., and Holtzer, H. (1969). *J. Cell Biol.* **41,** 188–200.
Bonner, P. H., and Hauschka, S. D. (1974). *Dev. Biol.* **37,** 317–328.
Buckley, P. A., and Konigsberg, I. R. (1977). *Proc. Natl. Acad. Sci. U.S.A.* **74,** 2031–2035.
Chevallier, A., Kieny, M., and Mauger, A. (1977). *J. Embryol. Exp. Morphol.* **41,** 245–258.
Christ, B., Jacob, H. L., and Jacob, M. (1977). *Anat. Embryol.* **150,** 171–186.
Coon, H. G. (1965). *Proc. Natl. Acad. Sci. U.S.A.* **55,** 66–73.
Coon, H. G., and Cahn, R. D. (1966). *Science* **153,** 1116–1119.
de la Haba, G., Kamali, H. M., and Tiede, D. M. (1975). *Proc. Natl. Acad. Sci. U.S.A.* **72,** 2729–2732.
Dienstman, S. R., Biehl, J., Holtzer, S., and Holtzer, H. (1974). *Dev. Biol.* **39,** 83–95.
Hall, B. K. (1970). *Biol. Rev.* **45,** 455–484.
Hardingham, T. E., and Muir, H. (1972). *Biochim. Biophys. Acta* **279,** 401–405.
Hascall, V. C., and Heinegard, D. (1974). *J. Biol. Chem.* **249,** 4232–4241.
Hauschka, S. D. (1974). *Dev. Biol.* **37,** 329–344.
Hauschka, S. D., and Konigsberg, I. R. (1966). *Proc. Natl. Acad. Sci. U.S.A.* **55,** 119–126.
Hay, E. D. (1983). *In* "Modern Cell Biology" (J. R. McIntosh, ed.), pp. 509–548. Liss, New York.
Hermann, H., Heywood, S. M., and Marchok, A. C. (1970). *Curr. Top. Dev. Biol.* **5,** 181–234.
Hilfer, S. R., Searls, R. L., and Fonte, V. G. (1973). *Dev. Biol.* **30,** 374–391.
Holtzer, H., Rubenstein, N., Dienstman, S., Chi, J., Biehl, J., and Somyle, A. (1974). *Biochimie* **53,** 1575–1580.
Holtzer, M., and Sanger, J. W. (1972). *In* "Research in Muscle Development and the Muscle Spindle" (B. Q. Banker *et al.,* eds.), pp. 122–132.
Kieny, M., Pautou, M.-P., and Chevallier, A. (1981). *Arch. Anat. Microsc. Morphol. Exp.* **70,** 81–90.
Konigsberg, I. R. (1963). *Science* **140,** 1273–1284.
Konigsberg, I. R. (1970). *In* "Chemistry and Molecular Biology of the Intercellular Matrix" (E. A. Balasz, ed.), Vol. 3, pp. 1779–1810. Academic Press, New York.
Konigsberg, I. R. (1971). *Dev. Biol.* **26,** 133–152.
Levitt, D., and Dorfman, A. (1973). *Proc. Natl. Acad. Sci. U.S.A.* **70,** 2201–2205.
McLachlan, V. C., and Hornbruch, A. (1979). *J. Embryol. Exp. Morphol.* **54,** 209–217.
Mizutani, H., and Urist, M. (1982). *Clin. Orthop.* **171,** 213–223.
Nathanson, M. A. (1979). *In* "Regeneration of Striated Muscle" (A. Mauro, ed.), pp. 99–107. Raven, New York.
Nathanson, M. A. (1983a). *In* "Limb Development and Regeneration" (R. O. Kelley, P. F. Goetnick, and J. A. MacCabe, eds.), Part B, pp. 215–227. Liss, New York.
Nathanson, M. A. (1983b). *Dev. Biol.* **96,** 42–62.
Nathanson, M. A. (1983c). *J. Biol. Chem.* **258,** 10325–10334.

Nathanson, M. A., and Hay, E. D. (1980a). *Dev. Biol.* **78**, 301–331.
Nathanson, M. A., and Hay, E. D. (1980b). *Dev. Biol.* **78**, 332–351.
Nathanson, M. A., Hilfer, S. R., and Searls, R. L. (1978). *Dev. Biol.* **64**, 99–117.
Nogami, H., and Urist, M. R. (1974). *J. Cell Biol.* **62**, 510–519.
Nogami, H., and Urist, M. R. (1975). *Calcif. Tissue Res.* **19**, 153–163.
Okada, T. S. (1980). *Curr. Top. Dev. Biol.* **16**, 349–380.
Okazaki, K., and Holtzer, H. (1966). *Proc. Natl. Acad. Sci. U.S.A.* **56**, 1484–1490.
O'Neill, M. C., and Stockdale, F. E. (1972a). *Dev. Biol.* **29**, 410–418.
O'Neill, M. C., and Stockdale, F. E. (1972b). *J. Cell Biol.* **52**, 52–65.
Palmoski, M. J., and Goetnick, P. F. (1972). *Proc. Natl. Acad. Sci. U.S.A.* **69**, 3385–3388.
Ramirez, O., and Aleman, V. (1972). *J. Embryol. Exp. Morphol.* **28**, 559–570.
Reddi, A. H., and Huggins, C. B. (1973). *Proc. Soc. Exp. Biol. Med.* **143**, 634–667.
Richler, C., and Yaffe, D. (1970). *Dev. Biol.* **23**, 1–22.
Saito, H., Yamagata, T., and Suzuki, S. (1968). *J. Biol. Chem.* **243**, 1536.
Sampath, T. K., and Reddi, A. H. (1981). *Proc. Natl. Acad. Sci., U.S.A.* **78**, 7599–7603.
Sampath, T. K., and Reddi, A. H. (1983). *Proc. Natl. Acad. Sci. U.S.A.* **80**, 6591–6595.
Sampath, T. K., Nathanson, M. A., and Reddi, A. H. (1984). *Proc. Natl. Acad. Sci. U.S.A.* **81**, 3419–3423.
Sato, K., and Urist, M. R. (1984). *Clin. Orthop. Relat. Res.* **183**, 180–187.
Searls, R. L., and Janners, M. Y. (1969). *J. Exp. Zool.* **170**, 356–376.
Seyedin, S. M., Thompson, A. Y., Rosen, D. M., and Piez, K. A. (1983). *J. Cell Biol.* **97**, 1950–1953.
Shainberg, A., Yagil, G., and Yaffe, D. (1969). *Exp. Cell Res.* **58**, 163–167.
Smith, G. N., Toole, B. P., and Gross, J. (1975). *Dev. Biol.* **43**, 221–232.
Solursh, M., Vaerewyck, S. A., and Reitner, R. S. (1974). *Dev. Biol.* **41**, 233–244.
Styftestad, G. T., and Caplan, A. H. (1984). *Dev. Biol.* **104**, 348–356.
Takaoka, K., Yoshikawa, H., Shimizu, N., Ono, K., Amitani, K., Nakata, Y., and Sakamoto, Y. (1981). *Biomed. Res.* **2**, 466–471.
Toole, B. P., and Gross, J. (1971). *Dev. Biol.* **25**, 57–77.
Toole, B. P., Jackson, G., and Gross, J. (1972). *Proc. Natl. Acad. Sci. U.S.A.* **69**, 1384–1386.
Urist, M. R. (1965). *Science* **150**, 893–895.
Urist, M. R., Granstein, R., Nogami, H., Svenson, L., and Murphy, R. (1977). *Arch. Surg.* **112**, 612–619.
Urist, M. R., Lietze, A., Mizutani, H., Takagi, K., Triffitt, J. T., Amstutz, J., DeLange, R., Termine, J., and Finerman, G. A. M. (1982). *Clin. Orthop.* **162**, 219–232.
Urist, M. R., DeLange, R. J., and Finerman, G. A. M. (1983). *Science* **20**, 680–686.
White, N. K., and Hauschka, S. D. (1971). *Exp. Cell Res.* **67**, 479–482.
Wolpert, L. (1969). *J. Theor. Biol.* **25**, 1–47.
Yaffe, D. (1971). *Exp. Cell Res.* **66**, 33–48.
Yoshikawa, H., Takaoka, K., Shimizu, N., and Ono, K. (1984). *Clin. Orthop.* **182**, 231–235.
Zwilling, E. (1966). *Ann. Med. Exp. Fenn.* **44**, 134–139.

CHAPTER 4

TRANSDIFFERENTIATED HEPATOCYTES IN RAT PANCREAS

M. Sambasiva Rao, Dante G. Scarpelli, and Janardan K. Reddy

DEPARTMENT OF PATHOLOGY
NORTHWESTERN UNIVERSITY MEDICAL SCHOOL
CHICAGO, ILLINOIS

I. Introduction

Cell differentiation involves the process by which progenitor cells having a common genotype give rise to a spectrum of cells with different morphological and biochemical properties that characterize the fully developed organism (Jacob and Monod, 1963). Once the cells are fully differentiated they acquire the ability to synthesize specific proteins and perform specialized functions. It is generally believed that cell differentiation is accomplished through the differential recruitment of genes and the expression of their products, since nuclei of all the cells in a multicellular organism possess the same complement of genes as the fertilized ovum (Schmid and Alder, 1984). Although earlier investigators (Grobstein, 1959) postulated that the process of differentiation is an irreversible phenomenon, recent evidence strongly suggests that fully differentiated cells can, with the proper stimulus, change their commitment and convert into an entirely different phenotype. The process of conversion of one differentiated cell type into a totally different cell type is termed "transdifferentiation" or "metaplasia" (Okada, 1980). Classical examples of transdifferentiation include conversion of pigmented epithelial cells of iris into lens in the newt, fish, and human fetal eye tissue (Yamada, 1982; Eguchi *et al.*,

63

1981; Yasuda *et al.*, 1978) and conversion of connective tissue or muscle into cartilage (Reddi and Huggins, 1975; Nathanson, this volume). A more frequently encountered example of *in vivo* transdifferentiation is the squamous metaplasia affecting the lining epithelium of mucous glands or that of the major branches of the respiratory tree which undergo squamous change in response to protracted physical or chemical injury (DiBerardino *et al.*, 1984). An unusual type of *in vivo* transdifferentiation in the pancreas of adult Syrian golden hamsters, leading to the emergence of cells resembling normal hepatocytes, was first described by Rao and Scarpelli (1980). These cells possess the characteristic morphology of liver cells and contain glycogen and albumin (Scarpelli and Rao, 1981). Rao and co-workers (1982) have shown that the hepatocyte-like cells induced in hamster pancreas also respond to the peroxisome proliferators by synthesis of peroxisomal enzymes and proliferation of peroxisomes. Further, these cells regenerate following partial hepatectomy and respond to phenobarbital by augmented synthesis of arylhydrocarbon hydroxylase and proliferation of smooth endoplasmic reticulum as do normal hepatocytes (Rao *et al.*, 1983).

The presence of hepatocyte-like cells in rat pancreas was first noted by Lalwani *et al.* (1981) in a rat treated with Wy 14643, i.e., [4-chloro-6-(2,3-xylidino)-2-pyrimidinylthio]acetic acid, a peroxisome proliferator. Recently, we observed that ciprofibrate, i.e., 2-[4-(2,2-dichlorocyclopropyl)phenoxy]2-methylpropionic acid, another peroxisome proliferator, also induces the development of hepatocytes in the pancreas of adult male rats (Reddy *et al.*, 1984). During the course of these studies, it was found that hepatocytes were induced in the pancreas of all rats during reversal of experimentally induced copper-depleted pancreatic acinar cell atrophy, following 4-hydroxyaminoquinoline 1-oxide injection (Rao *et al.*, 1985a). In this chapter, we briefly review the model of hepatocyte conversion in rat pancreas and present morphological and some functional aspects of these cells.

II. Induction of Pancreatic Hepatocytes in Rats

The induction of hepatocytes in the pancreas of adult rats has been achieved by using the following experimental procedures.

A. Ciprofibrate Model

Transdifferentiated hepatocytes in the pancreas were observed in rats fed ciprofibrate, a peroxisome proliferator (Reddy *et al.*, 1984).

Fig. 1. An island of hepatocytes (HC) in the pancreas of a male rat fed ciprofibrate for 60 weeks. The hepatocytes, each with a prominent central nucleus and large eosinophilic cytoplasm, are surrounded by exocrine acinar (AC) tissue. Hematoxylin–eosin. ×211.

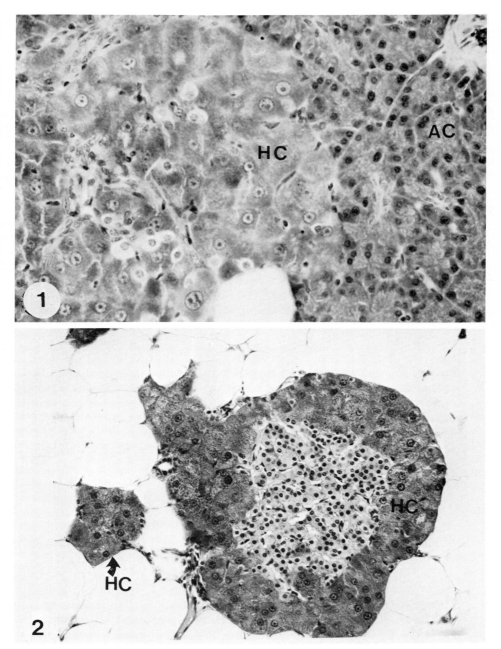

FIG. 2. Pancreas of a rat at 15 weeks of recovery from copper depletion. Atrophy of exocrine acinar tissue and fatty infiltration are prominent. The transdifferentiated hepatocytes (HC) form a two-to five-cell-layer-thick collar around an islet of Langerhans. Hematoxylin–eosin. ×182.

When ciprofibrate was added to the diet at a dosage of 10 mg/kg body weight, ~25% of the rats developed one or several foci of hepatocytes in the pancreas at 60 weeks (Fig. 1). The induced hepatocytes persisted in animals that were killed 12 weeks after the withdrawal of drug at 60 weeks. The hepatocytes were usually localized adjacent to islets of Langerhans with extensions into the surrounding acinar tissue. Unlike in the copper depletion–repletion model (see below), the remaining pancreas showed no evidence of acinar atrophy or fatty infiltration. The islands of hepatocytes in the pancreas can be easily identified by examination, under the dissecting microscope, of glutaraldehyde-fixed tissues incubated in alkaline 3,3'-diaminobenzidine (DAB) medium (Reddy et al., 1984).

B. COPPER DEPLETION–REPLETION MODEL

In the copper depletion–repletion model of hepatocyte trans-differentiation (Rao et al., 1985a), male rats were placed on a copper-deficient diet for 10 weeks, 22 weeks after a single iv injection of 4-hydroxyaminoquinoline 1-oxide (4-HAQO). The surviving rats were then placed on a normal diet. Groups of rats were sacrificed at 3-week intervals following their return to the normal diet for a subsequent 18-week period. The pancreas showed a marked loss of acinar tissue and severe fatty infiltration. Pancreata of all rats killed at 6 weeks or later during repletion contained at least several foci of hepatocytes (Fig. 2). At 6 weeks only a few foci of hepatocytes were present; however, the number and size of the foci increased with time. No hepatocytes were present in the pancreas of rats treated with 4-HAQO alone. A detailed characterization of this model may elucidate the role of copper deficiency in the development of pancreatic hepatocytes. Recent evidence clearly demonstrates that a simple copper depletion and repletion, without 4-HAQO administration, results in the induction of hepatocytes in rat pancreas (Rao et al., 1986).

III. Morphology of Pancreatic Hepatocytes

The hepatocytes observed in the pancreas of rats fed ciprofibrate or seen during recovery from copper depletion are morphologically indistinguishable from hepatic hepatocytes. They are usually arranged in groups of a few cells to several cells (Figs. 1 and 2). No obvious single cell trabecular pattern, such as that seen in adult rat liver, is discernible. The hepatocytes are polyhedral and measure about 30–35 μm in diameter. These cells display a centrally located round nucleus containing a prominent nucleolus. The cytoplasm is eosinophilic and finely granular. The majority of the hepatocytes stain strongly positive with the periodic acid–Schiff (PAS) stain for glycogen (Fig. 3). These

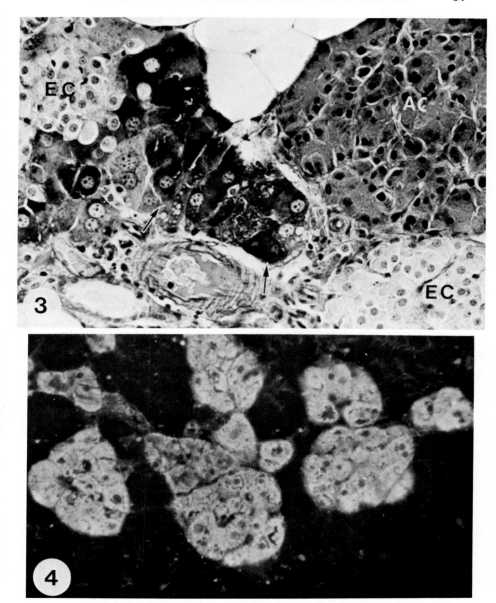

Fig. 3. Hepatocytes in the rat pancreas (arrows) are intensely stained by the periodic acid–Schiff (PAS) stain, whereas the endocrine cells (EC) and acinar cells (AC) are negative. Hematoxylin–eosin–PAS stain. ×198.

Fig. 4. Indirect immunofluorescence microscopy with antibody against albumin of rat pancreas at 15 weeks of recovery from copper depletion. Clusters of transdifferentiated hepatocytes show positive staining for albumin. ×139.

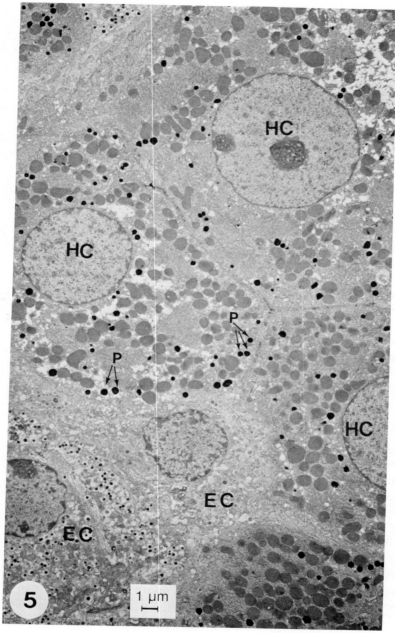

FIG. 5. Rat pancreas with transdifferentiated hepatocytes (HC). The animal was killed 12 weeks after the discontinuation of ciprofibrate treatment. The induced phenotype appears irreversible. The tissue was incubated in alkaline 3,3′-diaminobenzidine medium for the cytochemical localization of peroxisomal catalase. EC, Endocrine cell. Peroxisomes (P) reveal positive reaction product.

cells also contain albumin, as shown by strong immunofluorescence staining with anti-albumin antibodies (Fig. 4). Peroxisomal catalase is also demonstrable in these hepatocytes. Pancreatic secretory proteins, such as α-amylase and carboxypeptidase A, are not detectable in pancreatic hepatocytes by indirect immunofluorescence, whereas the adjacent acinar cells yield an intensely positive immunofluorescence staining.

By electron microscopy, the pancreatic hepatocytes show all the morphologic features characteristic of adult rat liver parenchymal cells. The nuclei contain evenly dispersed heterochromatin with a prominent centrally placed nucleolus. The cytoplasm contains short stacks of rough endoplasmic reticulum, a prominent Golgi complex, vesicles of smooth endoplasmic reticulum, many oval to round mitochondria, glycogen particles, lysosomes, and peroxisomes with the characteristic rat hepatocyte-specific nucleoids (Fig. 5). Well-developed bile canaliculi, with projections of microvilli, are seen between adjacent hepatocytes; the remaining intercellular surface is smooth without any complex cytoplasmic projections. However, the plasmalemmal surface exposed to vascular spaces has multiple microvilli. Between some vascular spaces and hepatocytes a prominent basement membrane is visible, while in others no basement membrane is noted.

Kupffer cells and other nonparenchymal cells of the liver such as Ito cells are not observed in the foci of pancreatic hepatocytes.

IV. Induction of Peroxisome Proliferation in Pancreatic Hepatocytes

Administration of peroxisome proliferators, a group of structurally unrelated xenobiotics, to rodents results in marked hepatomegaly, proliferation of peroxisomes, induction of peroxisome-associated enzymes, such as catalase, fatty acid β-oxidation enzymes, and carnitine acetyl transferase, and hepatocellular carcinomas (Reddy et al., 1980, 1982a; Reddy and Lalwani, 1983). Since the induction of peroxisome proliferation is a tissue-specific phenomenon (Reddy and Lalwani, 1983; Rao et al., 1984b), it appeared particularly relevant to assess the ability of pancreatic hepatocytes to recognize and respond to a peroxisome proliferator. Administration of ciprofibrate to rats containing pancreatic hepatocytes resulted in a 9-fold increase in the volume density of peroxisomes in pancreatic hepatocytes (Fig. 6). After discontinuation of ciprofibrate feeding the peroxisome volume density returned to normal level. Peroxisomes in pancreatic hepatocytes showed typical crystalloid nucleoids (inset, Fig. 6), characteristic of peroxisomes of hepatocytes of rat liver. Immunofluorescence staining with appropriate antibodies showed intense staining for catalase and peroxisomal enoyl-CoA hydratase (Fig. 7), signifying specific induction of perox-

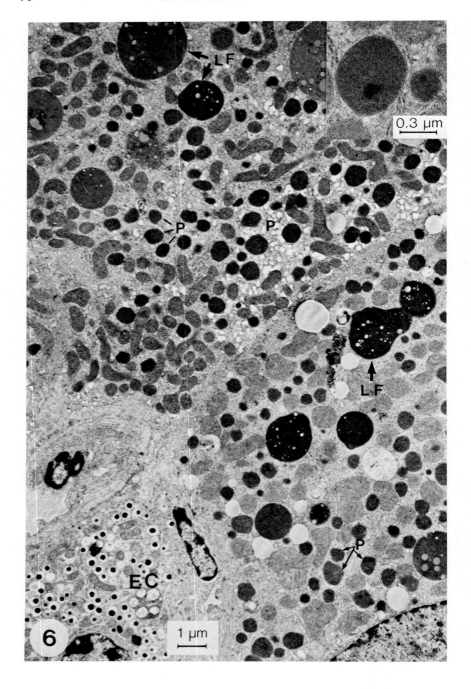

isome-associated enzymes. Rao *et al.* (1982) have previously demonstrated that pancreatic hepatocytes in hamsters also respond to the peroxisome-proliferative effect of methyl clofenapate. Pancreatic hepatocytes in rats continuously maintained on a ciprofibrate-containing diet (60 weeks) revealed an excessive amount of lipofuscin (Fig. 6) together with a significant increase in the volume density of peroxisomes, similar to that observed in the hepatocytes of liver of rats after chronic peroxisome proliferator treatment (Reddy *et al.,* 1982b).

V. Stability of Pancreatic Hepatocytes

Once the pancreatic hepatocytes develop, the resulting phenotype appears to be permanent and does not revert back to pancreatic cells. In experiments in which pancreatic hepatocytes were induced by ciprofibrate, the number and size of hepatic foci in pancreas persisted even after 12 weeks of drug withdrawl. Similarly, in the copper depletion–reversal model, hepatocytes in the pancreas emerge as early as 6 weeks after the rats are placed on a normal diet, and these hepatic foci continue to increase in number and size for up to several weeks. Based on these observations, it is reasonable to assume that the induced pancreatic hepatocytes in the rat are stable and apparently retain the induced phenotype indefinitely. However, lifelong animal studies are required to corroborate this assumption. In the hamster pancreas, the hepatocytes developed as early as 3 weeks after a single injection of *N*-nitrosobis(2-oxopropyl)amine during pancreatic regeneration and persisted up to 50 weeks (Scarpelli and Rao, 1981; M. S. Rao, unpublished results).

VI. Histogenesis of Pancreatic Hepatocytes

Transformation of one cell type to another type, although it seems to be a rare event, is a well-established phenomenon. The question whether the second cell type is arising from a stem cell or from a well-differentiated stable cell is not satisfactorily answered. The cells in a multicellular organism are divided into three types: "labile," "stable," and "permanent" cells, depending on their regenerative capacity. In organs such as gastrointestinal tract, bronchus, bone marrow, where stem cells have been demonstrated, it may be argued that a trans-differentiated cell may arise from a stem cell although "true" cellular

FIG. 6. Pancreatic hepatocyte in a rat fed ciprofibrate for 60 weeks. Note the presence of numerous peroxisomes (P) and accumulation of lipofuscin (LF) in these hepatocytes. The presence of lipofuscin in these cells reflects possible oxidative stress resulting from sustained proliferation of peroxisomes and the induction of H_2O_2-generating peroxisomal enzyme(s). EC, Endocrine cell.

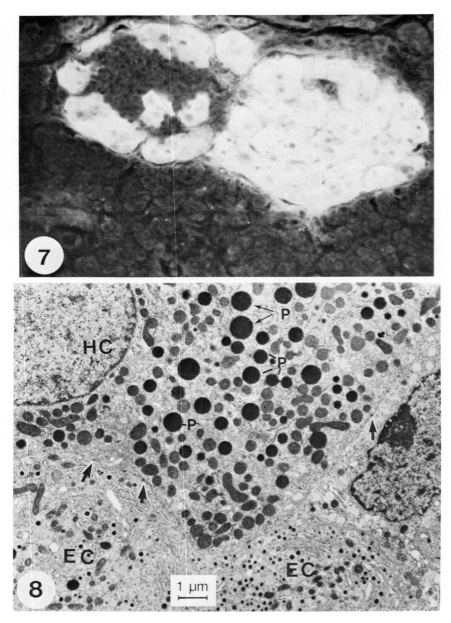

FIG. 7. Pancreatic hepatocytes induced by copper depletion–repletion protocol. Indirect immunofluorescence localization of peroxisome proliferation-associated 80,000-MW polypeptide (peroxisomal enoyl-CoA hydratase–dehydrogenase, the bifunctional enzyme of the peroxisomal fatty acid β-oxidation system) in the pancreatic hepatocytes of a rat fed ciprofibrate for 4 weeks to induce peroxisome proliferation. The intense fluorescence is localized exclusively to the hepatocyte cytoplasm. ×260.

transdifferentiation is generally envisioned as a direct conversion of cell types that are well differentiated (Okada, 1983). In an organ such as the pancreas which consists of only well-differentiated stable cells that undergo cell division only after experimental manipulation (Fitzgerald *et al.*, 1968; Reddy *et al.*, 1975; Longnecker *et al.*, 1975; Scarpelli *et al.*, 1981; McGuinness *et al.*, 1984) and which apparently lacks stem cells (Leblond, 1964), it is conceivable that hepatocytes arise as a result of conversion of one of the component differentiated cell types, i.e., acinar cells, centroacinar cells, ducts cells, or endocrine cells. Alternatively, the hepatocytes may be derived from a change in commitment of a rarely observed intermediate cell (Melmed *et al.*, 1972) or of a stem cell, as yet unidentified, in this organ. Whether the pancreatic hepatocytes arise from a differentiated cell or from an undifferentiated cell type, it is appropriate to consider this phenomenon as an example of transdifferentiation since the pancreatic rudiments never differentiate into hepatocytes during development (Rutter *et al.*, 1973). Intermediate cells in the pancreas are characterized by the presence of features of both the acinar cells and different types of islet cells. The intermediate cells have been identified in normal pancreas of different species (Melmed *et al.*, 1972). Rat pancreatic hepatocytes induced by ciprofibrate (Reddy *et al.*, 1984) are commonly seen in association with islets. Transmission electron microscopic examination reveals cell junctions between the hepatocytes and islet cells (Fig. 8). Some of the cells around the islets show transitional forms displaying features of both hepatocytes and acinar/endocrine cells (Fig. 9). The identifying marker for hepatocytes is the presence of peroxisomes with hepatocyte-specific urate oxidase-containing crystalloid nucleoids (inset, Fig. 9). In these transitional cells the matrix of peroxisomes yields a positive reaction for catalase when incubated in alkaline DAB cytochemical medium, whereas zymogen granules and islet cell granules are nonreactive. In addition, the immunochemical labeling for the hepatocyte-specific mitochondrial protein, carbamoyl-phosphate synthetase (ammonia), revealed specific labeling of some mitochondria in transitional cells (Fig. 10), suggesting that these cells contain hepatocyte-specific mitochondria (Rao *et al.*, 1985b). Carbamoyl-phosphate synthetase is a liver-specific protein that is localized strictly in the mitochondria of hepatic parenchymal cells (Clarke, 1976; Bendayan and Shore, 1982; Gaasbeek-Janzen *et al.*, 1984). The presence of liver cell markers in cells with transitional cell features provides a strong

---

FIG. 8. Portion of a rat pancreatic hepatocyte (HC) with numerous peroxisomes (P) and in contact with endocrine cells (EC). Note the presence of cell junctions (arrows) between a transdifferentiated hepatocyte and endocrine cells.

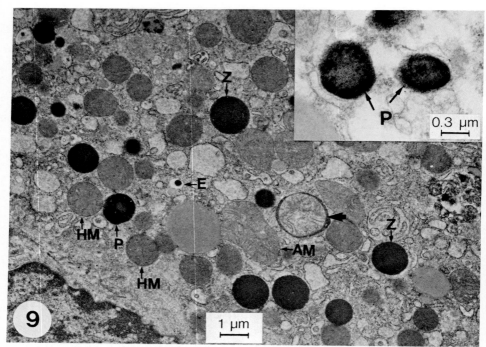

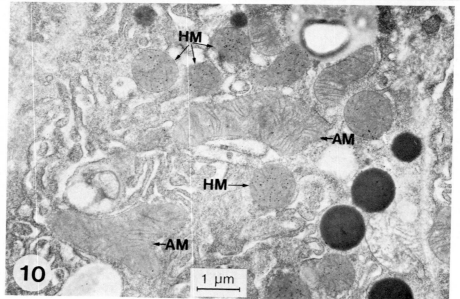

indication that acinar/intermediate cells are precursor cells for hepatocytes.

If the intermediate cell is the precursor for hepatocyte conversion in rat pancreas, the transdifferentiation from intermediate cell to hepatocyte could be a single-step process (i.e., it is direct and proceeds without an intervening cell cycle) (Fig. 11). However, the origin of intermediate cells is controversial. Becker *et al.* (1978) suggested that the intermediate cells result from abnormal differentiation of endocrine cells, whereas Kobayashi (1966) hypothesized that these cells result from cell fusion during fetal development. Melmed (1979) postulated that the "mass of exocrine pancreatic tissue contains an endocrine potentiality." In a recent study, Cossel *et al.* (1983) have observed intermediate cells (acinar A cell type) in diabetic patients and concluded that these cells are derived from acinar cells. It is conceivable that intermediate cells with exocrine and endocrine potential can arise either from cell division involving a putative stem cell or from cell division with the possible instability or destabilization of transcriptional controls of a fully differentiated pancreatic cell such as an acinar cell (Fig. 11). This multistep process involving DNA replication, intermediate cell formation, and transdifferentiation into hepatocytes may be operating in the adult rat pancreas during recovery from severe atrophy of exocrine acinar tissue resulting from copper deficiency. Additional studies are needed to determine whether the process of hepatocyte transdifferentiation involves a single step or multiple ones. Evidence indicates that the transdifferentiated hepatocytes are capable of undergoing cell division, thereby leading to the expansion in size of hepatic foci in the pancreas. Accordingly, the number of hepatocyte foci may depend upon a one-to-one conversion of a pancreatic cell (e.g., intermediate cell) into a hepatocyte, but an increase in the size of foci requires cell division of transdifferentiated hepatocytes.

In hamster pancreas the origin of hepatocytes is considered to be

---

Fig. 9. Transitional cell with features of hepatocyte and pancreatic acinar cell. Peroxisomes (P) reveal the 3,3'-diaminobenzidine reaction product indicative of catalase. This cell also shows zymogen granules (Z) and endocrine granules (E). The round mitochondria represent the hepatic mitochondria (HM) and elongated ones represent acinar cell mitochondria (AM). Some of the acinar cell mitochondria are degraded (arrowhead). Inset shows the urate oxidase-containing crystalloid nucleoids that are specific for peroxisomes of rat liver.

Fig. 10. Protein A–gold immunocytochemical procedure using the antibodies against carbamoyl-phosphate synthetase (ammonia), showing the localization of this enzyme (gold particles) in round to oval hepatocyte-type mitochondria (HM). The adjacent elongated mitochondria (AM) lack the gold particles, indicating that they are of pancreatic type. These features strongly indicate the "switching on" of dormant hepatocyte-specific genes and "switching off" of pancreatic genes in these transitional cells.

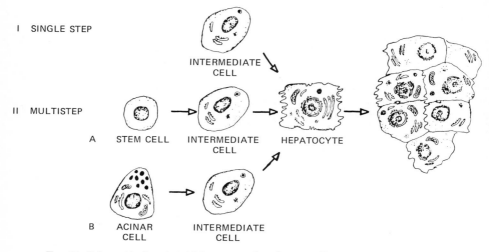

FIG. 11. Schematic drawing with suggested pathways of hepatocyte transdifferentiation in the rat pancreas. The conversion may be direct, requiring only a single step
without the necessity of DNA replication, if there is a substantial pool of "intermediate
cells" in the pancreas. The second, and most likely, possibility is that intermediate or
transitional cells are formed as a result of DNA replication involving a "pluripotential"
stem cell or a mature cell such as an acinar cell. Such intermediate cells as a result of
destabilization of transcriptional controls could become transitional cells and thus give
rise to hepatocytes. The expansion of hepatocyte colonies is the result of cell division of
transdifferentiated hepatocytes.

from acinar cells (Scarpelli and Rao, 1981) for two reasons: (1) the
carcinogen N-nitrosobis(2-oxopropyl)amine was administered when
acinar cells are in the DNA-replicative phase, at which stage they are
most susceptible to heritable changes (Holtzer *et al.,* 1972); and (2)
when acinar cell atrophy of the gastric segment of the pancreas was
induced by ligation of the duct draining that segment, hepatocytes
developed only in the splenic segments and duodenal, i.e., nonligated,
segments in which acinar cells were intact (Scarpelli and Rao, 1984).
Transitional cells showing hepatocyte and pancreatic acinar or endocrine elements have not yet been identified in the hamster pancreas
(Scarpelli *et al.,* 1984).

VII. Concluding Remarks

These studies clearly show that a specialized exocrine pancreatic
cell can transform into a different phenotypic cell, i.e., the hepatocyte.
Although in an adult animal the pancreatic and liver cells are functionally and morphologically different, both organs arise from gut
endoderm, thus sharing a common ancestry. Under proper conditions

repressed liver-specific genes in pancreatic cells are probably de-repressed. The way in which such dormant genes are activated is not clear. Although some carcinogens such as N-methyl-N-nitroso-guanidine were shown to induce transdifferentiation (Tsonis and Eguchi, 1981; Clayton and Patek, 1981), the possibility that 4-HAQO-induced genetic alterations may lead to hepatocyte conversion, when subjected to the added stress of copper depletion and repletion, can be excluded in view of the recent observation that a simple copper depletion–repletion regimen can lead to the development of pancreatic hepatocytes (Rao et al., 1986). As pointed out by DiBernardino et al. (1984), the stability of a differentiated cell could be due to various molecular mechanisms, such as DNA methylation, chromatin structure, DNA protein interactions, and DNA rearrangements. Modification of any of these could initiate transdifferentiation. Further studies are required to evaluate the mechanism(s) involved in the induction of pancreatic hepatocytes. By utilizing a variety of molecular techniques it is possible to analyze the control mechanisms of tissue-specific gene regulation in pancreatic hepatocytes. Studies with transdifferentiated hepatocytes may be expected to yield considerable new information about the mechanism(s) of cell differentiation and the attendant transcriptional controls.

ACKNOWLEDGMENTS

This research was supported by the National Institutes of Health Grants GM 23750 and CA 36130. We gratefully acknowledge the excellent secretarial assistance of Lowella Rivero and Karen McGhee.

REFERENCES

Becker, K., Wendel, U., Przyrembel, H., Tsotsalas, M., Muntefering, H., and Bremer, H. J. (1978). Eur. J. Pediatr. 127, 75–89.

Bendayan, M., and Shore, G. (1982). J. Histochem. Cytochem. 30, 139–147.

Clarke, S. (1976). J. Biol. Chem. 251, 950–961.

Clayton, R., and Patek, C. (1981). Adv. Exp. Med. Biol. 158, 229–238.

Cossel, L., Schade, J., Verlohren, H., Lohmann, D., and Mattig, H. (1983). Zentralbl. Allg. Pathol. Pathol. Anat. 128, 147–159.

DiBerardino, M. A., Hoffner, N. J., and Etkin, L. D. (1984). Science 224, 946–952.

Eguchi, G., Masuda, A., Karasawa, Y., Kodama, R., and Itoh, Y. (1981). Adv. Exp. Med. Biol. 158, 209–221.

Fitzgerald, P. J., Herman, L., Carol, B., Roque, A., Marsh, W. H., Rosenstock, L., Richardson, C., and Perl, D. (1968). Am. J. Pathol. 52, 983–1011.

Gaasbeek Janzen, J. W., Lamers, W. H., Moorman, A. F. M., DeGraaf, A., Los, J. A., and Charles, R. (1984). J. Histochem. Cytochem. 32, 557–564.

Grobstein, C. (1959). In "The Cell" (J. Brachet and A. E. Mirsky, eds.), Vol. 1, pp. 437–496. Academic Press, New York.

Holtzer, H., Weintraub, H., Mayne, R., and Mochan, R. (1972). Curr. Top. Dev. Biol. 7, 229–256.

78 M. SAMBASIVA RAO ET AL.

Jacob, F., and Monod, J. (1963). *In* "Cytodifferentiation and Macromolecular Synthesis" (M. Locke, ed.), pp. 30–64. Academic Press, New York.

Kobayashi, K. (1966). *Arch. Histochem. Jpn.* **26**, 439–482.

Lalwani, N. D., Reddy, M. K., Qureshi, S. A., and Reddy, J. K. (1981). *Carcinogenesis* **7**, 645–650.

Leblond, C. P. (1964). *Natl. Cancer Inst. Monogr.* **14**, 119–150.

Longnecker, D. S., Crawford, B. G., and Nadler, D. J. (1975). *Arch. Pathol.* **99**, 5–10.

McGuinness, E. E., Morgan, R. G. H., and Wormsley, K. G. (1984). *Environ. Health Perspect.* **56**, 205–212.

Melmed, R. N. (1979). *Gastroenterology* **76**, 196–201.

Melmed, R. N., Benitez, C. J., and Holt, S. J. (1972). *J. Cell Sci.* **11**, 449–475.

Okada, T. S. (1980). *Curr. Top. Dev. Biol.* **16**, 349–390.

Okada, T. S. (1983). *Cell Differ.* **13**, 177–183.

Pelc, S. R. (1964). *J. Cell Biol.* **22**, 21–28.

Rao, M. S., and Scarpelli, D. G. (1980). *Proc. Natl. Pancreat. Cancer Project Meet.* **5** (2).

Rao, M. S., Reddy, M. K., Reddy, J. K., and Scarpelli, D. G. (1982). *J. Cell Biol.* **95**, 50–56.

Rao, M. S., Subbarao, V., Luetteke, N., and Scarpelli, D. G. (1983). *Am. J. Pathol.* **110**, 89–94.

Rao, M. S., Thorgeirsson, S., and Reddy, J. K. (1984a). *J. Cell Biol.* **99**, 359a.

Rao, M. S., Lalwani, N. D., Watanabe, T. K., and Reddy, J. K. (1984b). *Cancer Res.* **44**, 1072–1076.

Rao, M. S., Subbarao, V., Scarpelli, D. G., and Reddy, J. K. (1985a). *Toxicologist* **5**, 637.

Rao, M. S., Bendayan, M., and Reddy, J. K. (1985b). *Fed. Proc. Fed. Am. Soc. Exp. Biol.* **44**, 740.

Rao, M. S., Subbarao, V., Scarpelli, D. G., and Reddy, J. K. (1986). *Cell Differ.*, in press.

Reddi, A. H., and Huggins, C. B. (1975). *Proc. Natl. Acad. Sci. U.S.A.* **72**, 2212–2216.

Reddy, J. K., and Lalwani, N. D. (1983). *CRC Crit. Rev. Toxicol.* **12**, 1–58.

Reddy, J. K., Rao, M. S., Svoboda, D. J., and Prasad, J. D. (1975). *Lab. Invest.* **32**, 98–104.

Reddy, J. K., Azarnoff, D. L., and Hignite, C. E. (1980). *Nature (London)* **283**, 397–398.

Reddy, J. K., Warren, J. R., Reddy, M. K., and Lalwani, N. D. (1982a). *Ann. N.Y. Acad. Sci.* **386**, 81–110.

Reddy, J. K., Lalwani, N. D., Reddy, M. K., and Qureshi, S. A. (1982b). *Cancer Res.* **42**, 259–266.

Reddy, J. K., Rao, M. S., Qureshi, S. A., Reddy, M. K., Scarpelli, D. G., and Lalwani, N. D. (1984). *J. Cell Biol.* **98**, 2082–2090.

Rutter, W. J., Pictet, R. L., and Morris, P. W. (1973). *Annu. Rev. Biochem.* **42**, 601–646.

Scarpelli, D. G., and Rao, M. S. (1981). *Proc. Natl. Acad. Sci. U.S.A.* **78**, 2577–2581.

Scarpelli, D. G., and Rao, M. S. (1984). *J. Cell Biol.* **99**, 341a.

Scarpelli, D. G., Rao, M. S., Subbarao, V., and Beversluis, M. (1981). *Cancer Res.* **41**, 1051–1057.

Scarpelli, D. G., Rao, M. S., and Reddy, J. K. (1984). *Environ. Health Perspect.* **56**, 219–227.

Schmid, V., and Alder, H. (1984). *Cell* **38**, 801–809.

Shnitka, T. S., and Youngman, M. M. (1966). *J. Ultrastruct. Res.* **16**, 598–625.

Svoboda, D. J., Grady, H., and Azarnoff, D. (1967). *J. Cell Biol.* **35**, 127–152.

Tsonis, D. A., and Eguchi, G. (1981). *Differentiation* **20**, 52.

Tsukada, H., Mochizuki, Y., and Fujiwara, S. (1966). *J. Cell Biol.* **28**, 449–460.

Yamada, T. (1982). *In* "Cell Biology of the Eye" (D. S. McDevitt, ed.), pp. 193–242. Academic Press, New York.

Yasuda, K., Okada, T. S., Eguchi, G., and Hayashi, M. (1978). *Exp. Eye Res.* **26**, 591–595.

CHAPTER 5

TRANSDIFFERENTIATION OF AMPHIBIAN CHROMATOPHORES

Hiroyuki Ide

BIOLOGICAL INSTITUTE
TÔHOKU UNIVERSITY
SENDAI, JAPAN

I. Introduction

Chromatophores are specialized cells for animal coloration. These cells are of neural crest origin, and the precursor cells of the chromatophores (chromatoblasts) migrate to various regions of the body surface and differentiate there to form characteristic pigment granules. Further, the chromatophores change the morphology and intracellular distribution of the pigment granules, a process corresponding to the background adaptation of lower vertebrates. Thus, the capabilities of changing cell morphology and of translocating pigment granules are also markers of differentiation.

Three types of chromatophores are known in amphibians. Melanophores contain melanosomes and disperse the granules in response to melanocyte-stimulating hormone (MSH). Iridophores, iridescent chromatophores, contain reflecting platelets (purine crystals) and contract the cell body and dendrites in response to the same hormonal stimulation. Xanthophores laden with pterinosomes and carotenoid vesicles show no response to MSH (Bagnara and Hadley, 1973).

We have succeeded in the cell culture of these chromatophores (Ide, 1974, 1978; Ide and Hama, 1976) and observed that conversions from

79

iridophores and xanthophores into melanophores occur during cell proliferation.

In this chapter, we will briefly review the conversion of cultured chromatophores of the bullfrog (*Rana catesbeiana*). This system is a good example of transdifferentiation and provides a good opportunity for further studies on the instability in cell differentiation.

II. Stability of Cell Commitment in Cultured Melanophores

The melanophores of bullfrog tadpoles were isolated by the methods of trypsin digestion and Ficoll density gradient centrifugation and cultured in a medium consisting of neuroretina-conditioned and diluted L-15 and fetal calf serum (Ide, 1973, 1974). Only chromatophores survived and proliferated under these conditions. The melanophores commenced cell proliferation within 3 weeks and doubled their number every 3–5 days. During the proliferation, the melanophores retained the activity of melanin formation and reactivity to MSH (Ide, 1974). Recently, we have established several cell lines of melanophores (Kondo and Ide, 1983). Some of them continued cell proliferation for over 6 years (more than 200 passages) without loss of melanin formation activity and MSH responsiveness. Thus, it was demonstrated that the commitment of melanophore was stable, at least under these culture conditions.

III. Transdifferentiation from Iridophores into Melanophores in Clonal Culture

The iridophores of bullfrog tadpoles were isolated and cultured under the same conditions mentioned above. By taking serial photographs of the iridophores in clonal culture, the conversion from the iridophores into melanophores was demonstrated (Ide and Hama, 1976). After cell proliferation, the melanized cells dispersed melanin in response to MSH. All the proliferated iridophores converted into melanophores, and no reverse conversion from the melanized cells into iridescent cells was observed during further cultivation.

Initially, the iridophores were filled with a large number of reflecting platelets (Fig. 1A), and neither melanosomes nor premelanosomes were observed in the cytoplasm. No melanized structures were de-

FIG. 1. Horizontal sections of iridophores before (A and B) and after (C and D) proliferation *in vitro*. (A) 1-Day culture; (B) 1-day culture, dopa reaction; (C) 45-day culture, melanized iridophore; (D) 45-day culture in the medium containing 10% tadpole serum; iridescent iridophore. R, Reflecting platelet; M, melanosome; P, premelanosome; D, electron-dense granule; N, nucleus. Bar, 1 μm. (D from Ide, 1984.)

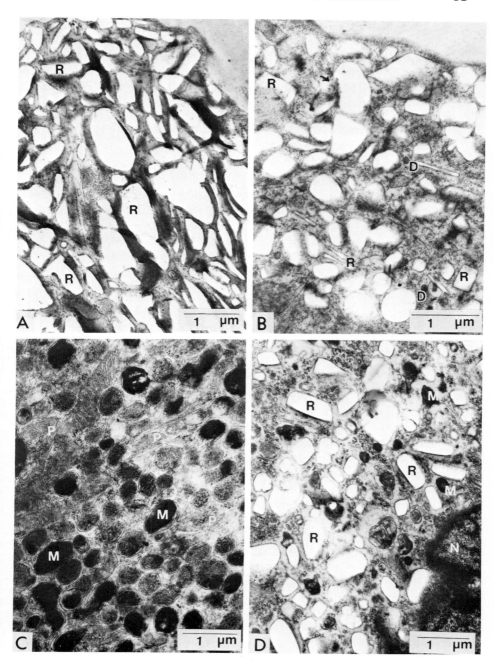

tected after dopa reaction, although a small number of small electron-dense granules were detected (Fig. 1B). After cell proliferation, the melanized cells were filled with melanosomes and premelanosomes (Fig. 1C). The reflecting platelets were completely lost from the melanized cells. Thus, transdifferentiation from differentiated irido-phores into melanophores was demonstrated.

IV. Transdifferentiation from Xanthophores into Melanophores in Clonal Culture

The xanthophores of bullfrog tadpoles were also isolated and clonally cultured as above. The xanthophores also converted into melanophores in clonal culture (Ide, 1978), and the resultant melanophores dispersed melanin in response to MSH. As in the case of iridophores, all the proliferated xanthophores converted into melanophores, and the conversion was irreversible. Nonproliferated xanthophores remained yellowish in color and unresponsive to MSH. Although the original xanthophores were filled with pterinosomes, the proliferated cells included a large number of melanosomes, indicating transdifferentiation into melanophores.

V. Proliferation of Iridophores without Transdifferentiation into Melanophores

As mentioned above, iridophores and xanthophores transdifferenti-ate into melanophores during *in vitro* proliferation. However, in the tadpole skin, there are many iridophores and xanthophores, in which no, or very few, melanosomes were detected. At least in the case of iridophores, we can trace cell proliferation in the tail skin *in vivo,* without conversion into melanized cells (Fig. 2). Thus, it seems that the conversion was induced in the present culture conditions.

We have attempted to identify the factor(s) in the tadpole skin *in vivo* that inhibit the conversion by modifying the components of the culture medium. First, we supplemented the culture medium with tad-pole serum (Ide, 1984). When the iridophores were cultured in a medi-um containing 5–20% tadpole serum, many colonies of iridescent cells appeared that never occurred in cultures with tadpole serum-free me-dium (Fig. 3). These cells were dendritic in shape and surrounded by melanophores, xanthophores, and other unidentified cells of skin ori-gin. Although it is difficult to count the iridophore number precisely, these cells seem to proliferate actively since cells in mitosis were fre-quently observed (Ide, 1984). These iridescent cells responded to MSH by contracting the cell bodies, as the iridophores *in vivo* in the skin and those immediately after *in vitro* cultivation did (Fig. 4). Thus, the

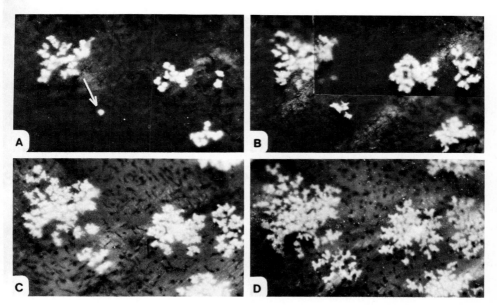

FIG. 2. Proliferation of iridophores in tadpole tail skin. (A) Day 0. An iridophore is indicated by the arrow. (B) Day 5. The iridophore in (A) proliferated into four cells. (C) Day 10. (D) Day 15. No melanization occurred in the proliferating iridophores. ×49.

iridescent cells were identified as iridophores on the basis of the responsiveness to MSH stimulation.

Electron microscopic observation of the iridophores revealed the presence of many reflecting platelets in the cytoplasm, although a considerable number of melanosomes were also observed (Fig. 1D). Xanthophores were frequently observed in the iridophore colony, although it was difficult to count the number of xanthophores in these mixed cultures. Further, in the xanthophores, immature pterinosomes (Yasutomi and Hama, 1972) were predominant, suggesting the formation of pterinosomes during xanthophore proliferation (Ide, 1984).

When pure cultures of iridophores or xanthophores were prepared at clonal cell density, they converted into melanophores even in the presence of tadpole serum. To retain the original phenotypes of these chromatophores, the cells must be surrounded by some unidentified cells of skin origin. Actually, these bright-colored chromatophores always coexisted with each other and with melanophores and the other unidentified cells (Ide, 1984). In our culture conditions without tadpole serum, only chromatophores survive and proliferate. Thus, it is possible to consider that the tadpole serum permits the survival and pro-

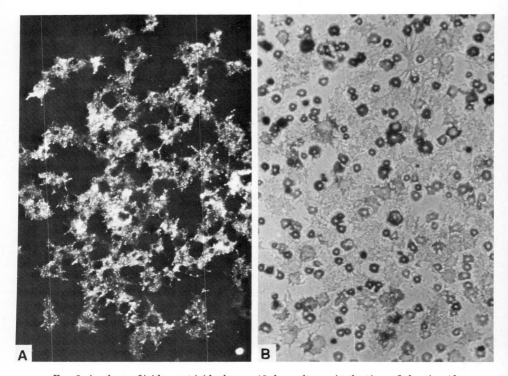

FIG. 3. A colony of iridescent iridophores, 40-day culture. At the time of planting (day 0), the culture contained two iridophores and a small number of melanophores. (A) Reflecting light. (B) Phase contrast. Many intermingling melanophores and other cells are observed in the iridophore colony. ×149.

liferation of nonpigment cells of skin origin and then supports the proliferation of bright-colored chromatophores without conversion into melanophores. FT cells, a fibroblastic cell line of bullfrog tongue (Wolf and Quimby, 1964), however, could not support the proliferation of iridophores without conversion, although the fibroblastic cells proliferated actively in tadpole serum-free medium.

Tadpole serum was effective for the maintenance of iridophore phenotypes at concentrations of 5–20%, and the effect was dose dependent. The serum factor was nondialyzable and heat labile, suggesting a protein nature (Ide, 1984).

VI. Reflecting Platelet Formation in Cultured Melanophores

As described above, melanophores continuously form melanin during proliferation. However, when guanosine, a precursor of platelet

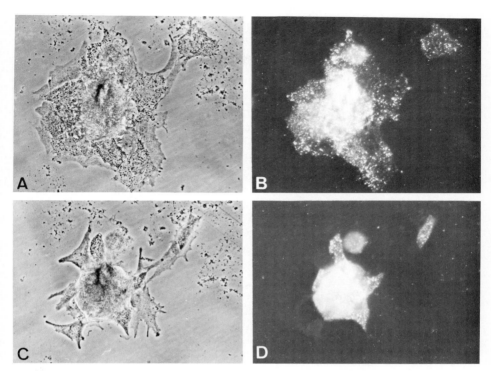

FIG. 4. Effects of α-MSH on proliferated iridescent iridophores. The iridophores were collected with a capillary pipette from the colony after trypsinization and inoculated onto a new dish, since dispersed melanosomes in contaminating melanophores prevented the observation of iridophore morphology after MSH stimulation. (A) Phase contrast. (B) Reflecting light. In C and D, the same material was used as in A, but treated with 0.1 μg/ml α-MSH for 60 minutes. The iridophores contracted the cell body. (C) Phase contrast. (D) Reflecting light. ×191.

component, was added in the culture medium, some clones of melanophores formed typical reflecting platelets other than melanosomes (Ide, 1979). These iridescent melanophores, however, showed melanin dispersion in response to MSH stimulation, and no cell body contraction was observed. Thus, the transdifferentiation from melanophores to iridophores is incomplete, at least in the present culture conditions.

VII. Concluding Remarks: Conversion between Different Chromatophore Types

The possibility of chromatophore metaplasia has been repeatedly indicated since the classical work of Niu (1954). However, due to the difficulty of tracing single chromatophores completely *in vivo* or in explant culture, definite evidence has not been demonstrated. We have succeeded in clearly demonstrating conversion between chromatophores. The commitment of iridophores and xanthophores is relatively labile, and these cells are convertible, at least in culture conditions. As mentioned above, original iridophores, filled with reflecting platelets, contained no or very few melanosomes and were tyrosinase negative. The converted cells included a large number of melanosomes and premelanosomes, indicating active melanin synthesis. Thus, the switching of the pathway of pigment formation was demonstrated to occur during the course of *in vitro* proliferation (Fig. 5).

Bagnara *et al.* (1979) proposed a model on the origin of various pigment granules, in which they hypothesized a primordial organelle derived from endoplasmic reticulum. This indicates that the various pigment granules were closely related to one another. Actually, the presence of tyrosinase activity has been suggested in amphibian iridophores (Frost *et al.*, 1984) and demonstrated in amphibian xanthophores (Yasutomi and Hama, 1976). The problem of the presence of a small number of tyrosinase molecules and tyrosinase mRNA in the iridophores and xanthophores remains to be solved.

Recently, the proliferation of fish iridophores has been reported (Yasutomi, 1984). The iridophores isolated from fry of *Gambusia affinis* proliferated *in vitro* without conversion into melanophores. Two explanations for the absence of conversion in fish iridophores are possible. First, fish iridophores, which show no response to MSH, are different from amphibian iridophores, which show cell body contraction in response to MSH, and the commitment of fish iridophores may be sta-

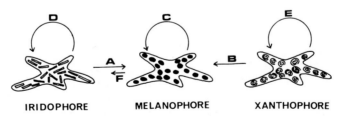

IRIDOPHORE   MELANOPHORE   XANTHOPHORE

FIG. 5. Conversion between different chromatophore types *in vitro*. Iridophores (A) and xanthophores (B) convert irreversibly into melanophores during proliferation in the medium without tadpole serum. The phenotypes of melanophores are stable during proliferation (C). In medium containing tadpole serum, the iridophores (D) and probably the xanthophores (E) retain their original phenotypes during proliferation. Some melanophores form reflecting platelets during the cultivation in medium containing guanosine (F).

ble in *in vitro* proliferation. Second, the fish iridophores proliferated only in the presence of nonpigmented cells of the fry. Thus, the coexisting cells may support iridophore proliferation without conversion into melanophores, as suggested in the present study.

ACKNOWLEDGMENTS

The author thanks Prof. T. S. Okada for suggestions in preparing the chapter. The author's work reviewed in this article was supported by a Grant-in-Aid for Special Project Research, Multicellular Organization, from the Ministry of Education, Science and Culture, Japan.

REFERENCES

Bagnara, J. T., and Hadley, M. E. (1973). "Chromatophores and Color Change." Prentice-Hall, New York.

Bagnara, J. T., Matsumoto, J., Ferris, W., Frost, S. K., Turner, W. A., Tchen, T. T., and Taylor, J. D. (1979). *Science* **203**, 410–415.

Frost, S. K., Epp, L. G., and Robinson, S. J. (1984). *J. Embryol. Exp. Morphol.* **81**, 127–142.

Ide, H. (1973). *Gen. Comp. Endocrinol.* **21**, 390–397.

Ide, H. (1974). *Dev. Biol.* **41**, 380–384.

Ide, H. (1978). *J. Exp. Zool.* **203**, 287–294.

Ide, H. (1979). *Pigment Cell* **4**, 28–34.

Ide, H. (1984). *Pigment Cell,* in press.

Ide, H., and Hama, T. (1976). *Dev. Biol.* **53**, 297–302.

Kondo, H., and Ide, H. (1983). *Exp. Cell Res.* **149**, 247–256.

Niu, M. C. (1954). *J. Exp. Zool.* **125**, 199–220.

Wolf, K., and Quimby, M. C. (1964). *Science* **144**, 1578–1580.

Yasutomi, M. (1984). *Zool. Sci.* **1**, 894.

Yasutomi, M., and Hama, T. (1972). *J. Ultrastruct. Res.* **38**, 421–432.

Yasutomi, M., and Hama, T. (1976). *Dev. Growth Differ.* **18**, 289–299.

CHAPTER 6

MULTIPOTENTIALITY IN DIFFERENTIATION OF THE PINEAL AS REVEALED BY CELL CULTURE

Kenji Watanabe

DEPARTMENT OF ANATOMY
FUKUI MEDICAL SCHOOL
MATSUOKA-CHO, YOSHIDA-GUN
FUKUI, JAPAN

I. Introduction

The pineal complex of lower vertebrates is a photoreceptive organ known as the "median eye" that furnishes function and morphology comparable with two lateral eyes (Eakin, 1973). In the course of phylogenic development, the pineal has changed its major function from photoreceptive to endocrine. A number of pinealogists have suggested repeatedly that photoreceptive cells in the pineal of lower vertebrates are direct ancestors of endocrine pinealocytes of higher vertebrates (Collin, 1971; Oksche and Hartwig, 1979).

It has been well demonstrated that nonlenticular cells of vertebrate eyes can convert their cell types, or transdifferentiate, into lens in cell culture conditions (reviews by Okada, 1980, 1983). Therefore, it seemed interesting to us to examine whether pineals of higher vertebrates still retain the potency to differentiate into their ancestral phenotypes, into lens and various other ocular cells, when cultured *in vitro*. Our recent results along this line indicate that pineals of higher vertebrates are provided with an unexpectedly wide repertoire of differentiation. This system may not only offer another interesting example of transdifferentiation but will also give us a unique opportunity to analyze the nature of an exogenous cue by which multipotent cells can select a given pathway of differentiation. In this chapter, studies on the multiplicity of differentiation of pineals will be reviewed.

89

II. "Oculopotency" in Pineal Cells

Cell components of the "median eye" in lower vertebrates are photoreceptor cells, neurons, pigment cells, and lens cells (Eakin, 1973).
These cells, which are relevant to photoreceptive function, have disappeared or changed remarkably in their characteristics during ontogeny and phylogeny. For example, photoreceptor cells of the median
eye in cold vertebrates have a stacked membrane structure, i.e., outer
segment, similar to the cone cells of retinae (Eakin, 1970). Rudimentary photoreceptor cells observed in the pineal of early stage avian and
mammalian animals have inner segments rich in mitochondria and a
cilium associated with swirled membrane (Zimmerman and Tso, 1975).
Pinealocytes of adult avian and mammalian animals have no such
appendices (Clabough, 1973). Pinealogists have often considered that
photoreceptor cells, rudimentary photoreceptor cells, and pinealocytes
are closely related cells, and all belong to a single category known as
the photoneuroendocrine cell line (Oksche and Hartwig, 1979). No one,
however, has explained what factors evoke interspecies diversification
of this cell line and determine the different differentiation states characteristic of each species and particular developmental stage.

The median eye is a vesicle pinched off from the brain wall during
embryonic development (Eakin, 1970). Lens cells constitute a part of
the vesicular wall, which faces the head skin. Lens cells of the median
eye of the American chameleon are of tall epithelium and contain
crystallins, which are usually considered to be specific proteins of lateral eyes (McDevitt, 1972). Though lens cells have never been observed
in the pineal of avian and mammalian animals *in situ,* its potentiality
for lens differentiation has recently been demonstrated by cell culture
experiments with avian pineal cells (Watanabe *et al.,* 1985). Pineal
cells of 8-day-old quail embryos were cell cultured for a prolonged
period. After about 4 weeks' cultivation, transparent cell masses were
sometimes observed (Fig. 1A), which resembled "lentoid bodies" from
cultured ocular cells (Okada, 1980). Immunological methods using specific antibodies against crystallins showed the presence of all three
classes of crystallins (α, β, and δ) in these lentoid bodies derived from
cultured pineal cells.

δ-Crystallin is detected in cell cultures of avian nonlenticular cells
such as neural and pigmented retinae (Okada, 1980; Eguchi and Itoh,
1981), pineal cells (Watanabe *et al.,* 1985), brain cells (Nomura, 1982;
Takagi *et al.,* 1983), and limb bud cells (Kodama and Eguchi, 1982,
1983), as well as in the hypophysis *in situ* (Barabanov, 1977, 1982).
However, cultured brain cells of older embryos (Takagi *et al.,* 1983)

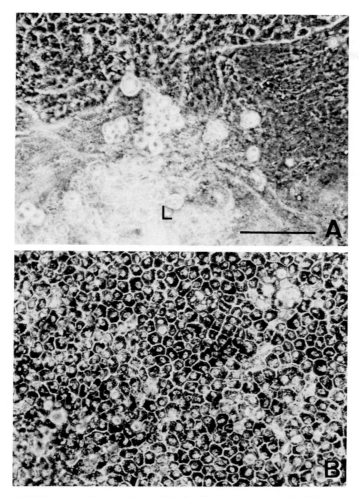

FIG. 1. (A) The formation of a lentoid body (L) after 30 days of cultivation of pineal cells. (B) Some of polygonal cells were pigmented around 14 days of cultivation of pineal cells and they resembled pigment cells of retinae of lateral eyes. Bar, 100 μm.

and limb bud cells neither form lentoid bodies nor produce α- or β-crystallins. The situation is the same in the hypophysis *in situ*. Therefore, pineals are the only example of nonocular tissues that perform lentoidogenesis accompanied by the synthesis of all three classes of crystallins.

Pigment cells have been found in the retina of the median eye and in the pineal of gulls (Wetzig, 1961), sheep (Jordan, 1921), cows (Santamarina, 1958), and humans (Quast, 1931). Wetzig (1961) investigated the cytology of the pineal of embryonic gulls and found that pigment cells increased once in number in the embryonic life, but tended to decrease around the time of hatching. He suggested that pigment cells are not functional components of the pineals in the gull, but rudimentary ones, related to the median eye of lower vertebrates. The *in situ* presence of pigment cells in the pineals of 8-day-old quail embryos has been discussed (K. Watanabe *et al.*, unpublished observation). Pigmentation of pineals in these materials was observed under stereomicroscopy at a frequency of roughly 50%. Cell cultures of pineals with and without pigments separately revealed that both were provided with a similar potentiality for melanogenesis under *in vitro* conditions. Pigment cells differentiated *in vitro* were polygonal epithelial cells (Fig. 1B) and similar to pigment cells found in cell cultures of the pigmented retina.

Tyrosinase is an enzyme that hydrolyzes tyrosin to dopa and then synthesizes melanin. Its activity is shown histochemically by applying dopa as a substrate to the cells. After dopa reaction, all the pineals of 8-day-old quail embryos were darkened as a result of producing dopa melanin (unpublished data). Therefore, all pineals at this stage possess cells containing tyrosinase, although the enzyme in about half of the pineals is inactivated by unknown factors.

III. Myogenic Potency in Pineal Cells

A few immature muscle fibers are very rarely observed in the *in situ* pineals of several mammalians (cf. Watanabe *et al.*, 1981). However, a number of well-developed muscle fibers were invariably differentiated from pineal cells of newborn rats when they were dissociated and cultured (Freschi, 1979; Watanabe *et al.*, 1981). This suggests that the expression of myogenic cells in the pineal is more or less suppressed *in situ*, but they can fully manifest myogenic potency if conditions surrounding the cells are changed, perhaps so as to derepress the suppressing mechanisms (Fig. 2).

Pineal cells of 8-day-old quail embryos are also able to differentiate into muscle fibers *in vitro* (Fig. 3; Watanabe *et al.*, 1984a). This information gave us the unique chance to investigate the origin of myogenic cells in pineal cultures and to determine the exogenous conditions favorable for the myogenesis of pineal cells as well. Pineals *in situ* consist of cells derived from two major different origins. The one is neural epithelium, from which pineal parenchyma is developed (Clabough, 1973; Calvo and Boya, 1978), and the other is mesoectoderm (neural crest) derived from the lower level of brain, which will

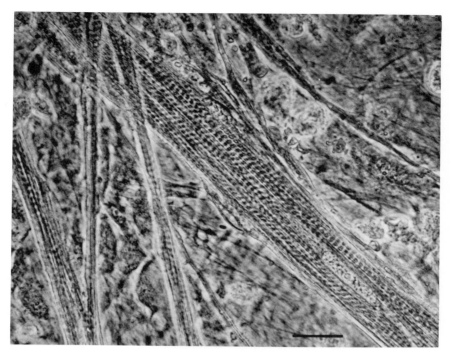

FIG. 2. Skeletal muscle fibers developed in a 5-week culture of pineal cells of newborn rats. They were multinuclear cells with regular striations. Bar, 20 μm.

form the pineal stroma (Weston, 1970). Cells derived from these two origins are intermingled in adulthood, but are separated in the embryonic period. It is possible to make embryonic pineals almost free from mesoectoderm by adequate treatment with collagenase and Dispase. Such tissue still retained myogenic potency when dissociated and cultured. The result shows that the neuroepithelium derived from pineal parenchyma contains myogenic cells.

A high ionic condition of the culture medium was critical for extensive myogenic differentiation of embryonic pineal cells. The standard culture condition was a basal medium (Eagles' MEM with several supplements) under 5% CO_2 atmosphere. The myogenic culture condition was a basal medium with an additional 25 mM NaCl under 12% CO_2 atmosphere. To quantify the extent of myogenecity of pineal cells under two different conditions, a specific antibody was raised against skeletal muscle type creatine kinase (MM-CK) and used in enzyme-linked immunoassay and immunohistochemistry for MM-CK. The

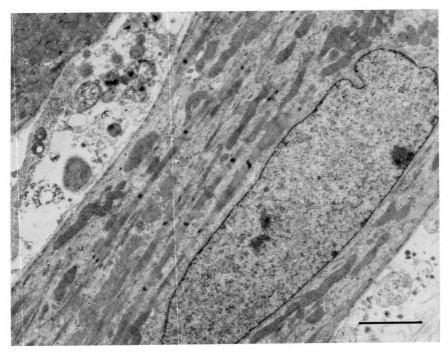

FIG. 3. An electron micrograph of multinucleated muscle fibers developed in 4 weeks of cultivation of pineal cells of quail embryos under high ionic condition. The characteristics of a mature muscle fiber, such as arrayed myofilaments with Z lines, elongated mitochondria, and an eccentric nucleus in the muscle fiber, are shown in this photograph. Bar, 2 μm.

quantity of MM-CK in soluble fractions of the cultured pineal cells remained low (about 1 ng/mg protein) up to 20 days under the standard condition, whereas it was much higher (30 ng/mg protein) under the "myogenic" condition. MM-CK-positive cells revealed by immunohistochemistry were epithelial at the early stage (Fig. 4) and elongated at a later stage of the culture period. Differentiation of myoblasts from pectoral muscle was not affected by the difference in these two conditions, and myoblasts stained by immunohistochemistry in the early stages of culturing were spindle shaped.

What types of cells are related to myogenic potency in pineals? The rare appearance of muscle fibers in pineals *in situ* may suggest that such cells are quite immature or are committed to other pathways of differentiation. Lenon and Peterson (1979) and Wier and Lenon (1981) reported that muscle fibers could be differentiated from a glial cell line

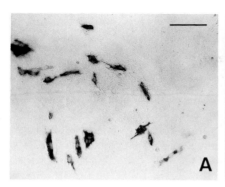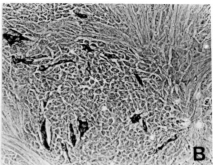

FIG. 4. Pineal cells after 14 days of cultivation, immunostained for MM-CK. (A) The appearance of positive cells in the epithelial island, which are in the initial stage of muscle differentiation. (B) A phase micrograph of the same field as A. Bar, 100 μm.

(B9) and also from cells of optic nerve, probably glia cells. As glia cells are also cell components of the pineal of avian and mammalian animals (Collin, 1971), these can be considered to be candidates for myogenic cells.

IV. Repertoire of Differentiation of Pineals

In this chapter the multipotential nature involved in the differentiation of pineals has been reviewed. In the avian and mammalian pineals, a major function of which is endocrine in normal development, two pathways fundamentally different from the normal course seem to be immanent when studied in cell culture. These are (1) the manifestation of oculopotency and (2) the expression of some repertoire of differentiation normally belonging to neural crest. The differentiation of pigmented epithelium as well as lentoid bodies in cultures of pineals is an example of the first, whereas the appearance of glia-like cells and skeletal muscle may be an example of the second pathway.

The pineal complex of lower vertebrates is a photoreceptive organ often known as the "median eye." The demonstration of oculopotency in avian and mammalian pineals seems to indicate that the ancestral potency is still retained in pineals of higher vertebrates. If so, transdifferentiation of pineals into ocular directions provides an interesting example of the manifestation of ancestral differentiation characters in culture.

The differentiation of skeletal muscle, possibly via glia-like cells, can be interpreted differently. We assume that pineals retain a developmental potentiality similar to that of the neural crest. Neural folds are a common origin in the development of both the neural crest and pineals (Le Douarin, 1982; Kamer, 1949). It has now been demonstrated that neural crest has a myogenic potency, as reviewed in another chapter in this volume (see Chapter 8, Nakamura and Ayer-Lièvre).

It should be noted here that both pathways have hardly been expressed in a single culture plate. As stated before, the differentiation of muscle fiber requires different culture conditions than those for ocular differentiation. The problem of cell commitment is to ask how a particular pathway of differentiation is selected and stabilized in multipotent cells. As indicated in the present case of pineals, there is little doubt that there is an exogenous cue that leads to cell commitment and to differential gene expression through cascade reactions occurring inside cells. The present situation of studies on avian and mammalian pineals seems to prime another promising system for understanding the instability of cell differentiation, together with studies on the transdifferentiation of eye tissues.

ACKNOWLEDGMENT

The author's study reviewed in this chapter was in part supported by a Grant-in-Aid for Special Project Research, Multicellular Organization, from the Ministry of Education, Science and Culture.

REFERENCES

Barabanov, V.M. (1977). *Dokl. Acad. Nauk SSSR* **234,** 195–198.
Barabanov, V. M. (1982). *Dokl. Acad. Nauk SSSR* **262,** 1491–1494.
Calvo, J., and Boya, J. (1978). *Acta Anat.* **101,** 289–303.
Clabough, J. W. (1973). *Am. J. Anat.* **137,** 215–230.
Collin, J. P. (1971). *In* "The Pineal" (G. E. W. Wolstenholme and J. Knight, eds.), pp. 79–125. Churchill-Livingstone, London.
Eakin, R. M. (1970). *Am. Sci.* **58,** 73–79.
Eakin, R. M. (1973). "The Third Eye." Univ. of California Press, Berkeley.
Eguchi, G., and Itoh, Y. (1981). *In* "Pigment Cell 1981" (M. Seiji, ed.), pp. 271–278. Univ. of Tokyo Press, Tokyo.
Freschi, J. E. (1979). *J. Physiol. (London)* **293,** 1–10.
Jordan, H. E. (1921). *Anat. Rec.* **22,** 275–287.
Kamer, J. C., van de (1949). Thesis, Faculty of Science, Arnhem.
Kodama, R., and Eguchi, G. (1982). *Dev. Biol.* **91,** 221–226.
Kodama, R., and Eguchi, G. (1983). *Dev. Growth Differ.* **25,** 261–270.
Le Douarin, N. M. (1982). "The Neural Crest." Cambridge Univ. Press, London and New York.
Lenon, V. A., and Peterson, S. (1979). *Nature (London)* **281,** 586–588.
McDevitt, D. S. (1972). *Science* **175,** 763–764.
Nomura, K. Y. (1982). *Differentiation* **22,** 179–184.
Okada, T. S. (1980). *Curr. Top. Dev. Biol.* **16,** 349–380.
Okada, T. S. (1983). *Cell Differ.* **13,** 177–183.
Oksche, A., and Hartwig, H. G. (1979). *Prog. Brain Res.* **52,** 113–130.

Quast, P. (1931). *Z. Mikrask.-Anat. Forsch.* **24,** 38–100.

Santamarina, E. (1958). *Can. J. Biochem. Physiol.* **36,** 227–235.

Takagi, S., Haruguchi, M., Agata, K., Araki, M., and Okada, T. S. (1983). *Dev. Growth Differ.* **25,** 421.

Watanabe, K., Aoyama, H., Nojyo, Y., Matsuura, T., and Okada, T. S. (1981). *Dev. Growth Differ.* **23,** 221–227.

Watanabe, K., Tamamaki, N., and Nojyo, Y. (1984a). *Dev. Growth Differ.* **26,** 386.

Watanabe, K., Aoyama, H., Tamamaki, N., Yasuzima, M., Nojyo, Y., Ueda, Y., and Okada, T. S. (1985). *Cell Differ.* **16,** 251–257.

Weston, J. A. (1970). *Adv. Morphol.* **8,** 41–114.

Wetzig, H. (1961). *Gegenbaurs. Morphol. Jb.* **101,** 406–431.

Wier, M. L., and Lenon, V. A. (1981). *J. Neuroimmunol.* **1,** 61–68.

Zimmerman, B. L., and Tso, M. O. M. (1975). *J. Cell Biol.* **66,** 60–75.

CHAPTER 7

TRANSDIFFERENTIATION OF ENDOCRINE CHROMAFFIN CELLS INTO NEURONAL CELLS

Masaharu Ogawa, Tomoichi Ishikawa,† and Hitoshi Ohta‡*

* DEPARTMENT OF PHYSIOLOGY
† DEPARTMENT OF ANATOMY
‡ DEPARTMENT OF NEUROPSYCHIATRY
KOCHI MEDICAL SCHOOL
KOCHI, JAPAN

I. Introduction

Most animal cells show striking stability in differentiated states once committed to a specific developmental program. However, a few instances of transformation from one differentiated state to another that is radically different are seen in some vertebrate eye tissues (Okada, 1980, 1983; chapters by Eguchi and Moscona, this volume). Such phenomena, called transdifferentiation or cellular metaplasia, are the result of experimental disruption of normal development, but can be related to the mechanisms of normal development. Thus, it is reasonable to expect that they might help to reveal developmental processes that are more difficult to investigate during normal development.

The adrenal gland is a composite organ. The cortex derived from coelomic epithelium develops first, and the primitive adrenal chromaffin cells enter the adrenal primordium. The adrenal medulla is derived from cells of the neural crest in association with the development of the rest of the sympathetic nervous system. Adrenal medullary chromaffin cells and sympathetic neurons have some common

*CURRENT TOPICS IN
DEVELOPMENTAL BIOLOGY, VOL. 20*

properties: both cell types synthesize the enzymes necessary for the formation of adrenergic transmitters, and the mechanism for releasing the transmitters is basically the same. However, their morphological and functional characteristics are very different. The adrenal medullary chromaffin cells are endocrine cells and do not have any processes.

Recently, a number of experimental data has been accumulated showing the conversion of adrenal chromaffin cells, even in the postmitotic state, from endocrine to neuronal phenotype both *in vivo* and *in vitro*. Such a system has great potential for analyzing the roles of environmental signals in the mechanism of commitment to either endocrine or neuronal cells of adrenomedullary neural crest-derived cells.

Conversion of adrenal chromaffin cells from endocrine to neuronal phenotype was first reported by Olson (Olson, 1970; Olson and Malmfors, 1970). He explanted the adrenal medulla to the anterior chamber of the eye and observed that the adrenal medullary cells themselves gave rise to nerve fibers that reinnervated the denervated host iris in the adult rat. Such transplantation experiments suggested that adrenal chromaffin cells can convert their cell phenotype from endocrine to neuronal cells because of imposed environmental disturbances and can maintain this ability until quite late in life. In 1978, Unsicker and colleagues first showed convincing evidence that isolated adrenal chromaffin cells produce fiber outgrowths and exhibit the structural features of neurons when cultured in nerve growth factor (NGF)-enriched medium (Unsicker *et al.*, 1978). The following year, Levi-Montalcini and colleague showed that NGF can convert immature adrenal chromaffin cells into sympathetic-like neurons *in situ* (Aloe and Levi-Montalcini, 1979).

In this chapter we will deal mainly with the phenomena occurring during neuronal transdifferentiation in adrenal chromaffin cells in culture. Transdifferentiation of adrenal chromaffin cells *in situ* and in ectopic sites will be briefly considered.

II. Transdifferentiation of Adrenal Chromaffin Cells in Culture

A. PROCESS OUTGROWTH FROM CULTURED CHROMAFFIN CELLS

When the adrenal medulla isolated from neonatal rats is dissociated into suspensions of single cells and then cultured under appropriate conditions, one can initially observe cells of different types. Chromaffin cells are distinctly different in morphology from several varieties of supportive cells that assume a fibroblastic or epithelial morphology, and from adrenal cortical cells. Even when neurons are present in adrenal medulla, at a rate of less than 1 per 200 chromaffin cells in young rats (Unsicker *et al.*, 1978; Aloe and Levi-Montalcini, 1979; Ogawa *et al.*, 1984), chromaffin cells are easily distinguished by their differences in morphology or by fluorochemical examination. Chromaffin cells are round and ovoid at plating and become partially polygonal after a few days in culture.

The chromaffin cells survived without NGF in medium, whereas the sympathetic neurons of superior cervical ganglia (SCG), isolated from the same animals, degenerated within a few days under the same experimental conditions. After individual or small groups of chromaffin cells settled onto the dish surface, fine unmyelinated processes grew out in response to NGF, as in the case of neurite outgrowth in sympathetic neurons. NGF is not the only medium capable of inducing process outgrowth from chromaffin cells. A conditioned medium of C6 glioma cells (Unsicker *et al.*, 1984) or media conditioned by such tissues as adrenal fibroblast-like cells (Unsicker and Hofmann, 1983) and heart cells (M. Ogawa, unpublished data) have a similar effect on chromaffin cells, even in the presence of anti-NGF antiserum. Most of the chromaffin cells isolated from neonatal rats are postmitotic and can elicit processes in the presence of cytosine arabinoside, which blocks DNA replication. Cell division is not required for process outgrowth, as in the case of neuronal cells.

In contrast to isolated sympathetic neurons of SCG, most of which elicited neurites with 1 or 2 days, the onset of process outgrowth from chromaffin cells was delayed, and only 10–20% of them elicited processes within 1 week of culture. After 1 week of culture, most chromaffin cells elicited processes. A possible interpretation for this delay in the onset of process outgrowth from chromaffin cells is that there is a quiescent period between the time of initial induction and the point at which they express structural features of the neuron. This may be the time required from the initial induction event to the synthesis of sufficient materials for visible expression as a neuronal phenotype. Treatment of adrenal chromaffin cells in culture with cycloheximide, which blocks protein synthesis, or actinomycin D, which blocks RNA synthesis, inhibited the process outgrowth. The fact that protein synthesis is not required for neurite regeneration in neuronal cells is thought to be due to the presence of presynthesized pools of the proteins required for growing neurites (Seeds *et al.*, 1970; Chan and Baxter, 1979). It seems likely that chromaffin cells require a new synthesis of RNA and proteins for the outgrowth of processes. It is interesting to note that cultured pheochromocytoma (PC12) cells, a clonal cell line derived from rat adrenal medullary tumor that responds to NGF to express properties characteristic of neuronal cells (Greene and

Tischler, 1976), require presynthesis of RNA before the outgrowth of processes (Burstein and Greene, 1978). Further studies on the molecular events leading to the expression of neuronal phenotype from chromaffin cells are awaited.

Although the sensitivity of cells to NGF depends on their age and species, it is noteworthy that the chromaffin cells isolated from normal adult humans can grow processes in NGF-containing medium (Tischler *et al.*, 1980). It suggests that at least human adrenal chromaffin cells maintain the potentiality to become neurons through adulthood. Withdrawal of NGF from the medium did not lead to reversion of the transdifferentiated chromaffin cells to that of the initial round endocrine phenotype. It can be inferred that once the adrenal chromaffin cells are committed as neuronal cells the genes that code for neuronal phenotype may be persistently active under positive control.

B. STRUCTURAL CHANGES

The tips of extended processes formed growth cones that had an essentially similar appearance and fine structure to those of neurons (Millar and Unsicker, 1982). Growing processes made contact with nearby nonchromaffin cells, but this connection was transient and the processes detached and extended further to other sites. These processes finally connected to other (sometimes their own) chromaffin cells and formed stable connections. Individual chromaffin cells and their processes showed a tendency to adhere to each other, and after 3 weeks in culture dense dendritic networks appeared, as shown in Fig. 1. Stimulation of one group of clustered chromaffin cells elicited compound excitatory postsynaptic potentials (EPSPs) in one of the cells of another cluster even 3 mm apart, indicating that most clustered cells are linked functionally. The diameter of the cell bodies increased from about 15 μm at plating to about 30 μm after 4 weeks of culture. The chromaffin cells showed strong catecholamine-specific histofluorescence in the early days of culture, and progressive process outgrowth was accompanied by decreasing fluorescence intensity in cell bodies and in processes. This may reflect a redistribution of catecholamines over a larger area in process-bearing cells.

Accompanying these changes at the light microscopic level, the fine structures also converted from the endocrine to the neuronal type. Rough endoplasmic reticulum, Golgi complex, and mitochondria inhabit a relatively sparse area in the cytoplasm of adrenal chromaffin cells *in situ* (compared with neuronal cells). The most characteristic feature of the chromaffin cell is the presence of a large number of membrane-bound granules that vary in size in the rat from 50 to 350

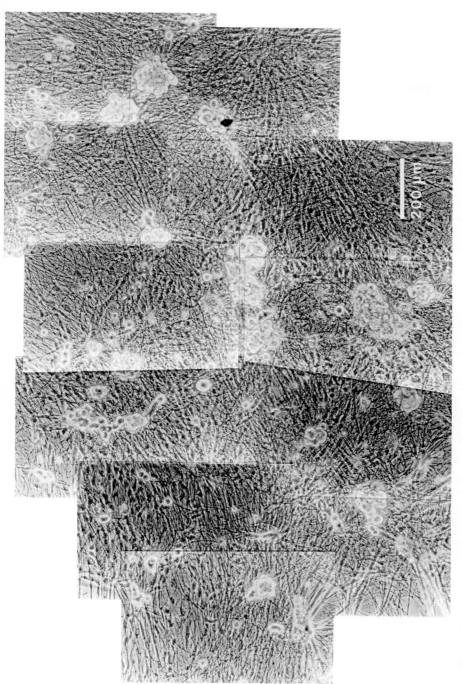

Fig. 1. Phase-contrast micrograph of adrenal chromaffin cells isolated from 10-day-old rats and cultured with NGF for 30 days. In this micrograph, proliferation of fibroblastic cells was substantially suppressed by treatment with cytosine arabinoside on days 21 and 22 after plating. Bar, 200 μm.

nm in diameter. An electron micrograph of the adrenal medulla of a
10-day-old rat (Fig. 2) reveals three types of vesicles: small clear vesicles and dense core vesicles in a preganglionic nerve ending between
two chromaffin cells and many electron-dense chromaffin granules in
chromaffin cells. In early days of culture, numerous chromaffin granules were dispersed, as *in situ,* throughout the cytoplasm. With increasing time in culture, the chromaffin granules decreased in the
central area and localized in the more peripheral areas of cytoplasm or
in the extended processes, and later disappeared from cells. During the
differentiation of the cultured chromaffin cells, there is a great increase in the amount of endoplasmic reticulum and mitochondria concomitant with increases in Golgi apparatus and microtubules. After 10
days of culture, two types of synaptic vesicles, clear and dense core
vesicles, appeared in extended processes (Fig. 3). The endings of processes swelled and increased in membrane density at the point of contact and formed so-called synapses among transformed chromaffin

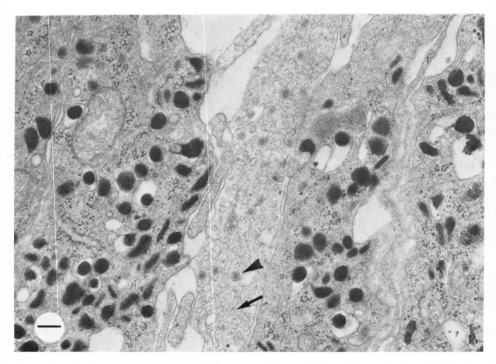

FIG. 2. Electron micrograph of adrenal medulla in 10-day-old rats. Both small clear
vesicles (arrow) and large dense core vesicles (arrowhead) are seen in preganglionic
nerve fibers between chromaffin cells. Bar, 200 nm.

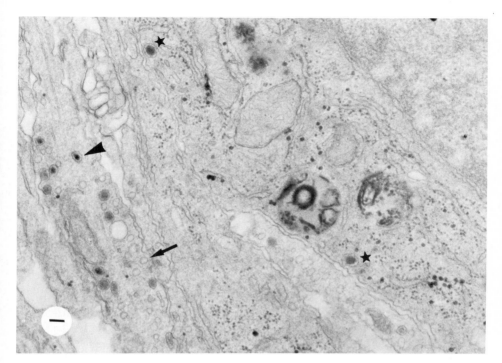

FIG. 3. Electron micrograph of adrenal chromaffin cells isolated from 10-day-old rats and cultured with NGF for 30 days. Associated with microtubules, small clear vesicles (arrow) and dense core (arrowhead) vesicles are seen in the extended process. In this figure, vesicles (stars) of around 110 nm in diameter are seen in the cell soma. These vesicles are larger and more dense than dense core vesicles, which are in the process. It might be argued whether these vesicles in the cell soma are survivors of small chromaffin granules present at plating or are newly synthesized dense core vesicles. Bar, 100 nm.

cells (Ogawa *et al.*, 1983). During the course, the content of heterochromatin decreased, the overall density of the nucleus was reduced, and the cells became clearly basophilic, which may reflect the increase of Nissl substance in the cytoplasm. It is noteworthy that these features of transdifferentiation of chromaffin cells are basically in accord with the developmental changes of neurons from immature cells *in vivo* and *in vitro* (see Jacobson, 1978).

C. FUNCTIONAL CHANGES

Accompanying these structural changes, synaptic transmission, which is the most characteristic property of neurons, appeared among transformed chromaffin cells (Ogawa *et al.*, 1983, 1984). Figure 4 pre-

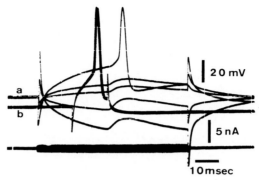

FIG. 4. Synaptic interaction at a newly formed synapse among cultured adrenal chromaffin cells. The recordings were made simultaneously from a pair of transformed chromaffin cells. Action potentials, elicited by depolarizing current injection through a micropipette, in the driver cell (cell b, middle trace) evoked EPSPs in the follower cell (cell a, top trace) after a delay of 5 milliseconds. In the follower cell, the membrane potential is set to different levels by passing current (bottom trace) through a micropipette.

sents an example of this. Action potentials in the driver cell (cell b in the figure, middle trace) evoked excitatory postsynaptic potentials (EPSPs) in the follower cell (cell a in the figure, upper trace). There was a delay of 5 milliseconds between the peak of the action potential in the driver cell and the beginning of the EPSP in the follower cell.

The following observations indicate that the synaptic transmission involved was chemically mediated: (1) a delay of a few milliseconds was observed between firing in the presynaptic cell and depolarization in the postsynaptic cell; (2) the variations in the amplitude of EPSPs suggested the quantum release of transmitter; (3) the amplitude of EPSPs was sensitive to the level of postsynaptic membrane potential and to Ca^{2+} concentration in the medium. Nicotinic cholinergic blocking agents, such as hexamethonium, curare, and tetraethylammonium, reversibly blocked these synaptic transmissions. These results collectively indicate that the adrenal chromaffin cells develop cholinergic synapses among themselves in culture. Furthermore, we observed electrophysiologically that the isolated adrenal chromaffin cells form cholinergic synapses with heart cells in culture (M. Ogawa, unpublished data).

These findings raised the possibility that the transdifferentiating chromaffin cells, which are adrenergic at plating, may also be plastic with respect to the transmitter choice, as previously reported for cultured sympathetic neurons. In the case of sympathetic neurons, the culture condition is known to determine neurotransmitter selection (see Patterson, 1978). When adrenergic sympathetic neurons are cocultured with appropriate nonneuronal cells, or even with conditioned medium prepared with nonneuronal cells from the same species, they develop striking cholinergic properties and form functional cholinergic synapses among themselves. On the contrary, when they are cultured under depolarizing conditions or cultured alone without nonneuronal cells, they are prevented from converting from an adrenergic to a cholinergic phenotype. The cultured adrenal chromaffin cells under depolarizing conditions also developed adrenergic synapses where norepinephrine was stored in synaptic vesicles when they were stained with permanganate for electron microscopy (Ogawa et al., 1983, 1984). We do not yet have sufficient data as to whether the phenotypic expression of the transmitter could be acquired before the chromaffin cells became committed to transdifferentiating as neuronal cells. We have observed that clear vesicles, which are storage vesicles for acetylcholine, are seen in extended processes or in their endings only and, further, that the chromaffin granules, which store catecholamines in situ, are destined to disappear after the cell becomes committed to its neuronal fate. These observations suggest that the neurochemical phenotype in chromaffin cells might be determined concomitantly in the course of neuronal transdifferentiation in culture.

III. Transdifferentiation of Adrenal Chromaffin Cells *in Situ* and in Ectopic Sites

On the assumption that neural transdifferentiation of adrenal chromaffin cells takes place *in situ* in a situation similar to that *in vitro,* Aloe and Levi-Montalcini examined the effects of NGF on embryonic and neonatal rats (Aloe and Levi-Montalcini, 1979). When NGF was injected into 16- to 17-day-old fetuses and daily after birth, they found dramatic morphologic changes in the adrenal gland: the whole chromaffin cell population underwent transformation into neuronal cells and the cortical section of the gland was in large part occupied with processes elongated from the chromaffin cells. The injection of NGF only after birth caused many chromaffin cells to transdifferentiate into neuronal cells and brought a decrease in the number of endocrine chromaffin cells.

Morphological evidence was accumulated to show that matured adrenal chromaffin cells can transdifferentiate into neuronal cells and form synapses with appropriate target cells in ectopic sites. Olson and colleagues showed histofluorochemically in adult rats that the transplanted adrenal chromaffin cells extend catecholamine-containing processes that invade the host iris (Olson, 1970; Olson and Malmfors, 1970) or the adjacent brain tissue grafts (Olson et al., 1980) in the anterior eye chamber. Such findings were corroborated by electron microscopic observations revealing that processes extending from chromaffin cells contain small clear and dense core vesicles in addition to chromaffin granules, and that the chromaffin cells form synapse-like contacts with each other (Unsicker *et al.*, 1981). The space between the capsule and the kidney parenchyma has also served as a useful site for grafts of adrenal medulla to study the outgrowth of their processes. When autologous transplantation of adrenal medulla was performed in adult guinea pigs, the cells developed processes that displayed longitudinally arranged microtubules and synaptic vesicles in addition to the chromaffin granules (Unsicker *et al.*, 1977, 1983). In transplantation experiments, the morphological cell changes were accompanied by a decrease in the size and number of chromaffin granules and by losses of epinephrine and the enzyme phenylethanolamine *N*-methyl transferase (PNMT), which converts norepinephrine into epinephrine.

These observations raise the possibility that the adrenal medulla can take on a variety of functions and serve as a potential replacement for various classes of nervous tissues. Experimental rats with unilateral lesions of substantia nigra pars compacta, a useful model of Parkinson's disease, showed significantly less abnormal rotational behavior with adrenal medulla grafts (Freed *et al.*, 1981; Herrera-Marschitz *et al.*, 1984). Although the implanted adrenal medullary cells could not innervate the host striatum, the tissue survived for some time and actively secreted catecholamines in the implantation site. At present, it seems premature even to speculate that the transplanted chromaffin cells might take the place of the damaged neural tissue and provide the lost input to the postsynaptic tissues. However, it is expected that transplantation of the adrenal medulla may provide corrective therapy for unmanageable neurological diseases such as Parkinson's, since the adrenal medulla represents a source of catecholaminergic cells in the body that can be used for autologous grafting.

IV. Concluding Remarks

There is little doubt that adrenal medullary chromaffin cells hold the potential to differentiate into neuronal cells. Why do the chromaffin cells differentiate into endocrine cells, and not neuronal cells, within the adrenal anlage in normal development? On their route to the final target site, the primordia of adrenal chromaffin cells are subjected to encounters with mesodermal cortical cells, whereas the sympathetic neurons are not. The influence of cortical cells on the differentiation of medullary cells has been suggested by the fact that when dexamethazone, a kind of glucocorticoid, was added to the culture medium, fiber outgrowth from chromaffin cells was completely abolished (Unsicker *et al.,* 1978). The adrenal medullae *in situ* are exposed to high concentrations of cortical hormones through a capillary plexus from the cortex. We might speculate that the adrenal chromaffin cells, throughout their lives, are repressed, as to their neuronal potentiality and are fated to differentiate into endocrine cells under the influence of cortical cells *in situ.* However, this speculation becomes controversial in the case of PC12 cells, which grow processes in the presence of dexamethazone (Tischler and Greene, 1980).

One could also anticipate that sympathetic neurons or their primordia might show reverse transdifferentiation to chromaffin cells under appropriate conditions. There is strong evidence for a conversion from ganglionic neurons to chromaffin cells *in vivo.* Le Douarin and colleagues have been doing ingenious transplantation experiments in bird embryos and have shown that immature ganglionic cells can redifferentiate into adrenal chromaffin cells when they are transplanted to a region where adrenal medulla normally arise (see Le Douarin, 1982). They have expressed the possibility that ciliary ganglionic cells, once differentiated as cholinergic neurons, can redifferentiate to adrenergic chromaffin cells (Le Douarin, 1980). From a different approach, Eränkö and colleagues showed the transdifferentiation of ganglionic neurons to chromaffin-like cells *in vitro* (Eränkö and Eränkö, 1980). They cultured the explants of SCG of newborn rats in hydrocortisone-enriched medium and observed the induction of sympathetic cells into small intensely fluorescent (SIF) cells that appeared intermediate in their properties between neurons and adrenal medullary cells (reviewed by Landis and Patterson, 1981).

In catecholaminergic cells, which are derived from part of the adrenomedullary neural crest, the differentiated structures are morphologically and functionally more closely related to the site where the crest cells become localized than to their origin in the neural primordium of their normal developmental fate. Environmental signals interact with their genomes and switch the gene activation pattern, though not irreversibly, from the neuronal to the endocrine type, or vice versa.

ACKNOWLEDGMENTS

We thank Drs. A. Inouye and A. Irimajiri for their many constructive discussions during the work. We thank Mr. C. Zagory, Jr. for his reading and Mrs. T. Ichinowatari for her typing of the manuscript. Original work of the authors was supported, in part, by a Grant-in-Aid for Special Project Research, "Multicellular Organization" (Project No. 59480023) and by No. 58570058 to M.O. from the Ministry of Education, Science and Culture of Japan.

REFERENCES

Aloe, L., and Levi-Montalcini, R. (1979). *Proc. Natl. Acad. Sci. U.S.A.* **76,** 1246–1250.

Burstein, D. E., and Greene, L. S. (1978). *Proc. Natl. Acad. Sci. U.S.A.* **75,** 6059–6063.

Chan, K. Y., and Baxter, C. F. (1979). *Brain Res.* **174,** 135–152.

Eränkö, O., and Eränkö, L. (1980). *In* "Histochemistry and Cell Biology of Autonomic Neurons, SIF Cells and Paraneurons" (O. Eränkö, S. Soinila, and H. Päivärinta, eds.), pp. 17–26. Raven, New York.

Freed, W. J., Morihisa, J. M., Spoor, E., Hoffer, B. J., Olson, L., Seiger, A., and Wyatt, R. J. (1981). *Nature (London)* **292,** 351–352.

Greene, L. A., and Tischler, A. S. (1976). *Proc. Natl. Acad. Sci. U.S.A.* **73,** 2424–2428.

Herrera-Marschitz, M., Stromberg, I., Olson, D., Ungerstedt, U., and Olson, L. (1984). *Brain Res.* **297,** 53–61.

Jacobson, M. (1978). "Developmental Neurobiology," 2nd Ed. Plenum, New York.

Landis, S. C., and Patterson, P. H. (1981). *Trends Neurosci.* **4,** 172–175.

Le Douarin, N. (1980). *Nature (London)* **286,** 663–669.

Le Douarin, N. (1982). "The Neural Crest." Cambridge Univ. Press, London and New York.

Millar, T. J., and Unsicker, K. (1982). *Dev. Brain Res.* **2,** 577–582.

Ogawa, M., and Ishikawa, T. (1982). *Adv. Biosci.* **36,** 103–110.

Ogawa, M., Ishikawa, T., and Ohta, H. (1983). *Tissue Cult.* **9,** 48–52 (in Japanese).

Ogawa, M., Ishikawa, T., and Irimajiri, A. (1984). *Nature (London)* **307,** 66–68.

Okada, T. S. (1980). *Curr. Top. Dev. Biol.* **16,** 349–379.

Okada, T. S. (1983). *Cell Differ.* **13,** 177–183.

Olson, L. (1970). *Histochemie* **22,** 1–7.

Olson, L., and Malmfors, T. (1970). *Acta Physiol. Scand. Suppl.* **348,** 1–112.

Olson, L., Seiger, A., Freedman, R., and Hoffer, B. (1980). *Exp. Neurol.* **70,** 414–426.

Patterson, P. H. (1978). *Annu. Rev. Neurosci.* **1,** 1–17.

Seeds, N., Gilman, A. G., Amano, T., and Nierenberg, M. W. (1970). *Proc. Natl. Acad. Sci. U.S.A.* **66,** 160–167.

Tischler, A. S., and Greene, L. A. (1980). *In* "Histochemistry and Cell Biology of Autonomic Neurons, SIF Cells and Paraneurons" (O. Eränkö, S. Soinila, and H. Päivärinta, eds.), pp. 61–68. Raven, New York.

Tischler, A. S., Delellis, R. A., Biales, B., Nunnemacher, G., Carabba, V., and Wolfe, H. J. (1980). *Lab. Invest.* **43,** 399–409.

Unsicker, K., and Hofmann, H. D. (1983). *Dev. Brain Res.* **7,** 41–52.

Unsicker, K., Zwarg, U., and Habura, O. (1977). *Brain Res.* **120,** 533–539.

Unsicker, K., Krisch, B., Otten, U., and Thoenen, H. (1978). *Proc. Natl. Acad. Sci. U.S.A.* **75,** 3498–3502.

Unsicker, K., Tschechne, B., and Tschechne, D. (1981). *Cell Tissue Res.* **215,** 341–367.

Unsicker, K., Vey, J., Hoffmann, H.-D., Muller, T. H., and Wilson, A. J. (1984). *Proc. Natl. Acad. Sci. U.S.A.* **81,** 2242–2246.

CHAPTER 8

NEURAL CREST AND THYMIC MYOID CELLS

Harukazu Nakamura

DEPARTMENT OF ANATOMY
HIROSHIMA UNIVERSITY SCHOOL OF MEDICINE
MINAMI-KU, HIROSHIMA, JAPAN

and

Christiane Ayer-Le Lièvre*

INSTITUT D'EMBRYOLOGIE
ANNEXE DU COLLÈGE DE FRANCE
CENTRE NATIONAL DE LA RECHERCHE SCIENTIFIQUE
NOGENT-SUR-MARNE, FRANCE

I. Introduction

Myoid cells commonly appear in the thymus of reptiles, birds, and mammals, including humans. The rather unexpected presence of these cells in this organ has attracted special attention, because it has been thought to be related with the autoimmune disease myasthenia gravis. Most myoid cells are situated in the medulla of the thymus (Raviola and Raviola, 1967; von Gaudecker and Müller-Hermelink, 1980), and they share common morphological and functional characteristics with skeletal muscle (Van de Velde and Friedman, 1970; Wekerle *et al.*, 1975; Kao and Drachman, 1977; Drenckhahn *et al.*, 1978, 1979, 1983).

The origin of the thymic myoid cells (TMC) has been a matter of dispute, and two alternative hypotheses have been proposed. One is that TMC occur as a result of transdifferentiation of thymic endodermal epithelial cells (Raviola and Raviola, 1967; Ito *et al.*, 1969). In the other hypothesis, TMC are considered to originate from the surrounding tissues (Cooper and Tochinai, 1982). Thus, the problem of seemingly ectopic differentiation of myoid cells in thymus is pertinent to review in this volume.

* Present address: Department of Histology, Karolinska Institute, Stockholm, Sweden.

*CURRENT TOPICS IN
DEVELOPMENTAL BIOLOGY, VOL. 20*

II. Hypotheses on the Origin of Thymic Myoid Cells (TMC)

Since TMC are surrounded by the epithelial cells and are connected to them by desmosomes, some authors have assumed that these cells arise from the thymic epithelial cells (Raviola and Raviola, 1967; Ito *et al.*, 1969).

Ito and Mori (1983) performed a clonal culture of myoid and epithelial cells of the thymus of the rat. In their cultures, myoid cells were not found in the clone of the epithelial cells, and no epithelial cells were found in the clone containing the myoid cells. They therefore suggested that thymic epithelial cells did not transdifferentiate into myoid cells.

Cooper and Tochinai (1982) cultured the thymus at various stages of *Xenopus* embryos. In cultures of young thymus, only epithelial cells are differentiated, though in cultures of older embryos, myoid cells, lymphoid cells, and epithelial cells are differentiated. Thus, these authors interpreted their results to support the hypothesis that TMC are of extrinsic origin.

III. Neural Crest and Thymus

The neural crest occurs transiently on the dorsal part of the neural tube, from which neural crest cells migrate and differentiate into many kinds of cells and tissues, including the peripheral nervous system, adrenal medulla, melanocytes, visceral skeleton of the face and neck, and others. Interspecific transplantation of the neural crest between quail and chick embryos has elucidated many aspects of neural crest migration and differentiation (Le Douarin, 1980, 1982) because quail cells are distinguishable from chick cells (Le Douarin, 1969, 1973).

It was shown that the head neural crest can produce mesectoderm, that is, mesenchyme from the neural crest. The connective tissue component of the pharyngeal derivatives, such as the thyroid and parathyroid glands, comes from the neural crest (Le Lièvre and Le Douarin, 1975). When the rhombencephalic neural crest was extirpated, the thymus and the parathyroid became hypoplastic or aplastic (Bockman and Kirby, 1984). If the TMC are of extrinsic origin, the possibility arises that myoid cells are differentiated from neural crest cells.

IV. Interspecific Transplantation

In order to test the possible origin of TMC from neural crest, an experiment in which the quail neural crest was transplanted into chick embryos was performed. In this experiment, a unilateral neural tube of

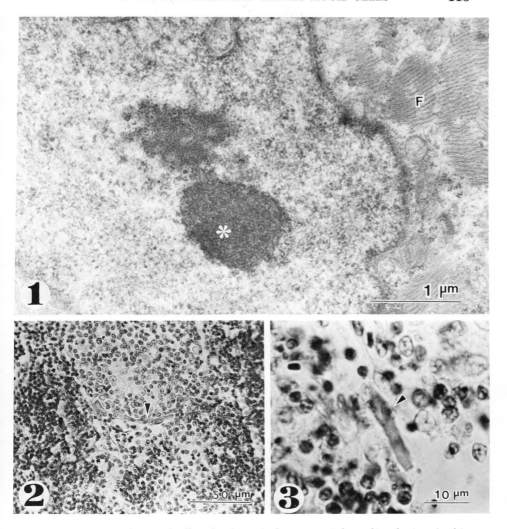

FIG. 1. A part of a myoid cell in the chimeric thymus at 19 days of incubation. As this cell has a large mass of DNA (*), it is thought to be a quail cell. Myofilaments (F) are beginning to be organized.

FIG. 2. Thymus of a newly hatched chick. The specimen was processed immunohistochemically using muscle-specific anti-creatine kinase antibody. An elongated form of myoid cell (arrowhead) is visible in the medulla. Myoid cells are rarely found in the cortex.

FIG. 3. Chimeric thymus after immunohistochemical reaction and Feulgen–Rossenbeck staining. Note the quail nucleus (arrowhead) in the myoid cell which reacted with antibody.

the quail embryo at the level of the rhombencephalon was transplanted into a chick embryo, since the thymic connective tissue comes from the neural crest at this level (Le Lièvre and Le Douarin, 1975). Both chick and quail embryos had 6–10 somites at the time of operation. The operated embryos were killed at 17–20 days of incubation because TMC become visible around hatching. The thymus of the operated side was processed for electron microscopy and immunohistochemistry. For immunohistochemistry with the PAP method, an anti-creatine kinase antibody that reacts specifically to creatine kinase of skeletal muscle was used as a primary antibody. The sections for this immunohistochemical preparation were stained according to the Feulgen–Rossenbeck method (Feulgen and Rossenbeck, 1924), making it possible to distinguish quail cells from chick cells.

Though the chimeric thymus formed at the operated side was somewhat smaller than the normal one, electron microscopic observations revealed morphologically well-differentiated TMC by 20 days of incubation. The quail and chick cells are distinguishable on electron micrograph. Quail nuclei have large masses of heterochromatin associated with nucleolar RNA, whereas heterochromatic DNA is scattered in the nucleus of the chick (Le Douarin, 1973). Quail-type myoid cells were found in the chimeric thymus (Fig. 1).

The anti-creatine kinase antibody reacts specifically to skeletal muscle. When this antibody was adopted for the detection of myoid cells, only the myoid cells reacted specifically in the thymus (Fig. 2). Immunohistochemistry using anti-creatine kinase antibody demonstrated that TMC are localized in the medulla. TMC were rarely found in the cortex, where lymphoid cells were predominant. The chimeric thymus stained by the Feulgen–Rossenbeck method after the immunohistochemical procedure contained myoid cells with nuclei of the quail type (Fig. 3). From the results of electron microscopic and immunohistochemical studies on the chimeric birds, we can conclude that TMC do not occur by transdifferentiation of the thymic epithelial cells, but derive from the neural crest.

ACKNOWLEDGMENTS

The authors thank Dr. K. Watanabe, Fukui Medical School, for kindly offering the antibody and Prof. M. Yasuda, Hiroshima University School of Medicine, for critical reading of the chapter. This work was supported in part by a Grant No. 83-01-32 from the National Center for Nervous, Mental and Muscular Disorders of the Ministry of Health and Welfare and by a Grant-in-Aid for Special Project Research from the Ministry of Education, Science and Culture (No. 59213011).

REFERENCES

Bockman, D. E., and Kirby, M. L. (1984). *Science* **223**, 498–500.

Cooper, E. L., and Tochinai, S. (1982). *In* "Developmental Immunology: Clinical Problems and Aging" (E. L. Cooper and M. A. B. Brazier, eds.), pp. 209–213. Academic Press, New York.

Drenckhahn, D., Unsicker, K., Griesser, G.-H., Schumacher, U., and Gröschel-Stewart, U. (1978). *Cell Tissue Res.* **187**, 97–103.

Drenckhahn, D., von Gaudecker, B., Müller-Hermelink, H. K., Unsicker, K., and Gröshcel-Stewart, U. (1979). *Virchows Arch. B Cell Pathol.* **32**, 33–45.

Drenckhahn, D., Gröschel-Stewart, U., Kendrick-Jones, J., and Scholey, J. M. (1983). *Eur. J. Cell Biol.* **30**, 100–111.

Feulgen, R., and Rossenbeck, H. (1924). *Hoppe-Seyler's Z. Physiol. Chem.* **135**, 203–252.

Ito, T., and Mori, T. (1983). *Acta Anat. Nippon.* **58**, 363.

Ito, T., Hoshino, T., and Abe, K. (1969). *Arch. Histol. Jpn.* **30**, 207–215.

Kao, I., and Drachman, D. B. (1977). *Science* **195**, 74–75.

Le Douarin, N. M. (1969). *Bull. Biol. Fr. Belg.* **103**, 436–452.

Le Douarin, N. M. (1973). *Dev. Biol.* **41**, 162–184.

Le Douarin, N. M. (1980). *Curr. Top. Dev. Biol.* **16**, 31–85.

Le Douarin, N. M. (1982). "The Neural Crest." Cambridge Univ. Press, London and New York.

Le Lièvre, C. S., and Le Douarin, N. M. (1975). *J. Embryol. Exp. Morphol.* **31**, 453–477.

Raviola, E., and Raviola, G. (1967). *Am. J. Anat.* **121**, 623–646.

Van de Velde, R. L., and Friedman, N. B. (1970). *Am. J. Pathol.* **59**, 347–368.

Von Gaudecker, B., and Müller-Hermelink, H. K. (1980). *Cell Tissue Res.* **207**, 287–306.

Wekerle, H., Paterson, B., Keterlsen, U.-P., and Feldman, M. (1975). *Nature (London)* **256**, 493–494.

CHAPTER 9

THE POTENTIAL FOR TRANSDIFFERENTIATION OF DIFFERENTIATED MEDUSA TISSUES *IN VITRO*

*Volker Schmid and Hansjürg Alder**

INSTITUTE OF ZOOLOGY
UNIVERSITY OF BASEL
BASEL, SWITZERLAND

I. Introduction

The question of how stable differential gene expression is achieved in ontogeny (i.e., cellular commitment) has been for a long time a field of interest in developmental biology. Is stability a consequence of loss of genomic information (Tobler *et al.,* 1972) or of irreversible alterations in the genome (Caplan and Ordahl, 1978)? Is a change in cellular commitment possible, and if so, to what extent and what are the mechanisms? In plants there is clear evidence that fully specialized adult cells can be totipotent (Vasil and Vasil, 1972); cellular commitment is therefore fully reversible and can extend even to the full morphogenetic potential necessary to shape the organism. In animals, however, gene expression in committed cells seems in general to be highly stable. Nuclear transplantation studies have indeed clearly demonstrated that the nuclei of adult somatic cells are still able to express pluripotency when transplanted into oocyte cytoplasm (for review see DiBerardino, 1980; DiBerardino *et al.,* 1984). Success in these studies is, however, rare and it is assumed that this could be the consequence of changes in the chromatin structure of the implanted nucleus

---

* Present address: Friday Harbor Laboratories, Friday Harbor, Washington.

CURRENT TOPICS IN
DEVELOPMENTAL BIOLOGY, VOL. 20

117

which prevent normal cooperation between the nucleus and the recipient cytoplasm (DiBerardino, 1980). Results are greatly improved when the implanted nuclei are conditioned prior to implantation by serial cloning. One has to conclude that the full repertoire of genes is still present, but reactivation in the foreign oocyte cytoplasm becomes increasingly difficult the further advanced in development the implanted nucleus is (DiBerardino, 1980).

An experimental alternative to nuclear transplantation is the study of transdifferentiation, where the commitment of fully differentiated cells is changed and the reactivation of previously silent genes can be analyzed. Such studies have the advantage of having the whole cell available as a reaction unit. Although several laboratories have looked for transdifferentiation processes *in vivo* (Oberpriller, 1967; Burnett, 1968; Hay, 1968), only the formation of lens cells from melanocytes in newts is convincingly documented (for review see Yamada, 1977). The reason for this failure probably lies in the fact that in all the examined systems tissue organization is too complex to allow the reliable identification of the original cell type throughout the transdifferentiation process.

Investigations using cell and organ cultures have demonstrated change in cellular commitment in the fruit fly *Drosophila* (Haynie and Bryant, 1976; Szabad *et al.*, 1979), in amphibians (for review see Yamada and McDevitt, 1984), and in the chicken eye (for review see Eguchi and Itoh, 1982; Okada, 1983). However, the number of well-documented cases of transdifferentiation is small, and the potency to express new phenotypes is restricted to one or two cell types, mostly related to the original germ layer. Previous to our recent work, cultured animal cells were never reported to have formed a functional organ by transdifferentiation.

We have recently developed an *in vitro* regeneration system which is entirely based on transdifferentiation (Schmid *et al.*, 1982; Schmid and Alder, 1984). The cells are isolated as tissue layers from anthomedusae, marine jelly fish which belong to the phylum Coelenterata. Medusae lack a mesoderm and are organized as bilayered animals. However, they have invented real organs and highly specialized tissues, nerve nets (Mackie, 1980), and a variety of sense organs, e.g., a visual system (Weber, 1981) or mechanical receptors (Tardent and Schmid, 1972). Additionally, they have two properties which have been of crucial importance for the establishment of this experimental system; first, the animals are able to regenerate most of their body parts (Schmid, 1974a) and second, the swimming organ (the umbrella or bell) is composed of three tissue layers, of which two consist of only

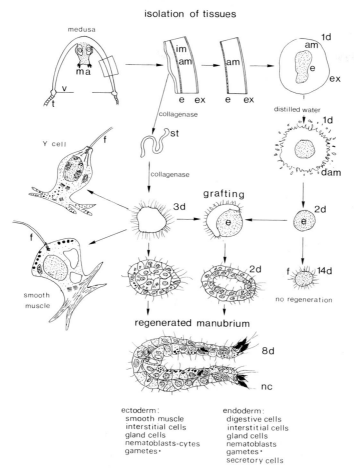

FIG. 1. Schematic drawing of the *in vitro* system used to regenerate manubria or tentacles. am, Outer mesoglea; d, days; dam, degenerating exumbrella and outer mesoglea; dst, destabilized striated muscle; e, endoderm of the subumbrellar plate; ex, exumbrella; f, flagella; im, inner mesoglea; ma, manubrium; nc, nematocytes; st, striated muscle; t, tentacles; v, velum; *, gametes observed in the grafting experiments.

one well-differentiated cell type (Schmid, 1972, 1980). Taking advantage of this unique anatomical situation, we have isolated these two tissues, mononucleated striated muscle and the endodermal plate of the subumbrella (Fig. 1), cultured them in artificial seawater, and tested the stability of their differentiated states, singly and in combination with each other. The results suggest that, on the tissue level, animals too have the potential to change cellular commitment to an extent almost comparable to that observed in plants.

II. The Isolation of Medusa Tissues

A. MONONUCLEATED STRIATED MUSCLE

Freshly hatched medusae, liberated from colonies of the polyps of *Podocoryne carnea* reared in the laboratory (Schmid, 1979), or medusae of *Polyorchis* sp., *Sarsia* sp., or *Stomotoca atra* collected in the plankton (Schmid, 1976) were washed several times in culture medium consisting of artificial Millipore-filtered seawater to which streptomycin (0.1 mg/ml) was added. The animals were then stained with DAPI (4',6-diamidino-2-phenylindole·2 HCl, Serva, Heidelberg, West Germany), 0.05 µg/ml, for 60 minutes. DAPI binds to AT-rich regions of the DNA and when applied *in vivo* for 1 hour stains brightly the nuclei of all cell types except those of striated muscle (Figs. 2–5; Schmid *et al.*, 1982). The medusae were then washed and incubated in collagenase (Millipore Corp., No. 5275, 150 units/ml culture medium) for 6–8 hours at 26°C. This treatment digests the inner mesoglea and separates the striated muscle from the subumbrellar plate endoderm (Schmid, 1980). Interradial fragments were excised with microscissors and watchmaker's forceps, the peripheral part with the velum and the

---

FIG. 2. Medusa of *Podocoryne carnea* after being treated with collagenase for 4 hours. The mononucleated striated muscle (st) has largely separated from the subumbrellar plate endoderm (sp). ma, Manubrium; te, tentacles. ×40.

FIG. 3. Enlarged view of a medusa treated with collagenase and stained with DAPI. The striated muscle (st) has detached from the subumbrellar plate endoderm (sp). ex, Exumbrella. Darkfield micrograph. ×227.

FIG. 4. Mirror image of Fig. 3. Nuclei (n) of exumbrella (ex) and subumbrellar plate endoderm (sp) stain brightly with DAPI, and nuclei of striated muscle (st) do not stain. Fluorescence micrograph. ×227.

FIG. 5. Fragment of isolated striated muscle stained with DAPI after isolation. Nuclei and cytoplasm of contaminating cells of the subumbrellar plate endoderm stain brightly. Fluorescence micrograph. ×114.

FIG. 6. Muscle fragment 10 days after isolation: smooth muscle cells and y cells are well differentiated. f, Flagellum; ny, nuclei of y cells; s, smooth muscle-like arrangement of myofilaments; sn, nucleus of smooth muscle cells; sv, small vesicles of y cells; solid arrow, bundles of striated myofilaments; open arrows, peripheral vacuoles typical for smooth muscle. ×4331.

FIG. 7. The basal part of smooth muscle and y cells of a 5-day-old isolate. s, Smooth muscle filaments; sv, small vesicles of y cells. ×14,413.

FIG. 8. Y cell adjacent to a smooth muscle cell with peripheral vacuoles (arrows); 10-day-old isolate. ny, Nucleus of y cell; sv, small vesicles. ×7455.

FIG. 9. Cytoplasmic bridge between two y cells in a 10-day-old muscle fragment. ny, Nuclei of y cells; sv, small vesicles. ×14,200.

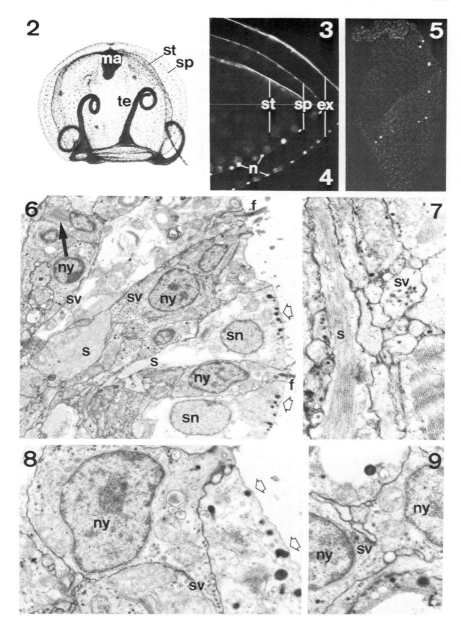

circular canal removed, and the loose striated muscle gently pulled off. Immediately after isolation the striated muscle was examined by fluorescence microscopy for contamination by DAPI-stained cells of the subumbrellar plate endoderm (Fig. 5; Schmid et al., 1982). Alternatively, striated muscle can be isolated mechanically without collagenase treatment by the same dissection procedure; this, however, is more difficult to do and many more isolated fragments are contaminated by endodermal cells.

B. SUBUMBRELLAR PLATE ENDODERM

After the striated muscle is excised, the remainder of the extirpated interradial medusa fragment consists of the subumbrellar plate endoderm, the bulky outer mesoglea (extracellular matrix), and the exumbrellar tissue (Fig. 1; Schmid et al., 1982). If replaced in culture this fragment rounds up within the next 12 hours so that the subumbrellar plate endoderm is concentrated in the center and is covered by the outer mesoglea, over which the exumbrella forms a continuous tissue layer. These rounded-up fragments are rinsed for about 10 seconds with distilled water and transferred back to culture medium (Fig. 1). The osmotic shock kills all the peripherally arranged exumbrella cells, whereas most of the endoderm, protected by the outer mesoglea, survives (Schmid et al., 1982, 1984). The remnants of the killed exumbrella cells (Fig. 1) are rinsed off by sucking the fragment repeatedly through a pipette, and the endoderm, together with remaining portions of the outer mesoglea, is transferred to culture at 12°C for further observation.

III. The Potential for Transdifferentiation

A. STRIATED MUSCLE

In situ striated muscle of all the different species investigated consists of mononucleated cells which are out of cell cycle (Schmid, 1972, 1978; Schmid et al., 1984), presumably in G_0 (resting stage). The cells form a thin, monolayered, epithelial tissue which lines the subumbrellar cavity and adheres to the inner mesoglea, an extracellular matrix (ECM) which is part of the bell (umbrella, Fig. 1; Schmid, 1980). Although the gross morphology is the same in all the species investigated, the bundles of myofibers vary in thickness, probably in direct relation to the swimming behavior of these animals.

1. Stability of the Differentiated State

Striated muscle mechanically isolated and without collagenase treatment contains variable amounts of the inner mesoglea. This can be directly observed in the large medusa species (*Polyorchis, Stomotoca*) where the inner mesoglea is sufficiently thick. When such fragments are cultured, the muscle tissue covers the mesoglea within a few hours. In the course of the next 2–3 days the volume of the isolated fragments increases, resulting in typical swollen isolates. DNA synthesis, flagellum formation, or the disappearance of the cross-striated myofibers is not observed; cellular commitment remains stable (Schmid, 1978; Schmid et al., 1984). Since these isolates can demonstrate the contractions typical of striated muscle tissue the cells must be functional in the original sense. No new cell types are formed and no medusa parts are regenerated. The isolates can live for months until they finally fall apart.

2. Destabilization

Striated muscle tissue isolated with collagenase and posttreated with collagenase (900 units/ml, 26°C, 4–8 hours) is in general not able to maintain the functional state of a mononucleated cross-striated muscle cell. In this case, the isolated muscle tissue shrinks shortly after isolation and forms an aggregate-like structure (Schmid, 1978; Schmid and Alder, 1984). By this process the flat muscle cells, originally arranged as a single-layered epithelium, become cuboidal and multilayered. One day after isolation the bundles of striated myofibers are somewhat disarranged in comparison to the other cytoplasmic components, but from the second day on, they start to fall apart at the Z bands (Schmid and Alder, 1984). The bundles of striated myofibers may eventually disappear completely, although there is considerable variation between different fragments and even within cells of the same fragment in this respect. The presence of bundles of striated myofibers seem to have no influence on the ability of the cells to transdifferentiate and/or to participate in regeneration. DNA synthesis at high synchrony starts at 52–55 hours postisolation (Schmid, 1975; Alder, 1982) followed by mitoses (after 3 days). *De novo* formation of flagella in all the peripheral cells of the isolate is observed first at day 2.5 (Fig. 6; Schmid, 1974b). In the following sections, isolated striated muscle cells exhibiting these features will be referred to as "destabilized."

3. Transdifferentiation to New Cell Types

The following observations refer exclusively to tissues isolated from *P. carnea*. After mitosis is completed (between 3 and 4 days after isolation) the destabilized isolates consist entirely of two new cell types (Figs. 1, 6–9; Schmid and Alder, 1984) present in about equal proportions. One cell type was identified as a smooth muscle cell; the other is not yet certainly identified and is therefore referred to as a y cell (Fig. 1).

The *smooth muscle cell* type (Figs. 1, 6, 7) is characterized by peripherally arranged medium-sized vacuoles filled with an electron-dense substance, large empty vacuoles in the mid part of the cell body, a large nucleus without prominent electron-dense material, and a flagellum; at the base of the cells contractile filaments are seen. Often these filaments are observed to form a continuum with bundles of striated myofibers. With the exception of the remnants of striated myofibril this cell type is indistinguishable at the ultrastructural level from smooth muscle cells seen in some medusa organs (e.g., tentacle and manubrium; Schmid and Alder, 1984).

The *y cell* type (Figs. 1, 8, 9) is distinctly smaller in size in comparison to smooth muscle cells. The cytoplasm bearing the base of the flagellum always forms a narrow bottle-shaped outgrowth at the periphery of the isolate. The cytoplasm contains many more ribosomes, and toward the base, numerous small vesicles filled with electron-dense material are seen (Figs. 7 and 8). The vesicles strongly resemble the dense-cored vesicles found in nerve cells of coelenterates (Westfall and Kinnamon, 1978). To date, however, no synaptic structures have been identified. Vacuoles or filament arrangements like those found in smooth muscle cells have never been observed in this cell type. Occasionally cytoplasmic bridges connect y cells (Fig. 9).

Fragmentary bundles of striated myofilaments are often seen in both cell types when young (Figs. 6 and 7), but they gradually disappear with increasing age. Both cell types are found in all isolates tested as described and remain present until the isolates fall apart after 3–12 weeks.

4. Formation of the Regenerate

When the culture temperature is reduced to 22°C or less after collagenase treatment, 20–30% of the isolates form; by the fifth day an inner, endodermal layer separated from the outer, ectodermal layer by a basal lamina (Fig. 10; Schmid and Alder, 1984). In some rare cases

---

FIG. 10. Muscle fragment 6 days after isolation. The fragment has formed an outer, ectodermal layer (ec) and an inner, endodermal layer (en) separated by a basal lamina (m); cells of both tissues contain bundles of striated myofilaments (solid arrows). ×8064.

FIG. 11. Muscle fragment 6 days after isolation (same specimen as in Fig. 10). The first nematoblasts differentiate in the endodermal layer; the thread (t) of the nematocyst is still outside the capsule (c). The same cell contains nonvacuolated bundles of striated myofilaments (solid arrow). n, Nucleus of nematoblast; sv, small vesicles of y cell. ×8136.

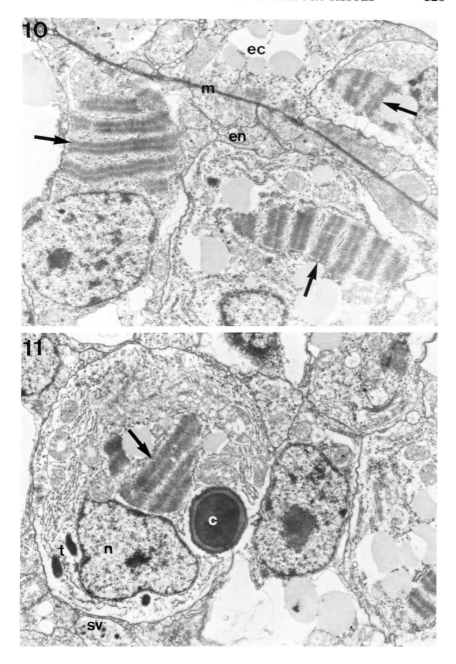

cells from both layers still contain portions of striated myofilaments (Fig. 10). Occasionally bundles of striated myofilaments are even found in diffentiating nematoblasts (Fig. 11). Between the fifth and sixth day directional swimming of the isolates indicates the expression of the proximodistal axis of the future regenerate. In the following days the normal inventory of at least seven to eight new cell types differentiates, perfectly arranged to form a tentacle or a manubrium (Figs. 1 and 12), the sexual and feeding organ of the medusa. At some occasions both manubrium and tentacle were expressed in the same regenerate.

The origin of most of the newly formed cell types from striated muscle cells is visually documented by the few rare cases in which the striated myofibers were not completely degraded and remained visible in the cytoplasm of the new cell types (Figs. 13–15). Usually the regenerates are very small (Fig. 12) and no attempts have been made so far to feed the regenerated manubrium and thus initiate gamete formation (see later).

B. The Endoderm

1. Stability of the Differentiated State

In vivo the endoderm forms a single layer of extremely flat cells beneath the striated muscle in the umbrella of the medusa (Schmid, 1980). It is covered on both sides by ECMs (inner and outer mesoglea, Fig. 1). Due to the isolation procedure, freshly isolated endoderm is covered by the outer mesoglea but might also contain remnants of the inner mesoglea (Schmid et al., 1984). In the following 12–24 hours the endoderm forms a sphere and reduces the gap junctions (visualized by lanthanium; Weber and Schmid, 1984), and the ECM material completely disappears from the surface of the isolates. Large patches of ECM can persist for many days between the cells (Schmid et al., 1984). The endoderm is extremely sensitive to temperature in the isolated state. When cultured like muscle isolates at 22°C, it dies within 2–4 days; however, when cultured at 10–12°C it will survive for weeks and

---

FIG. 12. Regenerated manubrium 12 days after isolation of the muscle. nc, Nematocytes of the mouth part (arrow). ×213.

FIG. 13. Figures 13–15 show the same specimen as Fig. 12. The electron micrographs of Figs. 13–15 demonstrate the presence of bundles of striated myofilaments (arrows) in the cytoplasm of nematocytes. c, Capsule of nematocyte; n, nucleus of nematocyte. ×11,928.

FIG. 14. Striated myofilaments in a gland cell. ×8023.

FIG. 15. Striated myofilaments in a digestive cell. nd, Nucleus of digestive cell, ng, nucleus of gland cell. ×6745.

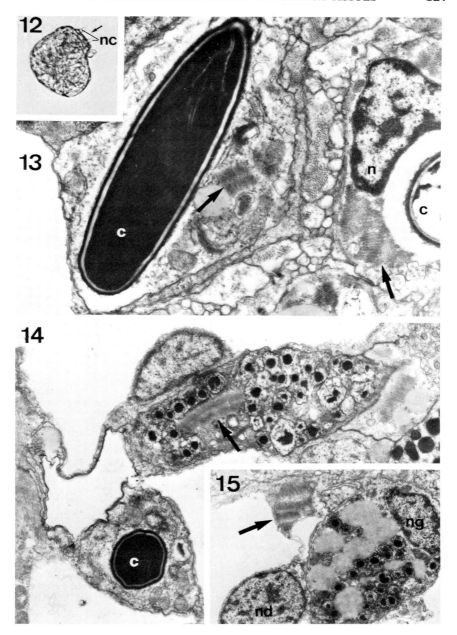

form functional flagella (Schmid et al., 1984). Gap junctions can be fully restored, at least in the first days, if the endoderm is allowed to reform a monolayer on ECM (Weber and Schmid, 1984).

Contrary to the situation in the isolated muscle, DNA synthesis and mitoses were never observed in the isolated endoderm, even after collagenase treatment (Schmid et al., 1984). Endoderm seems unable to form new cell types or a regenerate when submitted to the same procedures as are successful with muscle. However, the potential to transdifferentiate is present in endoderm, as demonstrated by the following experiments.

2. Destabilization of the Endoderm by Striated Muscle

Endoderm can be activated to synthesize DNA and to transdifferentiate to new cell types when combined with isolated striated muscle (Schmid et al., 1982). In this case regeneration leads to a fully functional manubrium, able to produce gametes, and occasionally to tentacles (Table I and Figs. 1, 16–18; Schmid et al., 1982). The ability to activate endoderm is present only in destabilized striated muscle (i.e., treated with collagenase prior to grafting; see above and Table I). Muscle isolated mechanically, without enzyme treatment, and then grafted to endoderm stably maintains its own differentiated state, and neither DNA synthesis nor formation of new cell types is observed in the endoderm (Table I and Fig. 19; Schmid et al., 1984); no regenerate is formed. Direct desmosomal contact between the muscle and the endoderm does not suffice for the activation of the endoderm (Fig. 20). The critical condition for activation is that the number of muscle cells must be large enough (20–40, Table II) to cover the endoderm. These few muscle cells have to stretch greatly in this process (Fig. 21; Schmid et al., 1984).

3. Transdifferentiation

As the isolated striated muscle alone can give rise to all the cell types found in complete regenerates (see above), the grafting experiment described in the previous section cannot conclusively demonstrate the transdifferentiation of the endoderm. More evidence is given by experiments in which the destabilized striated muscle was treated with mitomycin C, a drug which irreversibly inhibits DNA replication (Schmid et al. 1982). With the hypothesis that cell cycles are a prerequisite for cellular commitment, and therefore also for transdifferentiation (see below), one would predict destabilized muscle treated with mitomycin C cannot contribute any further new cell types to the regenerate, other than those already present or being initiated at the

TABLE I

The Effect of the Time of Collagenase Treatment on DNA Synthesis in the Regenerates[a]

Days of culture	Not treated with collagenase			Treated with collagenase for												
				4 hours			24 hours			48 hours			72 hours			
	n	sm(r)	en(r)	n	sm(r)	en(r)	n	sm(r)	en(r)	n	sm(r)	en(r)	n	sm(r)	en(r)	
1	4	0/369	0/388	4	0/483	0/222	4	0/230	0/335	3	2/302	0/131	4	18/152	4/213	
2	4	0/301	0/31	4	0/227	0/138	4	15/185	1/263	4	62/160	0/172	3	36/233	9/222	
3	4	1/372	0/205	3	0/219	0/122	4	54/228	22/230	4	17/37	3/131	4	30/263	45/194	
7–8	3	12/335	0/182	2	3/151	0/97	4	170/183	189/190	2	26/41	22/123	—	—	—	
Regeneration	0			0			+			+			+			

[a] The striated muscle (sm) was treated prior to grafting upon endoderm (en). Grafting occurred immediately after collagenase treatment. n, Number of experiments; r, labeling index (number of labeled cells/total number of cells).

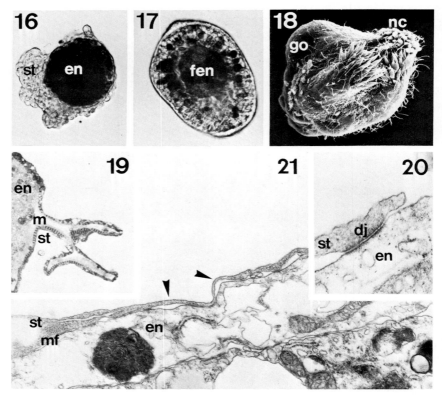

FIG. 16. Isolated destabilized striated muscle (st) grafted to endoderm (en), 4 hours after grafting. ×128.

FIG. 17. Regeneration 4 days after grafting; the endodermal cavity contains a ball-shaped fragment of endoderm (fen) not incorporated into the endodermal wall. ×129.

FIG. 18. Regenerated manubrium (4 weeks); nematocytes (nc) are well developed at the manubrial lips and eggs are maturing in the gonads (go). SEM. ×394.

FIG. 19. Semithin section of a 7-day-old graft consisting of endoderm (en) and muscle (st) not treated with collagenase. m, Mesoglea. ×1065.

FIG. 20. Frontal zone of a striated muscle cell in the process of covering the grafted endoderm. A desmosome junction (dj) has formed between the muscle cell (st) and the endoderm (en), ×35,500.

FIG. 21. Graft consisting of 5–10 destabilized muscle cells (st) and endoderm (en). The muscle cells have greatly stretched (solid arrows). mf, Cross section of striated myofibrils of striated muscle. ×21,300.

TABLE II

DNA SYNTHESIS IN GRAFTS CONSISTING OF 5–10 OR 20–40 DESTABILIZED
STRIATED MUSCLE CELLS AND 100–200 ENDODERM CELLS[a]

Days of culture	Number of muscle cells grafted					
	5–10			20–40		
	Muscle		Endoderm	Muscle		Endoderm
	n	r	r	n	r	r
2	3	1/9	0/272	4	9/54	0/371
3	—	—	—	3	3/45	4/191
4	3	1/4	0/228	3	52/147	35/100
Regeneration	0			Delayed		

[a] n, Number of grafts; r, labeling index (number of labeled cells/total number of cells).

time of drug treatment. This treatment was applied between day 3 and 4 after isolation, when only smooth muscle and y cells could have been present. Small but complete functional regenerates were formed, and the histodynamics and DNA synthesis pattern in both tissue layers clearly demonstrate transdifferentiation of the endoderm comparable to that of the muscle (Table III; Schmid et al., 1982). The reverse experiment, in which mitomycin C-treated endoderm is grafted onto destabilized muscle (Table III), confirmed the great transdifferentiation potential of the muscle. In both types of mitomycin C experiments, the regenerated manubria occasionally contained gametes (oocytes, Table III; Schmid et al., 1982). This indicates that both cell types can transdifferentiate to gametes.

IV. The Role of Cell Cycles

A general correlation seems to exist between the cell cycle event and stable differential gene expression. In ontogeny, DNA replication seems to be obligatory for specific gene expression in many animals (Alexandre et al., 1982; Satoh, 1982). The same seems to be true for transdetermination (Hadorn, 1966; Haynie and Bryant, 1976; Szabad et al., 1979) and for transdifferentiation (Yamada, 1977; Pritchard et al., 1978; Yamada and McDevitt, 1984) in other systems. In the medusa, too, neither regeneration nor transdifferentiation occur if DNA replication and mitoses are inhibited with aphidicolin, mitomycin C, or hydroxyurea (Schmid et al., 1982; Schmid and Alder, 1984).

TABLE III

FORMATION OF NEW CELL TYPES IN GRAFTS CONSISTING OF STRIATED MUSCLE AND ENDODERM[a]

Age (days)	Control			Treated with mitomycin C					
	n	ec	en	n	mec	en	n	ec	men
1	4	a (342)	a (219)						
2	5	a; b (390)	a (280)	2	a (144)	a (219)	2	a; b (389)	a (297)
3	11	a–c (893)	a; b; d (844)				3	a–c (571)	a (137)
4	7	a–d (623)	a–e; g (654)	2	a; b (105)	a; b (352)	4	a–c (861)	a (131)
5	3	b–d (346)	b–g (366)	1	a; b (75)	a; b; d; f (259)	5	b–d (1272)	a; b (167)
7				3	a; b; e (34)	a–g (654)	3	b–e (709)	a; b; c (75)
8–11	5	b–e; g (385)	b–g (516)				3	b–e; g (253)	a–d; f (54)

[a] ec, Striated muscle; en, endoderm. The tissues were alternately treated with mitomycin C before grafting. mec, Striated muscle treated with mitomycin C; men, endoderm treated with mitomycin C; a, original cell types of the ectoderm or the endoderm; b, interstitial cells; c, gland cells; d, nematoblasts; e, nematocytes; f, secretory cells; g, gametes; n, total of examined specimens; number in parentheses is the total number of examined nuclei. (From Schmid et al., 1982.)

The transdifferentiation of striated muscle to smooth muscle seems to be an exception to this general rule.

Most of the characteristics of the smooth muscle are already observed between the second and third day after isolation, when DNA replication has started but mitosis has not yet terminated. This indicates that transdifferentiation from striated to smooth muscle is direct; that is, it occurs without a preceding cell cycle. Earlier investigations have already shown that flagellum formation is independent of DNA synthesis (Schmid, 1975), but, as indicated by actinomycin D experiments (Bärtschi, 1977), it is dependent on transcription processes within the first 48 hours. The characteristics of y cells are seen first between the third and fourth day after isolation. Given a cell cycle time of 24 hours (H. Alder, unpublished results), y cells should require only one cell cycle. These observations have been confirmed recently in experiments using aphidicolin to inhibit mitosis (H. Alder, unpublished). On the basis of earlier studies (Schmid et al., 1982) we conclude that a maximum of two to three cell cycles are needed for the other cell types in the regenerate.

V. Concluding Remarks

The transdifferentiation processes observed in medusae clearly demonstrate that well-differentiated animal cells can retain the ability to change cellular commitment and to form various new cell types organized into a functional regenerate (Fig. 1). The only possible alternative explanation of our findings would be that the original material was contaminated by cells of other types. This hypothesis can be excluded on the following grounds:

1. Previous authors have concluded that tissue layers as prepared for these experiments are free of contaminating cell types (Kühn, 1914; Weiler-Stolt, 1960; Frey, 1968; Bölsterli, 1977).
2. Contaminating cells can be reliably identified and controlled for through the use of the DAPI method (Schmid et al., 1982; Schmid and Alder, 1984). This procedure is used throughout our experiments.
3. When contaminating cells were known to be present in the isolated muscle there was no effect on the rate of formation of endoderm and regenerate (Schmid and Alder, 1984).
4. In those rare cases in which degradation of striated myofibers was delayed, these can be found freely in the cytoplasm of the newly formed cell types (Figs. 6, 11, 13–15). Since the striated myofibers are not contained in vacuoles, and therefore not phagocytosed into the cell, the new cell types must originate from striated muscle. The contamination argument thus becomes irrelevant.

Cell cycles are necessary for this *in vitro* transdifferentiation to take place; this is in accordance with observations in most other systems (see above). However, recent results with mammalian tissue cultures indicate that functions associated with differential gene expression, while remaining reversible, can be gradually built up until stable gene expression is achieved (Bennett, 1983). In myogenic cell lines the switch from a genetic program controlling a proliferating tissue to one controlling a differentiating one occurs in the absence of DNA synthesis (Nadal-Ginard, 1978; Nguyen et al., 1983). Furthermore, in heterokaryons, muscle genes could be activated in nonmuscle cells without DNA replication (Chiu and Blau, 1984). The direct transformation from striated to smooth muscle demonstrated here is in accordance with these observations.

In contrast, the demonstrated transdifferentiation of striated muscle is unparalleled, and shows that a tissue layer composed of fully differentiated animal cells can undergo pluripotent transdifferentiation and retain the capacity for morphogenesis. This implies that the striated muscle can express, at least in part, the developmental potential of the fertilized egg. In many animals ontogenesis is understood as a time-dependent cascade of events, in which genome, cytoplasmatic determinants, and extracellular information finally shape a functional organism. It is difficult to understand how the functionally specialized somatic cell, the final product of such a complicated line of events, can maintain its developmental potential, but the experiments described here show that in medusae this is indeed the case.

In interpreting our observations the evolutionary background of the animal has to be considered. Although medusa belong to the "primitive" phylum Coelenterata, they are well-differentiated animals. They have developed a multitude of specialized cell types, in which gene products are stably expressed. Yet, these animals have a great ability to regenerate, very likely based on the transdifferentiation potential demonstrated here. The problem of maintaining the differentiated state of a cell might have been solved in a different way in the medusa in comparison to other animals with more limited powers of regeneration. How this is done, and what elements contribute to the *in vitro* regeneration process, are questions which can now be investigated under defined conditions.

ACKNOWLEDGMENTS

We thank Miss D. Keller for technical assistance and Dr. H. Rowell for reading and correcting the chapter. The work was supported by the Swiss National Science Foundation Grant 3.149-0.81.

REFERENCES

Alder, H. (1982). Thesis, University of Zurich, unpublished.
Alexandre, H., de Petrocellis, B., and Brachet, J. (1982). *Differentiation* **22**, 132–135.
Bärtschi, U. (1977). Thesis, University of Zurich, unpublished.
Bennett, D. C. (1983). *Cell* **34**, 445–453.
Bölsterli, U. (1977). *J. Morphol.* **154**, 259–290.
Burnett, A. L. (1968). "The Stability of the Differentiated State" (H. Ursprung, ed.), p. 109. Springer-Verlag, Berlin and New York.
Caplan, A. I., and Ordahl, C. P. (1978). *Science* **201**, 120–130.
Chiu, C. P., and Blau, H. M. (1984). *Cell* **37**, 879–887.
DiBerardino, M. A. (1980). *Differentiation* **17**, 17–30.
DiBerardino, M. A., Hoffner, J. N., and Etkin, L. D. (1984). *Science* **224**, 946–952.
Eguchi, G., and Itoh, Y. (1982). *Trans. Ophthalmol. Soc. U.K.* **102**, 374–378.
Frey, J. (1968). *Wilhelm Roux' Arch.* **160**, 428–464.
Hadorn, E. (1966). "Major Problems in Developmental Biology" (Locke, ed.), p. 85. Academic Press, New York.
Hay, E. D. (1968). "The Stability of the Differentiated State" (H. Ursprung, ed.), p. 85. Springer-Verlag, Berlin and New York.
Haynie, J. L., and Bryant, P. J. (1976). *Nature (London)* **259**, 659–662.

Kühn, A. (1914). *Ergebn. Fortschr. Zool.* **87.**
Mackie, G. O. (1980). *Trends NeuroSci.* **January,** 13–16.
Nadal-Ginard, B. (1978). *Cell* **15,** 858–864.
Nguyen, H. T., Russel, M. D., and Nadal-Ginard, B. (1983). *Cell* **34,** 281–293.
Oberpriller, J. (1967). *Growth* **31,** 251.
Okada, T. S. (1983). *Cell Differ.* **13,** 177–183.
Pritchard, D. J., Clayton, R. M., and DePomerai, D. I. (1978). *J. Embryol. Exp. Morphol.* **48,** 1–21.
Schmid, V. (1972). *Wilhelm Roux' Arch.* **169,** 281–307.
Schmid, V. (1974a). *Am. Zool.* **14,** 773–781.
Schmid, V. (1974b). *Exp. Cell Res.* **86,** 193–198.
Schmid, V. (1975). *Exp. Cell Res.* **94,** 401–408.
Schmid, V. (1976). *Dev. Biol.* **49,** 508–517.
Schmid, V. (1978). *Dev. Biol.* **64,** 48–59.
Schmid, V. (1979). *Ann. Soc. Fr. Biol. Dev.* pp. 35–38.
Schmid, V. (1980). "Invertebrate Tissue Culture" (E. Kurstak and K. Maramorosch, eds.), pp. 85–101. Academic Press, New York.
Schmid, V., and Alder, H. (1984). *Cell* **39,** 801–807.
Schmid, V., Wydler, M., and Alder, H. (1982). *Dev. Biol.* **92,** 476–488.
Schmid, V., Weber, Ch., and Keller, D. (1984). *Wilhelm Roux' Arch.* **193,** 36–41.
Szabad, J., Simpson, P., and Nöthiger, R. (1979). *J. Embryol. Exp. Morphol.* **49,** 229–241.
Tardent, P., and Schmid, V. (1972). *Exp. Cell Res.* **72,** 265–275.
Tobler, H., Smith, K. D., and Ursprung H. (1972). *Dev. Biol.* **27,** 190–203.
Vasil, J. K., and Vasil, V. (1972). *In Vitro* **8,** 117–127.
Weber, Ch. (1981). *J. Morphol.* **167,** 313–331.
Weber, Ch., and Schmid, V. (1984). *Exp. Cell Res.* **155,** 149–158.
Weiler-Stolt, B. (1960). *Wilhelm Roux' Arch.* **152,** 389–454.
Westfall, J. A., and Kinnamon, J. C. (1978). *Neurocytology* **7,** 365–379.
Yamada, T. (1977). *Monogr. Dev. Biol.* **13.**
Yamada, T., and McDevitt, D. S. (1984). *Differentiation* **27,** 1–12.

CHAPTER 10

THE PRESENCE OF EXTRALENTICULAR CRYSTALLINS AND ITS RELATIONSHIP WITH TRANSDIFFERENTIATION TO LENS

R. M. Clayton, J.-C. Jeanny,* D. J. Bower,† and L. H. Errington

INSTITUTE OF ANIMAL GENETICS
UNIVERSITY OF EDINBURGH
EDINBURGH, SCOTLAND

I. Introduction

The crystallins comprise some 80–90% of the proteins of the vertebrate lens. The times of appearance and the rates of increase of the α-, β-, and δ-crystallin classes in the developing chick lens, and of the individual polypeptide members of these classes, are temporally and spatially regulated (reviews in Clayton, 1970; Piatigorsky, 1981; McDevitt and Brahma, 1982). Several lines of evidence point to noncoordinate regulation of crystallin expression in lens cells. There is a wide range of differential levels of synthesis during ontogeny, and there are differential responses to long-term culture *in vitro* (Patek and Clayton, 1985), organ culture in contact with other tissues (McAvoy, 1980), cell position (McAvoy, 1980; Vermorken *et al.,* 1978), cell culture conditions (Clayton *et al.,* 1976), agents, such as insulin and eye-derived growth factor (EDGF) which affect *inter alia,* the rate of mitosis (de Pomerai and Clayton, 1978; Randall *et al.,* unpublished), and various pharmacological agents such as diphenylhydantoin, chlo-

---

* Present address: Unité de Recherches Gérontologiques, Paris, France.
† Present address: M. R. C. Clinical and Population Cytogenetics Unit, Edinburgh, Scotland.

CURRENT TOPICS IN
DEVELOPMENTAL BIOLOGY, VOL. 20

roquine, and chlordiazepoxide (Clayton and Zehir, 1982). These patterns of differential response were found to be further modulated by the genotype of the responding cell (for example Clayton et al., 1976; Clayton and Zehir, 1982).

Crystallins have also been found at low levels in several nonlens tissues, in both birds and amphibians (review in Clayton, 1982). These include the neural and pigmented retinas (NR and RPE), iris and cornea, and several extraocular embryonic tissues, including brain, adenohypophysis, and epiphysis (Barabanov 1977, 1982; Clayton, 1979; Ueda et al., 1983). The cornea and adenohypophysis are ectodermal in origin; the other tissues are of neuronal origin. The expression of ectopic crystallin antigenicity may be restricted during development in the quail and chick embryo (for example, cells of the adenohypophysis contain δ-crystallin at the beginning of the period of differentiation), but not when histodifferentiation begins (Barabanov, 1977, 1982). Many of these tissues are able, under appropriate conditions, to transdifferentiate to lens cells expressing high levels of crystallins (reviews in Eguchi, 1979; Okada, 1980, 1983; Clayton, 1982; Yamada, 1984). Transdifferentiation potential has been demonstrated in all vertebrates so far tested. Although an apparent relationship between such potential and crystallin expression has mainly been investigated in the chick, extralenticular antigenicity has been found in Xenopus (Clayton et al., 1968), Anolis carolinensis (McDevitt, 1972), and Lacerta viridis (Clayton, 1974); in these last two cases it is clearly a functional presence in the pineal lens.

In the chick, δ-crystallin is the major constituent of the early embryo lens, but its contribution falls during development (reviewed Clayton, 1974) and synthesis completely ceases between 3 and 4 months posthatch (Treton et al., 1982).

Evidently fiber cell morphology does not require the presence of δ-crystallin. However there is some evidence suggesting that δ-crystallin alone, in the absence of α- and β-crystallins, may not lead to fiber-like morphology. Cells in the adenohypophysis and epiphysis which express only δ-crystallin in vivo have an epithelial morphology (Barabanov, 1977, 1983), and cells in the limb bud, which after experimental manipulation synthesize δ-crystallin only, are rounded in shape (Kodama and Eguchi, 1982). Cells in the diencephalon which are positive for δ-crystallin during the fourth and fifth days of incubation in the quail embryo have a characteristic neuronal morphology, both the cell bodies and axons fluorescing strongly with anti-δ-crystallin antiserum (Barabanov, 1982).

The crystallins that appear during transdifferentiation are indistinguishable by the criteria of antigenicity and size and charge from crystallins in the lens itself. It would appear, however, that the precise qualitative and quantitative array of crystallins in a transdifferenti-ated lentoid is governed by a number of factors including the age of the donor embryo, the tissue of origin, the genotype of the cells, and growth conditions (Clayton, 1982). The relative contribution of δ-crystallin to lens and to NR-derived lentoids is very high early in development and falls with increasing embryo age to zero (Clayton *et al.*, 1979). Thus the ontogenic progression is the same in these two tissues, although the expression is not synchronous. Lentoids from embryo brain express high δ-crystallin levels but some α-crystallin is also present (Nomura, 1982). In the present chapter, recent studies on the extralenticular distribution of crystallins will be reviewed and its possible role in the transdifferentiation of nonlenticular tissues into lens will be discussed.

II. Crystallin RNAs in Retina and Other Nonlens Tissues

Crystallin RNAs were found at moderate levels in 8-day chick em-bryo NR and RPE, using solution hybridization to cDNA made to crystallin RNAs, but no hybridization was obtained with RNA from headless embryo bodies (Jackson *et al.*, 1978). These sequences in-creased during transdifferentiation to lentoids (Thomson *et al.*, 1979, 1981). Since the frequency and rate of transdifferentiation and the levels of hybridizable crystallin RNA were found to fall together dur-ing development, we suggested (Clayton *et al.*, 1979) that the capacity for transdifferentiation to lens was related to the expression of crystal-lin RNA in that tissue. It was not possible at the time to determine whether this decline affected all crystallins equally, and whether all crystallin RNAs were relevant to transdifferentiation potential. Yamada (1984) suggested that the relationship is indirect and that the expression of crystallin RNA implies a chromosomal state which is also conducive to transdifferentiation.

Although 8-day embryo RPE had a level of hybridization inter-mediate between 3.5- and 8-day NR, it took much longer for it to transdifferentiate and the final levels of crystallin RNA were lower (Thomson *et al.*, 1981). To account for this difference we suggested that only some cells in retina tissues express crystallin RNA, and that the numbers of such cells and the levels per cell vary independently. Cell heterogeneity is also implied by the results of clonal cultures (cf. Okada, 1980). We describe in the next section evidence which confirms this heterogeneity in terms of δ-crystallin gene transcription.

δ-Crystallin sequences have been detected in the poly(A)$^+$ RNA of brain and NR by Northern transfer, using a cDNA probe to coding sequences of δ-crystallin RNA, by Agata *et al.* (1983) and Bower *et al.* (1983a). A large proportion consists of high-molecular-weight precursors. Agata *et al.* (1983) also found such sequences in embryo limb bud, but failed to detect any in liver or heart. However, using larger amounts of RNA we were able to detect very low levels of δ-crystallin sequences in brain, lung, kidney, and heart. Traces of crystallin-like antigencity had previously been detected in liver and kidney (Clayton *et al.*, 1968). It is evident from the investigations of Bower *et al.* (1983a) and Agata *et al.* (1983) that there is a considerable difference in the levels of δ-crystallin sequences in tissues able to transdifferentiate, such as NR, and those not shown to be able to do so, such as heart.

III. Detection of δ-Crystallin RNA by *in Situ* Hybridization

We examined several tissues from embryos of 3.5, 4.5, and 6 days of incubation by the technique of *in situ* hybridization (Jeanny *et al.*, 1985). The results show the presence of δ-crystallin RNA sequences in all lens cells, but in only a subpopulation of cells, forming groups or clusters, in all other tissues examined, including embryonic NR, RPE, otic vesicle, epiphysis, adenohypophysis, and heart (Fig. 1). Controls with the plasmid vector were negative. The δ-crystallin clone used hybridizes only to the two δ-crystallin genes and is highly specific (Bower *et al.*, 1981). It does not hybridize to ribosomal RNA under any conditions (Bower *et al.*, 1983b) and there are no repetitive sequences. The plasmid vector sequences give no signal either in Northern transfers or in *in situ* hybridization. Finally, the nuclear labeling cannot be

---

FIG. 1. *In situ* hybridization. Each tissue was removed rapidly from the chick embryo, squashed on sterile slides, and hybridized to either of two δ-crystallin cDNA probes (Bower *et al.*, 1981, 1983a) by the method of Maitland *et al.* (1981). Further treatment including thermal washes and autoradiography was as described in Bower *et al.* (1983b). The specific activity of the probes was 3×10^6 dpm/μg and the final concentration 2×10^6 dpm/ml. In order to avoid possible processing or degradation during the preparation of tissues for *in situ* hybridization, we processed samples rapidly, using each embryo for one tissue and taking no more that 1–3 minutes, according to the tissue, from opening the egg to freezing the squash in liquid nitrogen. The speed of dissection deemed necessary precluded the use of CMF or Dispase. (A) 3.5-day lens (the processing of this sample has led to lighter labeling overall than that obtained in other preparations). ×66. Inset, 3.5-day lens control hybridized to the ³H-labeled plasmid vector. ×27. (B) 4-day NR. ×66. (C) 6-day NR. ×27. (D) 6-day RPE. ×66. Inset, 6-day RPE hybridized to a genomic probe to nuclear sequences corresponding to mitochondrial DNA. ×27. (E) 6-day adenohypophysis. ×27. (F) 6-day epiphysis. ×66. (G) 4-day otic vesicle. ×27. (H) 5-day heart. ×27.

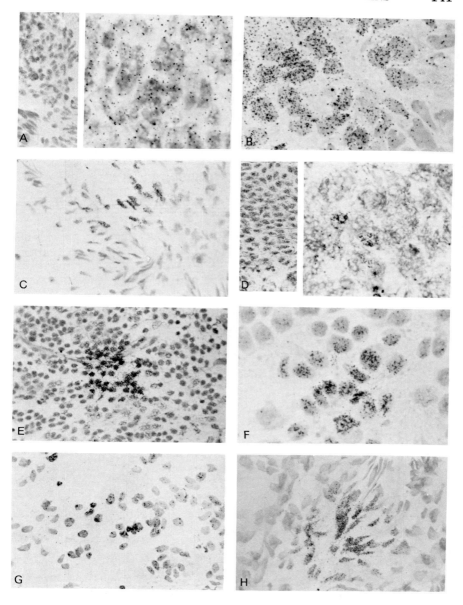

due to DNA δ-crystallin sequences, since there are only two copies of each of the two genes per genome, and the level of label obtained, given the methods employed, implies several thousands of copies of the sequences in positive cells (cf. Hafen *et al.*, 1983). Squashes cannot show whether these groups of positive cells are distributed nonrandomly, but at the stages examined, there were no obvious differences in morphology between labeled cells and their unlabeled neighbors. Lens cells were labeled over the nucleus and cytoplasm, but most of the label in the other cell types is nuclear. Some degree of cytoplasmic localization appeared in NR and adenohypophysis (Fig. 1B and E). This would be compatible with observations of Northern transfers of NR (Agata *et al.*, 1983; Bower *et al.*, 1983a) and immunofluorescence of adenohypophysis with anti-δ antisera (Barabanov, 1977; Ueda *et al.*, 1983). This immunofluorescence points to nonrandom cell expression of δ-crystallin, as does the restricted localization of transcribing cells, seen in sections of 3.5-day embryo NR (Bower *et al.*, 1983b).

The quantitative estimates indicated that NR expresses about 2% and heart and mid-brain about 0.02–0.04% of the amount of δ-crystallin RNA in the day-old chick lens (Bower *et al.*, unpublished). About 0.1% of the cells in the heart are transcribing δ-crystallin sequences, which indicates that individual heart cells may be transcribing at very high levels. There is an average of 27 labeled cells per cluster in heart and 44 in NR, which also has more clusters of transcribing cells. The separation of neural and pigmented retinas avoided treatment with CMF (Ca^{2+} Mg^{2+} free) or enzymes, and fragments which were not clearly separated were discarded, which would affect estimates if positive cells are nonrandomly distributed. In sections of 3.5-day embryo eye, the positive cells were found only on the vitreal border of the NR (Bower *et al.*, 1983a) in a position corresponding to putative future glial cells (Bhatacharjee and Sanyal, 1975), and we estaimated that 15% of NR cells were labeled. The difference between tissues with and those without transdifferentiation potential is so considerable that it seems likely to be significant, but characteristics other than the number of transcribing cells will certainly require investigation.

In the Northern blot the relative intensity of the bands which hybridized to a δ-crystallin cDNA probe in a given tissue was the same for all ages from 7 to 21 days of development, indicating no major change in amount (Figs. 2 and 3). If the amount of δ-crystallin RNA does not change significantly between 7 and 21 days of development, it would suggest that the proportion of transcribing cells remains steady,

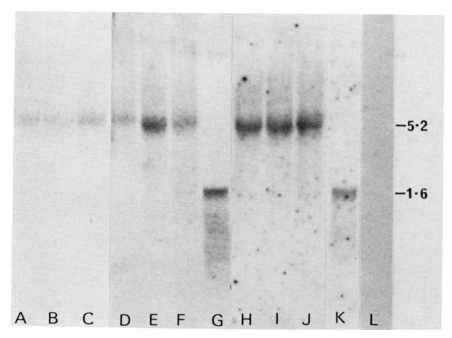

FIG. 2. δ-Crystallin RNA in embryonic chick tissues. RNA from freshly excised tissue was prepared according to Girard (1967). Electrophoresis, transfer to nitrocellulose, conditions of hybridization, subsequent washes under stringent conditions, and autoradiography were all as described in Bower *et al.* (1983b). (A,B,C) 10-μg RNA from 7-, 17-, and 21-day embryo midbrain. (D,E,F) 10-μg RNA from 7-, 17-, and 21-day embryo NR. (G) 100-ng RNA from 21-day lens. (H,I,J) 7-, 17-, and 21-day embryo heart. (K) 100-ng RNA from 21-day embryo lens. (L) 10-μg lens RNA, hybridized to the plasmid vector. Total RNA for lens and NR samples, and poly(A)+ RNA for all the samples.

either by recruitment or by a growth rate equal to that of other cells in the tissue.

It would seem that there are several types of cellular expression of crystallin sequences: cells which do not transcribe, cells which transcribe but have little or no processing, cells which transcribe and have some processing capacity, and cells in the lens itself, which transcribe rapidly and process commensurately. There are two δ-crystallin genes (Bhat *et al.*, 1980; Bower *et al.*, 1981). Whether ectopic expression can involve either one, according to the site, or whether there are alternative promoter sites or splice sites which are used in different cell types remains to be investigated.

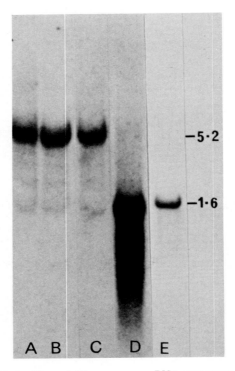

Fig. 3. Quantitation of δ-crystallin sequences. RNA was prepared from tissues put immediately into extraction buffer, using the guanidinium thiocyanate method (Chirgwin *et al.*, 1979), electrophoresed through formaldehyde agarose gels, and transferred to nitrocellulose. RNA was hybridized to a ^{32}P-labeled δ-crystallin probe, pM56 (Bower *et al.*, 1981). (A,B) 10-μg NR RNA from 17- and 21-day embryos, respectively. (C) 10-μg 15-day embryo heart poly(A)+ RNA. (D) 1 μg of 21-day lens RNA. (E) 15 μg of 8-day embryo NR RNA, incubated before extraction according to Girard (1967), for 15 minutes in ice-cold CMF medium. (Errington and Clayton, 1986).

IV. Noncoordinate Regulation of Crystallin RNAs during Transdifferentiation

Lens cells and NR-derived lentoids both show an increase in β-crystallins and a decrease in δ-crystallin with increasing embryo age (Clayton *et al.*, 1979). Both tissues express increased δ-crystallin under all conditions that accelerate growth rate *in vivo* and *in vitro* (review; Clayton, 1982). There are also striking differences between them. The order of appearance and the rate of increase of δ-, αA_2-, and 25-kDa β-crystallin RNAs during transdifferentiation differ between NR and RPE, and both differ from the lens itself. Table I shows the results of the increase in these three RNAs in transdifferentiating cultures of 8-day embryo NR and RPE. Direct comparisons between signal strengths of the three probes may not be made, but the differences between stages or tissues may be safely compared for each probe separately. In the lens, δ-crystallin appears before the α- and β-crystallins (Clayton, 1974; Piatigorsky, 1981; McDevitt and Brahma, 1982). Figure 4 and Table I show that αA-crystallin RNA appears first in transdifferentiating 8-day embryo NR, and α- and β- before δ-crystallin in transdifferentiating 8-day embryo RPE. In the lens itself, the regulation of the order of appearance and the rates of synthesis have presumably been under the strong selective pressure exerted by the requirement to produce cells with the refractive index appropriate to their position, and by the requirements of vision of the species (Clayton, 1974), but no such requirements operate in the transdifferentiating systems. Only δ-crystallin and a faint trace of αA_2-crystallin were detected in freshly isolated 8-day embryo NR, and none of the three crystallin RNAs were detected in the early culture period. Appreciable αA_2-crystallin RNA was detected in NR cell cultures at 7 days and δ- and β-crystallin RNAs by 21 days, although a trace of δ-crystallin may be detected at 14 days. However, the rise in α-crystallin RNA levels is slow, while that of δ-crystallin RNA is very rapid and it is the major component at later stages in transdifferentiation. This is in reasonable agreement with the data on the crystallin proteins in this system (de Pomerai et al., 1977).

In transdifferentiating RPE, the 25-kDa β-crystallin RNA is the first to appear, at 14 days, and neither αA_2- nor δ-crystallin sequences were detected before 28 days. The β-crystallin appears to be a particularly important component at later stages of transdifferentiation of RPE. None of these sequences were found in freshly isolated RPE. Since solution hybridization showed that 8-day embryo RPE expressed more total crystallin RNA than NR of the same age (Jackson et al., 1978; Clayton et al., 1979), it is now necessary to examine the expression of αB-crystallin RNA and of sequences of the other β-crystallins, since some of these may account for the earlier results.

αA-Crystallin RNA was detected in the 3.5-day embryo lens, and it rose rapidly during development, although protein levels rise more slowly (Errington et al., 1985b). Precursor RNAs were not seen in the early embryo, although large amounts of precursor δ-crystallin RNAs were found at these stages (Bower et al., 1983a). In contrast, a precursor α-crystallin RNA was observed in 7-day transdifferentiating NR cultures, although it was not seen in subsequent stages.

Rapid handling was necessary in our investigations of crystallin RNAs. When NR was homogenized in guanidinium thiocyanate buffer

Days of culture

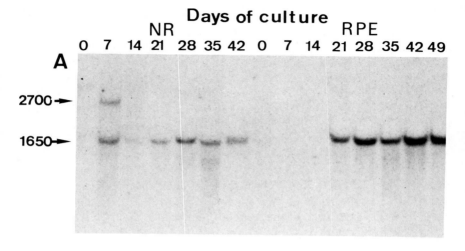

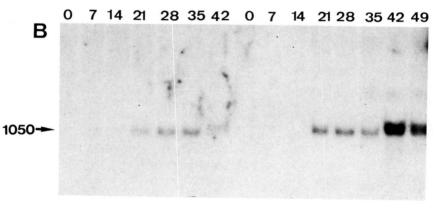

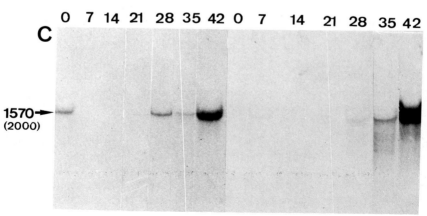

TABLE I

THE INCREASE IN THREE CRYSTALLIN RNAs DURING TRANSDIFFERENTIATION IN
SPREADING CULTURES OF 8-DAY CHICK EMBRYO NR OR RPE CELLS

Tissue	Crystallin RNA species	Days of culture							
		0	7	14	21	28	35	42	49
8-day embryo NR	δ-Crystallin	+	−	TR	++	++	+++	NT^a	
	αA$_2$-Crystallin	TR	+	+	+	+	+	+	
	25-kDA β-Crystallin	−	−	−	+	+	+	+	
8-day embryo RPE	δ-Crystallin	−	−	−	−	+	+	++	NT
	αA$_2$-Crystallin	TR	−	−	++	++	++	++	++
	25-kDA β-Crystallin	−	−	−	+	+	+	+++	+++

[a] NT, Not tested; TR, trace. (Not seen on all fluorographs.)

immediately on isolation, the bulk of the δ-crystallin RNA was of high molecular weight, but if the retinas were first incubated in ice-cold CMF medium for 15 minutes, although there was no evidence of degradation, all the δ-crystallin RNA had been processed to the 2-kb species, the main component in lens (Fig. 3). This implies that this RNA in the NR is capable of being processed and that NR has processing capacity which is suppressed *in vivo*.

V. Evolutionary Considerations

The appearance of δ-crystallin in moderate amounts in such embryonic structures as the adenohypophysis or in the putative glial cell precursor layer of NR implies a regulatory process which is cell and stage specific, and which has presumably been subject to evolutionary constraints. Kondoh *et al.* (1983) found that δ-crystallin DNA injected

FIG. 4. Crystallin RNAs in transdifferentiating cultures. 15-μg RNA was extracted from pooled samples of 8-day embryo NR or RPE cultures, taken at 7-day intervals. Extraction, electrophoresis, and hybridization were as described in Bower *et al.* (1983b). Day 0 represents freshly excited tissue. Series on the left, NR. Series on the right, RPE. (A) α-Crystallin probe; (B) β-crystallin probe; (C) δ-crystallin probe. The nucleotide sequence of δ-crystallin RNA gives a size of 1570 bp plus a poly(A)$^+$ tail of 50–100 nucleotides (cf. Yasuda *et al.*, 1984), but runs in gel electrophoresis to give an apparent size of 2000 bp. The α-crystallin geonomic probe was obtained by screening a λ charon 4A chicken genomic library, by the method of Maniatis *et al.* (1975), with minor modifications. The characterization of the 25-kDa β-crystallin cDNA probe is described in Errington *et al.* (1985). The probe cross-hybridizes with 35-kDa β-crystallin RNA in conditions of low stringency, confirming the observation of Hejtmancik and Piatigorsky (1983). The characterization of the δ-crystallin cDNA probe is described in Bower *et al.* (1981).

into mouse cells was well expressed in lens cells, slightly so in epithelial cells, and not expressed in fibroblasts. This demonstrates that the regulation was neither species specific nor crystallin class specific, but that the regulation of cell-specific expression is very conservative in evolution.

The adenohypophysis and epiphysis are topologically homologous to the retina. All three tissues have some responsiveness to light fluctuations and all produce growth regulatory substances (reviewed Clayton, 1982), while conditions such as trilateral retinoblastoma (Bader *et al.*, 1982) suggest a shared susceptibility of pineal and eye which may have a molecular basis. It may be that the δ-crystallin genes are near other genes which are essential to the development of neural and ectodermal structures, and are merely activated together with them. Or, it may be that the crystallins have an extralenticular function related to their molecular structure. A specialized organ, expressing a particular molecular species at abundant levels, has presumably evolved to its current function by virtue of some of the properties of the molecules as expressed in precursor cell types which evolved earlier, and these properties may still be relevant to the function on the less specialized cells. For example, many nonmuscle cells require the contractile properties of myosin, and express very low levels, but the muscle cell has evolved to specialize in contractility and has considerable amounts. This general problem is discussed elsewhere (Clayton, 1982).

A homology between small heat-shock proteins and bovine αB-crystallin was reported by Ingolia and Craig (1982). A consideration of all known αA- and αB-crystallin sequences makes the degree of homology even greater, and it is perhaps interesting that heat-shock proteins and α-crystallins are both able to bind to various structural components of the cell, suggesting a possible evolutionary modus for the development of α-crystallin-like from heat-shock protein-like molecules. We cannot yet dismiss the possibility that a search for homologies for other crystallin classes (for example, Kodama and Eguchi, 1983) may point not only to evolutionary relationships but also to possible cellular roles for crystallin-like molecules expressed in lower abundance than in the lens.

VI. Regulation of δ-Crystallin Gene Expression: Concluding Remarks

It is evident that crystallin gene expression is regulated at many levels. Regulation of transcription accounts for the increase of δ-crystallin RNA during embryo fiber cell differentiation (Piatigorsky *et al.*, 1976) and αA-crystallin during development (Errington *et al.*,

1985a; Clayton *et al.*, 1985). Regulation of transcription must also account for the increase in crystallin RNAs overall during transdifferentiation (e.g., Thomson *et al.*, 1979, 1981; Clayton *et al.*, 1979) and of δ-crystallin RNA during transdifferentiation (Bower *et al.*, 1983b; Yasuda *et al.*, 1983; Errington *et al.*, 1985a,b; Clayton *et al.*, 1985). The degree of methylation does not appear to be an important regulatory factor for δ-crystallin RNA transcription (Errington *et al.*, 1983; Grainger *et al.*, 1983). There is also evidence for selective post-transcriptional regulation: for example, while all δ-crystallin RNA is fully processed in day-old lens, 6-day embryo lens contains a notable proportion of high-molecular-weight precursors (Bower *et al.*, 1983b) and a preponderance of such precursors is seen in nonlens tissues (Agata *et al.*, 1983; Bower *et al.*, 1983a). This is in agreement with *in situ* hybridization, which shows most or virtually all the label in extralenticular tissues as supranuclear, according to the tissue examined. No precursors of αA_2-crystallin RNA were found in the developing lens but they were observed early in NR transdifferentiation (Errington *et al.*, 1985a). *In situ* hybridization shows that transcription outruns processing both during the early stages of fiber cell differentiation and in the early stages of transdifferentiation.

mRNA stability changes during fiber differentiation (Yoshida and Katoh, 1972; Delcour *et al.*, 1976) and changes in stability during development are differential, as are rates of translation (Clayton, 1974). Ontogenic changes in translational efficiencies were also found for δ-crystallin (Beebe and Piatigorsky, 1977) and αA_2-crystallin (Errington *et al.*, 1985a,b). There is also differential distribution of RNAs between polysomal and postpolysomal fractions, and with respect to poly(A)$^+$ binding (Thomson *et al.*, 1978). Such multiple levels of regulation, taken together with the possible variations of the primary transcript referred to above, presumably are sufficient to account for the wide range of differential changes in crystallin expression at different stages of lens ontogeny, for the differences between crystallin gene expression in lens and in ectopic sites, and for the differences between nonexpressing and expressing ectopic cells.

ACKNOWLEDGMENTS

This work was supported by a grant from the Medical Research Council to R. M. Clayton, who also thanks Professor T. S. Okada for making it possible to present these results at the Yamada Conference. J.-C. Jeanny was supported by a Research Fellowship from the Royal Society.

REFERENCES

Agata, K., Yasuda, K., and Okada, T. S. (1983). *Dev. Biol.* **100**, 222–226.
Barabanov, V. M. (1977). *Dokl. Akad. Nauk SSSR* **234**, 195–198.

Barabanov, V. M. (1982). *Dokl. Akad. Nauk SSSR* **262**, 22–24.

Beebe, D. C., and Piatigorsky, J. (1977). *Dev. Biol.* **59**, 174–182.

Bhat, S. Pl., Jones, R. E., Sullivan, M. A., and Piatigorsky, J. (1980). *Nature (London)* **284**, 234–238.

Bhattacharjee, J., and Sanyal, S. (1975). *J. Anat.* **120**, 367–372.

Bower, D. J., Errington, L. H., Wainwright, N. R., Sime, C., Morris, S., and Clayton, R. M. (1981). *Biochem. J.* **201**, 339–344.

Bower, D. J., Errington, L. H., Cooper, D. N., Morris, S., and Clayton, R. M. (1983a). *Nucleic Acids Res.* **11**, 2513–2527.

Bower, D. J., Errington, L. H., Pollock, B. J., Morris, S., and Clayton, R. M. (1983b). *EMBO J.* **2**, 333–338.

Chirgwin, J. M., Przybyla, A. E., MacDonald, R. J., and Rutter, W. J. (1979). *Biochemistry* **18**, 5294–5299.

Clayton, R. M. (1970). *Curr. Top. Dev. Biol.* **5**, 115–180.

Clayton, R. M. (1974). *In* "The Eye" (H. Davson and L. T. Graham, eds.), Vol. 5, pp. 399–494. Academic Press, New York.

Clayton, R. M. (1979). *Ophthalmic. Res.* **11**, 324–328.

Clayton, R. M. (1982). *In* "Differentiation *in Vitro*" (M. M. Yeoman and D. E. S. Truman, eds.), pp. 83–120. Cambridge Univ. Press, London and New York.

Clayton, R. M., and Zehir, A. (1982). *In* "Developmental Toxicology" (K. Snell, ed.), pp. 59–92. Croom Helm, London.

Clayton, R. M., Campbell, J. C., and Truman, D. E. S. (1968). *Exp. Eye Res.* **7**, 11–29.

Clayton, R. M., Odeigah, P. G., de Pomerai, D. I., Pritchard, D. J., Thomson, I., and Truman, D. E. S. (1976). *Colloq. INSERM* **60**, 123–136.

Clayton, R. M., Thomson, I., and de Pomerai, D. I. (1979). *Nature (London)* **282**, 628–629.

Clayton, R. M., Errington, L. H., Bower, D. J., and Morris, S. (1985). *In* "EURAGE Symp. Transparency of the Lens" (G. Duncan and C. Slingsby, ed.). In press.

Delcour, J., Odaert, S., and Bouchet, H. (1976). *Colloq. INSERM* **60**, 39–47.

Eguchi, G. (1979). In "Mechanisms of Cell Change" (J. D. Ebert and T. S. Okada, eds.), pp. 273–291. Wiley, New York.

Errington, L. H., and Clayton, R. M. (1986). In preparation.

Errington, L. H., Cooper, D. N., and Clayton, R. M. (1983). *Differentiation* **24**, 33–38.

Errington, L. H., Bower, J. D., and Clayton, R. M. (1985a). *In* "Coordinated Regulation of Gene Expression" (R. M. Clayton and D. E. S. Truman, eds.). Plenum, New York, in press.

Errington, L. H., Bower, D. J., Cuthbert, J., and Clayton, R. M. (1985b). *Biology of the Cell* **54**, 1–8.

Girard, M. (1967). *In* "Methods in Enzymology" (L. Grossman and K. Moldave, eds.), Vol. 12A, pp. 581–588. Academic Press, New York.

Grainger, R. M., Hazard-Leonards, R. M., Samaha, F., Hougan, L. M., Lesk, M. R., and Thomson, G. H. (1983). *Nature (London)* **306**, 88–91.

Hafen, E., Levine, M., Garber, R. L., and Gehring, W. J. (1983). *EMBO J.* **2**, 617–623.

Hejtmancik, F., and Piatigorsky, J. (1983). *J. Biol. Chem.* **258**, 3382–3387.

Ingolia, T. D., and Craig, E. A. (1982). *Proc. Natl. Acad. Sci. U.S.A.* **79**, 2360–2364.

Jackson, J. F., Clayton, R. M., Williamson, R., Thomson, I., Truman, D. E. S., and de Pomerai, D. I. (1978). *Dev. Biol.* **65**, 383–395.

Jeanny, J.-C., Bower, D. J., Errington, L. H., Morris, S., and Clayton, R. M. (1985). *Dev. Biol.* **111**, 1–6.

Kodama, R., and Eguchi, G. (1982). *Dev. Biol.* **91**, 221–226.

Kodama, R., and Eguchi, G. (1983). *Dev. Growth Differ.* **25**, 261–270.

Kondoh, H., Yasuda, K., and Okada, T. S. (1983). *Nature (London)* **301,** 440–442.

McAvoy, J. W. (1980). *Differentiation* **17,** 85–91.

McDevitt, D. S. (1972). *Science* **175,** 763–764.

McDevitt, D. S., and Brahma, S. K. (1982). In "Cell Biology of the Eye" (D. S. McDevitt, ed.), pp. 143–191. Academic Press, New York.

Maitland, N. J., Kinross, J. H., Busuttil, A., Ludgate, S. M., Smart, G. E., and Jones, K. W. (1981). *J. Gen. Virol.* **55,** 123–137.

Maniatis, T., Jeffrey, A., and Kleid, D. G. (1975). *Proc. Natl. Acad. Sci. U.S.A.* **72,** 1184–1188.

Nomura, K. (1982). *Differentiation* **22,** 179–184.

Okada, T. S. (1980). *Curr. Top. Dev. Biol.* **16,** 349–390.

Okada, T. S. (1983). *Cell Differ.* **13,** 177–183.

Patek, C. E., and Clayton, R. M. (1985). *Exp. Eye Res.* **40,** 357–378.

Piatigorsky, J. (1981). *Differentiation* **19,** 134–153.

Piatigorsky, J., Beebe, D., Zelenka, P., Milstone, L. M., and Shinohara, T. (1976). *Colloq. INSERM* **60,** 85–112.

de Pomerai, D. I., and Clayton, R. M. (1978). *J. Embryol. Exp. Morphol.* **147,** 179–193.

de Pomerai, D. I., Pritchard, D. J., and Clayton, R. M. (1977). *Dev. Biol.* **60,** 416–427.

Thomson, I., Wilkinson, C. E., Jackson, J. F., de Pomerai, D. I., Clayton, R. M., Truman, D. E. S., and Williamson, R. (1978). *Dev. Biol.* **65,** 372–382.

Thomson, I., de Pomerai, D. I., Jackson, J. F., and Clayton, R. M. (1979). *Exp. Cell Res.* **122,** 73–81.

Thomson, I., Yasuda, K., de Pomerai, D. I., Clayton, R. M., and Okada, T. S. (1981). *Exp. Cell Res.* **135,** 445–449.

Treton, J. A., Shinohara, T., and Piatigorsky, J. (1982). *Dev. Biol.* **100,** 222–226.

Ueda, Y., Takeichi, M., Okada, T. S., and Hino, F. (1983). *Dev. Growth Differ.* **25,** 421 (Abstr.).

Yasuda, K., Okuyama, K., and Okada, T. S. (1983). *Cell Differ.* **12,** 177–183.

CHAPTER 11

DUAL REGULATION OF EXPRESSION OF EXOGENOUS δ-CRYSTALLIN GENE IN MAMMALIAN CELLS: A SEARCH FOR MOLECULAR BACKGROUND OF INSTABILITY IN DIFFERENTIATION

*Hisato Kondoh and T. S. Okada**

INSTITUTE FOR BIOPHYSICS
FACULTY OF SCIENCE
KYOTO UNIVERSITY
KYOTO, JAPAN

I. Introduction

Recently, it has been shown that cloned animal genes are well regulated after reintroduction into living cells and mimic, to a great extent, normal regulation. Differentiation-dependent genes are of course no exception. Results of the large number of experiments along this line in the last few years have convinced us that experiments using gene transfer techniques provide a promising approach to uncovering the mechanisms of differential gene expression. As an example, we have previously demonstrated that the lens-specific δ-crystallin gene of the chicken is expressed, when introduced into mouse cells in primary culture, preferentially into homologous lens cells (Kondoh *et al.*, 1983).

* Present address: The National Institute for Basic Biology, Myodaiji, Okazaki, Japan.

*CURRENT TOPICS IN
DEVELOPMENTAL BIOLOGY, VOL. 20*

Crystallins are a group of soluble proteins specific to vertebrate lens. Among the crystallins, δ-crystallin is found only in avians and reptilians; in the chicken it is the earliest one to be expressed during lens development. It represents the most abundant crystallin in the lens (Piatigorsky, 1984).

Crystallins can be used as molecular markers of lens differentiation. We have repeatedly demonstrated that certain ocular tissues other than lens, such as neural retina and retinal pigmented epithelium, also transdifferentiate to form lens cells in appropriate culture conditions (for reviews, see Okada, 1980, 1983). One of the criteria, other than morphology, used to demonstrate this overt cell change was the expression of all the crystallin classes (Table I).

In addition to crystallin expression associated with lens differentiation, we came across expression of the δ-crystallin in certain ectopic (including nonocular) tissues as well, either in embryos (Barabanov, 1977; Clayton, 1982; Okada, 1983; Agata et al., 1983; Ueda et al., 1983) or in tissue cultures in vitro (Kodama and Eguchi, 1982; Takagi, 1985). Ectopic δ-crystallin expression is characterized not only by its very low level but also by the absence of any other crystallins (Table I). Ectopic δ-crystallin expression has been discussed in terms of the idea of its representing a rudimental lens differentiation (cf. Clayton, 1982). However, in this chapter we will present evidence for another explanation.

It is important to investigate the molecular mechanism as to why the gene coding for δ-crystallin is "leaky" (Moscona and Linser, 1983). Such studies may not only uncover the mechanism of transdifferentiation into lens of nonlenticular tissues both in regeneration and in cell cultures, but may also contribute toward understanding the molecular background of the unstable nature of cell differentiation in general.

TABLE I

CRYSTALLIN EXPRESSION IN CHICKEN EMBRYONIC TISSUES

Category of expression	Crystallins		
	α	β	δ
Normal lens phenotype	+ +	+ +	+ + +
Ectopic expression (e.g., adenohypophysis)	−	−	+
Expression in ordinary tissues including retina	−/±[a]	−	−
Expression in transdifferentiated retina	+ +	+ +	+ + +

[a] Compare to the chapter by Moscona, this volume.

Recent experiments on transfer of the chicken δ-crystallin gene into mouse cells offer interesting results with respect to this problem.

II. The Experimental System

We have previously cloned a continuous stretch of chicken δ-crystallin gene (Yasuda *et al.*, 1982, 1984; Ohno *et al.*, 1985). We injected the δ-crystallin gene on plasmid vectors directly into the nucleus of mouse cells in primary culture (Kondoh *et al.*, 1983). In the mouse, δ-crystallin is totally absent and is replaced by γ-crystallin. Thus, the expression of chicken δ-crystallin in mouse cells, if it takes place, can be detected either by radiolabeled polynucleotides complementary to δ-crystallin mRNA sequences, or by specific antibodies to δ-crystallin. We found that the chicken δ-crystallin gene was expressed in the mouse lens epithelial cells as efficiently as in the homologous chicken cells. Synthesis of the δ-crystallin took place 2–3 days after injection of the gene (Kondoh *et al.*, 1983) and increased linearly parallel to the injected DNA dose at least up to 500 copies per nucleus (Hayashi *et al.*, 1985). The δ-crystallin polypeptide synthesized in mouse lens epithelium had the authentic molecular weight (50,000) and antigenicity. The correct processing of the transcript was indicated by the presence of normal-sized (2-kb) cytoplasmic mRNA, as revealed by Northern blot analysis, and the translation into normal polypeptide (Kondoh *et al.*, 1983, and unpublished results).

In other nonlens cells in the initial survey, there was some detectable expression, but the level was far below that found in lens epithelial cells in terms of both mRNA and polypeptide (Kondoh *et al.*, 1983, and unpublished results). This difference in the level of δ-crystallin expression between the cell types is not primarily due to differences in the efficiency of injection or in the stability of the injected DNA, because an equivalent amount of the injected DNA was recovered as extrachromosomal copies (i.e., almost exclusively in Hirt's supernatant) from different types of cells (Hayashi *et al.*, 1985). Thus, the observed difference between cell types must reflect differential gene expression. The observations also suggested that there are regulatory signals in the δ-crystallin and associated DNA sequences which interact with tissue-specific regulatory mechanisms.

The cloned δ-crystallin gene is approximately 9 kb long, consists of 17 exons (see Fig. 2) (Ohno *et al.*, 1985, and references therein), and is associated with 2.2- and 0.5-kb-long 5'- and 3'-flanking sequences, respectively. For convenience, we divided the transcribed region into two segments, the 5' and 3' segments, at a unique *Kpn*I site located in the second intron. The 5' segment, 677 bp from the cap site, includes the initial two exons, I and II. Exon I, 35 bp long, is not translated and is thus probably required mainly for transcriptional initiation. Exon II includes the authentic initiator methionine codon and codes for four amino acids. The 3' segment contains most of the exons and accounts for the antigenic determinants. Although the 3' segment by itself appeared to contain no transcriptional initiation activity, it became active to synthesize δ-crystallin-like polypeptide (MW 48,000) when attached to viral or other promoters. This occurs by virtue of an alternative initiator methionine codon in exon III that appears to function only in the absence of exon II (Hayashi et al., 1985). In order to take advantage of this situation, a chimeric gene, Mo-δ, was constructed from a promoter complex of Moloney murine leukemia virus long terminal repeating segment (LTR) and from the 3' segment of the δ-crystallin gene and its 3'-flanking sequence. The level of the expression of this chimeric gene appeared fairly constant between the cell types (see below), indicating that the 3' segment of the δ-crystallin gene is not substantially involved in tissue-specific regulation. In contrast, the large part of the 5' fragment including the exon II could be deleted without changing the specificity of the gene expression. Thus, the 5'-flanking sequences are left as the strong candidate for including the putative tissue-specific regulatory signals.

III. Dual Regulation of δ-Crystallin Gene

In the 5'-flanking sequences, only the 100 bp from the cap site appear essential for the tissue-specific regulation, because removal of the sequences upstream from position −100 did not affect expression in any of the mouse cells so far tested. This region of course contains the TATAA box homology and other characteristic features of the transcriptional promoter.

We performed progressive deletion of the 100-bp 5' flanking region by exonucleolytic digestion of the sequences from upstream and assessed the gene activity in the fibroblastic cell line L801 and in primary mouse lens epithelial cells (Hayashi et al., 1985). The results are summarized in Fig. 1. We found a dramatic difference between the results of lens epithelium and fibroblasts. In lens epithelial cells, deletion to −80 did not affect gene activity to any significant extent, continued deletion to −65 reduced the activity to 3% of the original, and further deletions totally abolished the activity. In contrast, gene activity in the fibroblasts was already reduced by deletion to −88 and was completely lost by deletion to −80. These results indicate that the low-level expression of the δ-crystallin gene in the fibroblasts requires

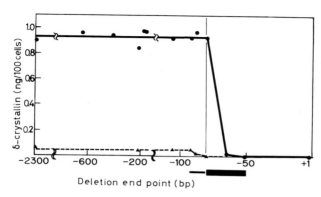

FIG. 1. Effect of deletions in the 5'-flanking sequence starting from upstream. Activity of the δ-crystallin gene in the lens epithelial cells (●) and in the fibroblast cell line L801 (▲) is expressed as the δ-crystallin synthesis in 2 days in 100 cells injected at 125 copies/cell. The regions where progressive deletions reduced gene activity in the fibroblast and the lens epithelial cells are indicated by thin and thick bars, respectively. [Data are taken from Hayashi *et al.* (1985).]

the region −92 to −80, but lens-characteristic high-level expression does not.

When the results are viewed from another aspect, the deletion mutant gene with the 80-bp flanking sequence represents all-or-none regulation with respect to lens specificity, and inclusion of an additional 12-bp upstream sequence brings about the residual activity in the fibroblasts. The results clearly demonstrated that expression in lens epithelial cells and in fibroblasts is distinct from each other not only in terms of level but also in terms of regulation.

IV. Expression of Exogenous δ-Crystallin Gene in Various Cell Types

A. EXPRESSION IN PRIMARY CULTURES AND CELL LINES

By comparing lens epithelial cells and a fibroblast line, we could distinguish two distinct types of regulation of δ-crystallin gene expression that require different regulatory sequences. We carried out an extensive survey of mouse cells to investigate in which type the δ-crystallin gene is expressed. Gene expression was assessed by comparison with a coinjected nonspecific *Mo-δ* gene (Fig. 2). Expression of the δ-crystallin gene results in the synthesis of a 50K polypeptide, whereas expression of the *Mo-δ* gene results in the synthesis of a 48K polypeptide as the major component and of the 51K polypeptide as a

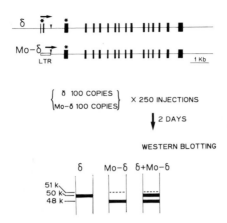

FIG. 2. Strategy of the analysis of δ-crystallin expression in various cell types after injection of the gene. Top: Schematic presentation of δ-crystallin gene and its derivative *Mo*-δ. Rectangles on the map indicate exons, arrows the transcriptional units, vertical wedges the *Kpn*I sites, and asterisks the Met codons for translational initiation. Middle: Experimental protocol. Bottom: Schematic drawing of a typical Western blot to distinguish expression of δ and *Mo*-δ genes.

minor component. Therefore, the expression of the two genes on different plasmids in the same mixture can be measured separately. The results are shown in Fig. 3 and summarized in Table II.

The primary cultures were injected within 3 days after preparation to ensure maintenance of their original differentiated state. We found that among a number of mouse tissue cells in primary cultures, not only lens but also epidermal cells supported a high level of expression of the introduced δ-crystallin gene. In epidermal cells, the deletion gene with the 80-bp-long 5'-flanking sequence exhibited full activity, but the deletion gene with the 50-bp-long flanking sequence exhibited no activity. Thus, the expression of the δ-crystallin gene in epidermal cells requires the same 5'-flanking sequences as in lens epithelial cells; its level was also comparable to the lens cells. If we consider the ontogeny of the lens and the epidermis, it is obvious that these two cell types are closely related in that both originate from the embryonic ectoderm and become distinct only after the induction of the lens by the optic cup (see Fig. 5).

Of particular interest among other cell types were retinal pigment epithelium, retinal glia, brain glia, and adenohypophysis. The former two have the potency to transdifferentiate into lens in a long-term culture under appropriate conditions (reviews by Okada, 1980, 1983; and Clayton, 1982), whereas the latter two are characterized by ectopic

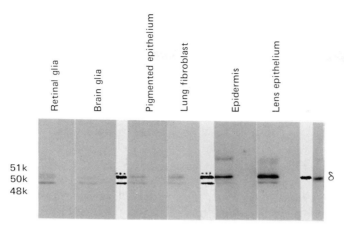

FIG. 3. Western blot analysis of the δ-crystallin synthesis in various cell types injected with the mixture of the δ-crystallin and the *Mo*-δ genes. For each cell type, injected (left) and uninjected (right) samples were compared.

δ-crystallin expression in chicken embryonic tissues (Barabanov, 1977; Ueda *et al.*, 1983; Takagi, 1985).

None of these cells allowed high-level expression of the δ-crystallin gene. In addition, low-level expression in these cells required the 5'-flanking sequences longer than 80 bp from the cap site, indicating that regulation is probably of the same type as that found in the fibroblasts.

B. EXPRESSION IN LONG-TERM CULTURES

It is well known that transdifferentiation into lens occurs in long-term cultures of embryonic neural retina and pigment epithelium of

TABLE II

EXPRESSION OF THE δ-CRYSTALLIN GENE IN VARIOUS CELL TYPES AS
NORMALIZED TO THE *Mo*-δ EXPRESSION

Cell type	δ (52K)/*Mo*-δ (50K)
Lens epithelium; epidermis	7.5–8.0
Kidney epithelium; adenohypophysis; adenocortical tumor (Y1); brain glia; kidney fibroblast	1.5–1.9
Retinal-pigmented epithelium; visceral endoderm (PSA5E); cardiac muscle; fibroblast (STO)	1.0–1.4
Liver fibroblast; lung fibroblast; retinal glia; neuroblastoma (NB2A); fibroblast (L801)	0.3–0.9

some birds and mammals. It is thus expected that cultures of retinal glia, for instance, would eventually acquire the capacity to express the δ-crystallin gene at the high level characteristic of the lens. It is interesting to see from what point this capacity is present. We set up cultures of neural retina from mouse fetuses and injected glial cells at two different time points, 2 and 25 days after the start of the cultures, to determine the effects of prolonged culturing.

After 2 days, the retinal glial cells expressed δ-crystallin only at the low level, but after 4 weeks, the retinal glial cells expressed the exogenous δ-crystallin gene at the high level characteristic of the lens (Fig. 4), although neither lentoidogenesis nor the expression of any lens phenotypes occurred in these cultures. This shift in the type of δ-crystallin expression is probably not due to replacement of glial subpopulations, but rather to changes in cell characteristics in the same glial population, because there is only one glial cell type, the Müller cell, in these retinal cultures.

In contrast, brain glial cells, as well as dermal fibroblasts in long-term cultures, continued to express the δ-crystallin gene at a low level (Fig. 4). Therefore, high-level expression is not a general tendency of cells in long-term cultures, but probably reflects the transdifferentiation ability of retinal glial cells. It should be reemphasized that the expression of endogenous lens-specific genes did not take place.

Thus, we observed lens-type expression of the δ-crystallin gene in three kinds of cell, i.e., authentic lens epithelial cells, epidermal cells, and retinal glial cells in long-term culture (Fig. 5). The latter two do not express the endogenous lens-specific markers, but are respectively

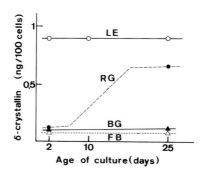

FIG. 4. Effect of the culture age on the expression of the injected δ-crystallin gene. Lens epithelial cells (LE), retinal glia (RG), brain glia (BG), and fibroblasts (FB) were studied. At the culture ages indicated, the δ-crystallin gene was injected and after 2 days the samples were recovered. The gene activity is expressed as in Fig. 1. The lines are merely conceptual.

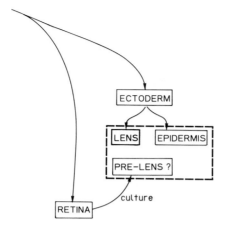

Fig. 5. The pathways of lens differentiation from different sources in both normal development and the transdifferentiation in cell cultures. The cell types in the broken rectangle supported the high-level expression of the exogenous δ-crystallin gene characteristic to the lens.

closely related to lens in the pathway of cell differentiation in normal development and in the event of transdifferentiation. It is thus tempting to speculate that high-level expression of the exogenous δ-crystallin gene reflects differentiated states of cells that are very close to lens cells and that such cells share crucial regulatory mechanisms with the lens cells.

V. Correspondence to the Natural Situation

Taking advantage of the present gene transfer system, we localized the DNA sequences responsible for the lens-specific regulation of the δ-crystallin gene within the 80-bp region upstream from the transcriptional initiation site. However, the expression of δ-crystallin in this experimental system deviated from its normal expression in the chicken in two respects. First, a low but significant level of expression took place in virtually all the cell types we examined. Second, efficient lens-type expression of δ-crystallin was not confined to the authentic lens cells but was also observed in other lens-related cells.

A. ECTOPIC EXPRESSION in Vivo

Low-level expression of the δ-crystallin gene was observed in most of the nonlens tissues after nuclear injection under the condition that the majority of the injected gene copies remained unintegrated into the chromosome (see Section II). However, we have shown, using stably transformed teratocarcinoma cell lines, that following the integration of the exogenous δ-crystallin gene into mouse chromosomes, expression was largely suppressed, and the low level of expression was restricted to certain defined cell types specific to each integration site (Kondoh *et al.*, 1984). Thus, the chromosomal position effect appears dominant over the intrinsic potential of the δ-crystallin gene, which alone may lead to ubiquitous low-level expression. It should also be emphasized that the expression of exogenous δ-crystallin in nonlens cells has never provoked other lens characteristics, including expression of different crystallin classes (Kondoh *et al.*, 1984, and unpublished observations). However, in the normal development of the chicken embryo and in *in vitro* cultures, the transcription (and possibly the translation in some cases) of the δ-crystallin gene has been documented in a variety of nonlens tissues (Clayton, 1982; Kodama and Eguchi, 1982; Okada, 1983; Agata *et al.*, 1983; Ueda *et al.*, 1983; Takagi, 1985). This kind of ectopic δ-crystallin expression in chicken tissues is characterized by its low level and by the absence of the expression of other crystallins. We thus feel it very likely that ectopic δ-crystallin expression in the developing chicken is brought about by the chromosomal activity of the gene locus, which elicits the otherwise cryptic δ-crystallin expression of the nonlens type, and is not necessarily related to lens differentiation.

B. Expression in Lens-Related Cells after Gene Transfer

Lens-type expression of δ-crystallin was also observed in related nonlens cell types, a situation that never occurs normally. This suggests multistep regulation of δ-crystallin expression. Our experimental system probably succeeded in mimicking normal regulation to a large extent, but lacked the mechanism to confine the expressing tissue to the lens. Unsuccessful regulation in the latter part may be due to a lack of the chromosomal effect or to a xenogenic combination of the gene and the recipient cells. In any event, the missing part must involve a kind of negative regulation. However, if we take advantage of the relaxed tissue specificity, the xenogenic gene transfer system described here should provide a means to assay the differentiated state of the cell in terms of its closeness to the lens.

VI. Concluding Remarks

The present analysis of exogenous chicken δ-crystallin expression in mouse cells provides an example to show how a gene transfer system can be useful in studying the differentiated state of the cells.

There are three distinct cell groups in respect to the level of expression of the transferred δ-crystallin gene: (1) very high expression in lens cells and epidermal cells that are ontogenically closely related; (2) low expression at the beginning and high expression after long-term culture as in retinal glia and possibly some other cell types that have been known to transdifferentiate into lens in culture; (3) low expression in fibroblasts and others. It was indicated that cells of the first group utilize a nontranscribed regulatory sequence of the gene different from the sequence required for the low expression in the second and third groups. When retinal glial cells were placed in long-term culture, the very high level of translation of the injected δ-crystallin gene was attained probably because the sequence utilized for the expression of the injected δ-crystallin gene switched from that for low expression to that for high expression. Such a switch is considered to be a key for the occurrence of transdifferentiation of retinal glial cells to lens in cell cultures. Thus, we have probably succeeded in detecting change of differentiated state by analysis of expression of an exogenous tissue-specific gene before cells endogenously express analogous genes. This kind of analysis should be applicable to a wide variety of experimental systems and provide an important approach to uncovering the molecular background of the instability in differentiated cells.

ACKNOWLEDGMENT

The original research described here was supported by Special Research Grant "Multicellular Organization" (No. 59113004) and Grant for Basic Cancer Research (No. 59010034) from the Japanese Ministry of Education, Science and Culture.

REFERENCES

Agata, K., Yasuda, K., and Okada, T. S. (1983). *Dev. Biol.* **100**, 222–226.
Barabanov, V. M. (1977). *Dokl. Akad. Nauk SSSR* **234**, 195–198.
Clayton, R. M. (1982). *In* "Differentiation *in Vitro*" (M. M. Yeoman and D. E. S. Truman eds.). Cambridge Univ. Press, London and New York.
Hayashi, S., Kondoh, H., Yasuda, K., Soma, G., Ikawa, Y., and Okada, T. S. (1985). *EMBO J.* **4**, in press.
Kodama, R., and Eguchi, G. (1982). *Dev. Biol.* **91**, 221–226.
Kondoh, H., Yasuda, K., and Okada, T. S. (1983). *Nature (London)* **301**, 440–442.
Kondoh, H., Takahashi, Y., and Okada, T. S. (1984). *EMBO J.* **3**, 2009–2014.
Moscona, A. A., and Linser, P. (1983). *Curr. Top. Dev. Biol.* **18**, 155–188.
Ohno, M., Sakamoto, H., Yasuda, K., Okada, T. S., and Shimura, Y. (1985). *Nucleic Acids Res.* **13**, 1593–1606.
Okada, T. S. (1980). *Curr. Top. Dev. Biol.* **16**, 349–380.
Okada, T. S. (1983). *Cell Differ.* **13**, 177–183.

Piatigorsky, J. (1984). *Mol. Cell. Biochem.* **59,** 33–56.

Takagi, S. (1985). *Wilhelm Roux's Arch. Dev. Biol.,* in press.

Ueda, Y., Hino, F., Takeichi, M., and Okada, T. S. (1983). *Dev. Growth Differ.* **25,** 422.

Yasuda, K., Kondoh, H., Okada, T. S., Nakajima, N., and Shimura, Y. (1982). *Nucleic Acids Res.* **10,** 2879–2891.

Yasuda, K., Nakajima, N., Isobe, T., Okada, T. S., and Shimura, Y. (1984). *EMBO J.* **3,** 1397–1402.

CHAPTER 12

NEUROTRANSMITTER PHENOTYPIC PLASTICITY IN THE MAMMALIAN EMBRYO

G. Miller Jonakait* and Ira B. Black

DIVISION OF DEVELOPMENTAL NEUROLOGY
CORNELL UNIVERSITY MEDICAL COLLEGE
NEW YORK, NEW YORK

I. Introduction

Expression of specific neurotransmitter characteristics was once considered an immutable hallmark of the fully differentiated neuron. More recent work in the neurosciences, however, has revealed that developing neurons are considerably more labile with respect to neurotransmitter traits than was once thought. Moreover, such phenotypic malleability extends to the adult neuron as well (Wakshull *et al.,* 1979; Johnson, 1983; Adler and Black, 1984). Both neonatal and adult sympathetic neurons *in vivo* are replete with noradrenergic phenotypic traits. These include the catecholamine (CA) biosynthetic enzymes tyrosine hydroxylase (T-OH) and dopamine β-hydroxylase (DBH), as well as the CA neurotransmitter norepinephrine (NE). In addition, these neurons exhibit a high-affinity uptake process for CAs which facilitates inactivation of the transmitter following release (Iversen, 1967). When grown under appropriate culture conditions, however,

* Present address: Department of Zoology and Physiology, Rutgers University, Newark, New Jersey.

CURRENT TOPICS IN
DEVELOPMENTAL BIOLOGY, VOL. 20

principal sympathetic ganglion cells acquire detectable levels of choline acetyltransferase (ChAT; Johnson *et al.*, 1976; Iacovitti *et al.*, 1981; Adler and Black, 1985; Kessler, 1984) and synthesize acetylcholine (Patterson and Chun, 1974). In culture the cells may lose the ability to maintain high levels of NE (Patterson and Chun, 1974), but increase production of the peptide transmitters substance P (Kessler *et al.*, 1981, 1983, 1984; Adler and Black, 1984) and somatostatin (Kessler *et al.*, 1983). Interestingly, even electrophysiologically identified cholinergic cells retain, in some cases, immunoreactive T-OH (Higgins *et al.*, 1981) as well as the ability to take up and store CAs (Landis, 1976; Reichardt and Patterson, 1977). Thus, sympathetic ganglion neurons display remarkable phenotypic plasticity in response to appropriate environmental cues.

However, neurotransmitter plasticity is not restricted in the peripheral nervous system to sympathetic ganglia. Chick ciliary ganglion neurons, most of which are cholinergic *in vivo*, display an analogous neurotransmitter plasticity when grown *in vitro*: after a week in culture they express immunoreactivity to both T-OH and phenylethanolamine *N*-methyltransferase (PNMT), the enzyme responsible for the conversion of NE to epinephrine. Interestingly, ciliary ganglion cell cultures retain detectable levels of ChAT, but lack CA histofluorescence and an uptake process for CAs (Teitelman *et al.*, 1985; Iacovitti *et al.*, 1985), suggesting that expression of one (or even two) catecholaminergic traits does not necessarily presuppose the expression of the entire range of neurotransmitter characteristics.

These examples from *in vitro* systems reveal the neurotransmitter phenotypic potential of certain peripheral nervous system (PNS) derivatives of the neural crest, raising questions regarding the range of phenotypic plasticity utilized by normal neurons developing *in vivo*.

II. Transient Expression of Neurotransmitter Phenotype during Normal Development

A. NORADRENERGIC EXPRESSION IN SYMPATHETIC AND ENTERIC CELLS

The embryonic microenvironment plays a potent role in determining and/or stabilizing phenotypic expression of PNS derivative of avian neural crest (see Le Douarin, 1980; Jonakait and Black, 1986, for reviews). To begin to examine this relationship in mammals, efforts were made to document the initial expression of the noradrenergic phenotype in the peripheral nervous system of rats (Cochard *et al.*, 1978, 1979). Phenotypic markers for noradrenergic neurons included immunoreactivity to both T-OH and DBH as well as CA histofluores-

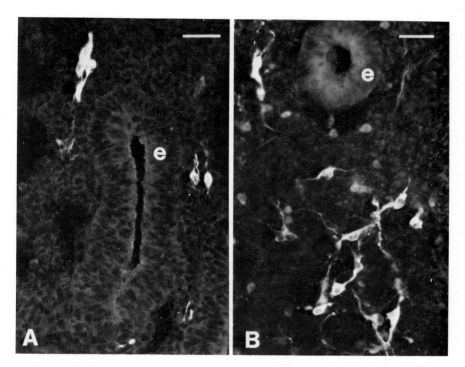

FIG. 1. T-OH immunoreactivity in cells of the embryonic rat intestine. (A) T-OH-positive cells are first detected at E11.5 (27 somites). (B) At E12.5 (36 somites) many T-OH-positive cells are present. e, Epithelium. Bar, 40 μm. [Reprinted from Black *et al.*, (1984) with permission from Plenum Press.]

cence. Primitive sympathetic ganglion cells simultaneously expressed these traits on day 11.5 of gestation (E11.5; 27 somites). Unexpectedly, noradrenergic markers also appeared at E11.5 in a population of cells in the embryonic intestinal mesenchyme (Fig. 1A). Since noradrenergic cells are not known constituents of the adult rat gut, the subsequent fate of these noradrenergic cells was of interest. By E12.5 T-OH-positive cells of the gut had increased in number (Fig. 1B), but by E13.5 CA histofluorescence had virtually disappeared, and by E15.5 T-OH immunoreactivity was undetectable. The presence of neurofilament protein in these cells confirmed their neuronal nature (Cochard, 1984; G. M. Jonakait, unpublished observation). These findings were confirmed by Teitelman *et al.* (1979) who found further that T-OH-positive cells of the gut took up [^{3}H]thymidine, suggesting continuing cell division (Teitelman *et al.*, 1981a).

The loss of noradrenergic characteristics from enteric neurons contrasted sharply with the maintenance of these traits in sympathetic

ganglia. The failure of enteric mesenchyme to support catecholaminergic traits in avian sympathetic ganglion precursors *in vitro* (Le Douarin *et al.,* 1977) suggested that the intestinal microenvironment
might suppress the expression of noradrenergic characteristics. However, programmed loss of noradrenergic traits by neural crest cells
destined to colonize the enteric ganglia could not be excluded.

B. Do Transiently Noradrenergic Cells of the Gut Survive? Analysis Following Uptake of [³H]Norepinephrine

We sought to determine the fate of transiently noradrenergic cells
of the gut. In search of a more long-lived marker for these cells, we
asked whether, in addition to their other noradrenergic traits, these
cells possessed a specific, high-affinity uptake process for CAs. If so,
exposure to radiolabeled CAs would label the cells, allowing subsequent detection.

In order to determine whether specific, high-affinity uptake of
[³H]CA was a property of the gut during the period of T-OH expression
(E12.5–E13.5), intestines were incubated with micromolar concentrations of [³H]NE. Label accumulated in a tightly bound compartment
(Jonakait *et al.,* 1985). Incubation with [³H]NE in the presence of
desmethylimipramine (DMI), a specific uptake inhibitor, blocked NE
accumulation by this compartment at all ages examined. Moreover,
kinetic analysis of uptake revealed that the uptake process had a high
affinity ($K_m = 1.4$ μM), was saturable, and was blocked during incubation at 4°C. These data in aggregate suggested that E12.5–E13.5 intestines possessed a high-affinity, energy-requiring uptake process, specific for NE.

To determine whether T-OH-positive perikarya were responsible
for the observed uptake, tissue sections prepared for radioautography
were examined simultaneously for T-OH immunoreactivity. At E12.5
and E13.5, silver grains marking areas of [³H]NE accumulation were
detected only over cells that also expressed T-OH (Fig. 2), suggesting
that T-OH-positive cells in the gut were responsible for the observed
uptake. The initial expression of CA biosynthetic machinery preceded
the initial appearance of the uptake process, since T-OH-positive cells
at E11.5 lacked detectable silver grain accumulations. More critically,
the *disappearance* of CA biosynthetic machinery preceded the loss of
the uptake process, since cells lacking T-OH immunoreactivity at
E15.5 retained the ability to concentrate ³H-labeled amines.

These data suggested that cells in the gut mesenchyme that transiently express T-OH possess a complete repertoire of noradrenergic
traits including biosynthetic machinery, the presence of NE itself, as

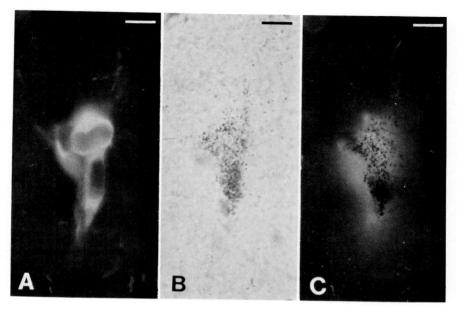

Fɪɢ. 2. High-power view of T-OH-positive cells in E13.5 intestine. (A) Cells photographed using fluorescent epi-illumination to show T-OH immunoreactivity. (B) The same field photographed with bright-field optics shows silver grains accumulated over the T-OH-positive area. (C) Combined use of epi-illumination and bright-field transillumination reveals silver grains (plane of focus) over T-OH-positive cell bodies (below plane of focus). Bar, 10 μm. [Reprinted from Jonakait et al., (1985) by permission of Academic Press].

well as a neurotransmitter inactivation mechanism (uptake$_1$; Iversen, 1967). Moreover, the noradrenergic cell population probably does not die, but remains in gut, selectively losing its detectable catecholaminergic characters while acquiring new, as yet unknown, phenotypic traits (Jonakait et al., 1979, 1985).

An analogous conversion of phenotype in vivo characterizes embryonic mammalian pancreas cells which, though T-OH-positive during early stages of development, acquire detectable levels of glucagon (Teitelman et al., 1981b) as they lose catecholaminergic traits.

Moreover, the apparent loss of certain noradrenergic traits with the concomitant retention of high-affinity uptake is analogous to the phenomenon that characterizes dissociated sympathetic ganglion cells which have acquired cholinergic properties (Landis, 1976; Reichardt and Patterson, 1977; see above). Furthermore, maintenance of uptake characterizes cholinergic sympathetic terminals that innervate the rat eccrine sweat gland in vivo (Landis and Keefe, 1983). The developmental loss by these terminals of CA histofluorescence proceeds together with the acquisition of acetylcholinesterase staining and maintenance of the CA-specific uptake mechanism.

C. Noradrenergic Expression in Cranial Sensory and Dorsal Root Ganglia

We have more recently described in the rat other populations of cells, located in some embryonic cranial sensory and dorsal root ganglia, which transiently express T-OH immunoreactivity (Jonakait et al., 1984). At E10.5, T-OH immunoreactivity appears in a small population of bipolar cells in the trigeminal ganglion anlage. Its appearance then follows a rostral–caudal sequence. By E11.5, T-OH immunoreactivity is evident in both the petrosal and nodose ganglia, serving the IXth and Xth cranial nerves, respectively. At this time, isolated cells expressing T-OH are also found in the most rostral dorsal root ganglion. By E13.5, however, T-OH is not evident in any cranial sensory or dorsal root ganglia. Similar to T-OH-positive ciliary ganglion cells in culture (see above), transiently T-OH-positive cells in embryonic sensory ganglia lack detectable CA fluorescence as well as the uptake process for CAs. They differ significantly, then, from their enteric counterparts in lacking a complete noradrenergic repertoire. This suggests further that not all noradrenergic traits are coordinately regulated, but can be expressed and regulated independently.

III. Factors Affecting Neurotransmitter Phenotype

A. Glucocorticoid Hormones

In studies designed to characterize transiently noradrenergic gut cells more fully, we asked whether vesicular storage was an additional catecholaminergic property. Since reserpine depletes vesicular NE (Carlsson, 1966) and crosses the placenta (Kovacic and Robinson, 1966), pregnant rats were treated with reserpine on E11.5, the day that cells first express noradrenergic traits. Unexpectedly, reserpine treatment resulted not in depletion of CAs, but rather in the persistence of noradrenergic traits in embryonic intestines beyond the time of normal disappearance (Jonakait et al., 1980, 1981a). In comparison to E13.5-day controls, CA histofluorescence and T-OH catalytic activity were significantly increased. T-OH and DBH immunoreactivity persisted as long as day 14.5. Recruitment from hitherto noncatecholaminergic populations seemed unlikely since uptake of [³H]NE was unchanged following reserpine treatment (unpublished observation).

This paradoxical result, achieved with a drug expected to deplete CAs, prompted us to investigate other mechanisms of reserpine action.

Since reserpine treatment results in a dramatic rise in circulating levels of plasma glucocorticoid hormones, (Maickel *et al.*, 1961; Carr and Moore, 1968), we examined the possible involvement of glucocorticoid hormones in mediating the action of reserpine. Several experiments suggested that this might be the case. First, treatment of pregnant rats with pure cortisol mimicked the action of reserpine in raising levels of T-OH activity and in prolonging the appearance of CA fluorescence. Second, when pregnant rats were pretreated with low doses of the synthetic glucocorticoid hormone dexamethasone to inhibit a stress-induced rise in plasma glucocorticoid hormones, the action of reserpine in prolonging noradrenergic traits was inhibited. Third, pretreatment of animals with mitotane, a drug cytotoxic to the maternal adrenal cortex, diminished the effectiveness of reserpine treatment. Finally, pretreatment of pregnant rats with pargyline, known to block the reserpine-induced increase in glucocorticoid hormones (Maickel and Martel, 1983), suppressed the action of reserpine in prolonging T-OH expression (Fig. 3). These data taken together suggested that the effect of reserpine was mediated, at least in part, by its action on the maternal pituitary–adrenal axis.

Glucocorticoid hormones play a similar role in other instances of

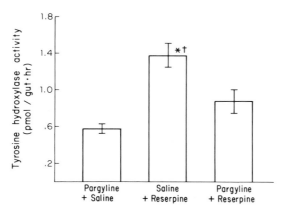

FIG. 3. Pregnant rats were treated on E10.5 with pargyline (25 mg/kg) or saline; 18 hours later, they received an injection of reserpine (5 mg/kg) or saline. At E13.5, embryonic intestines were dissected and assayed for T-OH enzymatic activity. Pargyline pretreatment blocked the reserpine-induced increase in T-OH. $*p < 0.0025$ when compared to pargyline + saline-treated control; $\dagger p < 0.025$ when compared to pargyline + reserpine-treated group.

noradrenergic phenotypic plasticity. The addition of dexamethasone to cultures of sympathetic ganglion cells inhibits the acquisition of cholinergic characteristics and promotes the retention of noradrenergic traits (McLennan *et al.*, 1980; Fukada, 1981). Though glucocorticoid hormones act directly on pheochromocytoma cells (PC12) to induce mRNA specific for T-OH (Baetge *et al.*, 1981), the effect of glucocorticoid hormones in dissociated sympathetic ganglion cell cultures is secondary to an action on the nonneuronal cells included in the culture (Fukada, 1981).

Whatever the mechanism of glucocorticoid hormone action on transiently noradrenergic cells of the gut, it is clear that maternal drug experience dramatically alters the developmental course of immature autonomic neurons.

B. NERVE GROWTH FACTOR

Nerve growth factor (NGF) is required for the normal survival and maturation of many neural crest-derived neuronal populations (see Greene and Shooter, 1980; Harper and Thoenen, 1981, for reviews). Moreover, the action of NGF in eliciting a neuronal phenotype from the adrenal medullary chromaffin cell and its malignant counterpart, the pheochromocytoma cell line PC12, is well documented (Tischler and Greene, 1975; Doupe *et al.*, 1982; Ogawa *et al.*, 1984, and chapter by Ogawa *et al.*, this volume).

Transiently noradrenergic cells of the embryonic intestine, too, respond to a transuterine injection of NGF on day E11.5 with prolonged expression of CA histofluorescence (Kessler *et al.*, 1979) and increased levels of T-OH catalytic activity measured at E13.5 (Jonakait *et al.*, 1981b).

IV. Concluding Remarks

It has become increasingly clear that transient neurotransmitter phenotypic expression is not a phenomenon limited to unrepresentative or idiosyncratic cell populations. Documented instances of transient neurotransmitter expression have increased dramatically since 1979 (Table I) and include examples from both peripheral and central nervous systems in a variety of vertebrates.

It is not clear, however, the level at which regulation of transient neurotransmitter expression occurs. Loss of specific gene products associated with specific neurotransmitters may involve steps ranging from extinction of a gene through a host of posttranscriptional or posttranslational processes more proximal to expression of a phenotypic trait. Therefore, it is not clear whether transient expression involves a

TABLE I

NEUROTRANSMITTER PHENOTYPIC PLASTICITY

In vitro			
Sympathetic ganglia	Rat	NE → ACh	Patterson and Chun (1974); Bunge *et al.* (1974)
Sympathetic ganglia	Rat	NE + SP	Kessler *et al.* (1981); Adler and Black (1985)
Dorsal root ganglia	Rat	? → T-OH	Price and Mudge (1983)
Ciliary ganglia	Chick	ACh → {T-OH / PNMT	Iacovitti *et al.* (1983); Teitelman *et al.* (1984)
In vivo			
Embryonic gut	Rat	T-OH } → ? / DBH	Cochard *et al.* (1978, 1979); Teitelman *et al.* (1979); Jonakait *et al.* (1979)
Neural crest	Chick	ChAT → ?	Smith *et al.* (1980)
Embryonic pancreas	Mouse	T-OH → glucagon	Teitelman *et al.* (1981b)
Ventral neural tube	Mouse	T-OH → ?	Teitelman *et al.* (1981a); Jonakait *et al.* (1985)
Lower brain stem	Rat	Somatostatin → ?	Shiosaka *et al.* (1981)
Upper brain stem	Rat	SP → ?	Inagaki *et al.* (1982)
Dorsal medulla	Rat	β-Endorphin → ?	Baetge *et al.* (1982)
Embryonic CNS	Rat	Enkephalin → ?	Palmer *et al.* (1982)
Sweat gland innervation	Rat	NE → AChE	Landis and Keefe (1983)
Neural crest	Chick	AChE → ?	Cochard and Coltey (1983)
Inferior colliculus	Rat	T-OH → ?	Jaeger and Joh (1983)
Spinal cord	Opossum	5-HT → ?	DiTirro *et al.* (1983)
Medial dorsal nucleus of thalamus	Monkey, human	AChE → ?	Kostovic and Goldman-Rakic (1983)
Cranial sensory dorsal root ganglia	Rat	T-OH → ?	Jonakait *et al.* (1984)
Deep cerebellar nucleus	Rat	AChE → ?	Martin-MacKinnon and Kristt (1984)

qualitative change at the genomic level or whether it reflects a quantitative change in transmitter levels. Moreover, it is not clear where or how factors such as NGF or hormones interact with these processes. It is also not clear whether the capacity for reexpression of the transiently appearing neurotransmitter traits remains a latent property of more mature descendants throughout life, allowing the organism increased flexibility in responding to disease or injury.

It is clear, however, that neurotransmitter phenotypic plasticity is not restricted to the developing nervous system, but is exhibited by adult neurons as well. This suggests that a certain developmental potential exists throughout the life of the organism. Determination of the factors that restrict that potential may reveal the fundamental processes involved in the maintenance and/or aging of the mature nervous system.

REFERENCES

Adler, J. E., and Black, I. B. (1984). *Science* **225,** 1499–1500.

Adler, J. E., and Black, I. B. (1985). *Proc. Natl. Acad. Sci. U.S.A.* **82,** 4296–4300.

Baetge, E., Kaplan, B. B., Reis, D., and Joh, T. (1981). *Proc. Natl. Acad. Sci. U.S.A.* **78,** 1269–1273.

Baetge, G., Shoemaker, W. J., Bayon, A., Azad, R., and Bloom, F. E. (1982). *Soc. Neurosci. Abstr.* **8,** 636.

Black, I. B., Adler, J. E., Bohn, M. C., Jonakait, G. M., Kessler, J. A., and Markey, K. A. (1984). *In* "Cellular Regulation and Molecular Biology of Neuronal Development" I. B. Black, ed.), pp. 117–130. Plenum, New York.

Bunge, R. P., Rees, R., Wood, P., Burton, H., and Ko, C.-P. (1974). *Brain Res.* **66,** 401–412.

Carlsson, A. (1966). *Handb. Exp. Pharmacol.* **19,** 529–592.

Carr, L. A., and Moore, K. E. (1968). *Neuroendocrinology* **3,** 285–302.

Cochard, P. (1984). Thèse L'L.E.R. de Medicine et Biologie Humaine de L'Université Paris-Nord.

Cochard, P., and Coltey, P. (1983). *Dev. Biol.* **98,** 221–238.

Cochard, P., Goldstein, M., and Black, I. B. (1978). *Proc. Natl. Acad. Sci. U.S.A.* **75,** 2986–2990.

Cochard, P., Goldstein, M., and Black, I. B. (1979). *Dev. Biol.* **71,** 100–114.

DiTirro, F. J., Martin, G. F., and Ho, R. H. (1983). *J. Comp. Neurol.* **213,** 241–261.

Doupe, A. J., Patterson, P. H., and Landis, S. C. (1982). *Abstr. Soc. Neurosci.* **8,** 257.

Fukada, K. (1981). *Nature (London)* **287,** 553–554.

Greene, L. A., and Shooter, E. (1980). *Annu. Rev. Neurosci.* **3,** 353–402.

Harper, G. P., and Thoenen, H. (1981). *Annu. Rev. Pharmacol. Toxicol.* **21,** 205–209.

Higgins, D., Iacovitti, L., Joh, T. H., and Burton, H. (1981). *J. Neurosci.* **1,** 126–131.

Iacovitti, L., Joh, T. H., Park, D. H., and Bunge, R. P. (1981). *J. Neurosci.* **1,** 685–690.

Iacovitti, L., Teitelman, G., Grayson, L., Joh, T. H., and Reis, D. J. (1985). *Dev. Biol.* **110,** 402–412.

Inagaki, S., Sakanaka, M., Shiosaka, S., Senba, E., Takatsuki, K., Takagi, H., Kawai, Y., Minagawa, H., and Tohyama, M. (1982). *Neuroscience* **7,** 251–277.

Iversen, L. L. (1967). The Uptake and Storage of Noradrenaline in Sympathetic Nerves. Cambridge Univ. Press, London and New York.

Jaeger, C. B., and Joh, T. H. (1983). *Dev. Brain Res.* **11,** 128–132.

Johnson, M. I. (1983). *Abstr. Soc. Neurosci.* **9,** 846.

Johnson, M., Ross, D., Meyers, M., Rees, R., Bunge, R., Wakshull, E., and Burton, H. (1976). *Nature (London)* **262,** 308–310.

Jonakait, G. M., and Black, I. B. (1986). *Handb. Exp. Pharmacol.,* in press.

Jonakait, G. M., Wolf, J., Cochard, P., Goldstein, M., and Black, I. B. (1979). *Proc. Natl. Acad. Sci. U.S.A.* **76,** 4683–4686.

Jonakait, G. M., Bohn, M. C., and Black, I. B. (1980). *Science* **210,** 551–553.

Jonakait, G. M., Bohn, M. C., Goldstein, M., Markey, K., and Black, I. B. (1981a). *Dev. Biol.* **88,** 288–296.

Jonakait, G. M., Kessler, J. A., Goldstein, M., Markey, K., and Black, I. B. (1981b). *Abstr. Soc. Neurosci.* **7,** 289.

Jonakait, G. M., Markey, K. A., Goldstein, M., and Black, I. B. (1984). *Dev. Biol.* **101,** 51–60.

Jonakait, G. M., Markey, K. A., Goldstein, M., Dreyfus, C. F., and Black, I. B. (1985). *Dev. Biol.* **108,** 6–17.

Kessler, J. A. (1984). *Abstr. Soc. Neurosci.* **10,** 13.

Kessler, J. A., Cochard, P., and Black, I. B. (1979). *Nature (London)* **280,** 141–142.

Kessler, J. A., Adler, J. E., Bohn, M. C., and Black, I. B. (1981). *Science* **214,** 335–336.

Kessler, J. A., Adler, J. E., Bell, W. O., and Black, I. B. (1983). *Neuroscience* **10,** 309–318.

Kessler, J. A., Adler, J. E., Jonakait, G. M., and Black, I. B. (1984). *Dev. Biol.* **103,** 71–79.

Kostovic, I., and Goldman-Rakic, P. S. (1983). *J. Comp. Neurol.* **219,** 431–447.

Kovacic, B., and Robinson, R. L. (1966). *J. Pharmacol. Exp. Ther.* **152,** 37–41.

Landis, S. C. (1976). *Proc. Natl. Acad. Sci. U.S.A.* **73,** 4220–4224.

Landis, S. C., and Keefe, D. (1983). *Dev. Biol.* **98,** 349–372.

Le Douarin, N. M. (1980). *Nature (London)* **286,** 663–669.

Le Douarin, N. M., Teillet, M.-A., and Le Lievre, C. (1977). *In* "Cell and Tissue Interactions" (J. W. Lash and M. M. Burger, eds.), pp. 11–27. Raven, New York.

McLennan, I. S., Hill, C. E., and Hendry, I. A. (1980). *Nature (London)* **283,** 206–207.

Maickel, R. P., and Martel, R. R. (1983). *Pharmacol. Biochem. Behav.* **19,** 321–325.

Maickel, R. P., Westerman, E. O., and Brodie, B. B. (1961). *J. Pharmacol. Exp. Ther.* **134,** 167–175.

Martin-MacKinnon, N. A., and Kristt, D. A. (1984). *Abstr. Soc. Neurosci.* **10,** 43.

Ogawa, M., Ishikawa, T., and Irimajiri, A. (1984). *Nature (London)* **307,** 66–68.

Palmer, M. R., Miller, R. J., Olson, L., and Seiger, A. (1982). *Med. Biol.* **60,** 61–88.

Patterson, P. H., and Chun, L. L. Y. (1974). *Proc. Natl. Acad. Sci. U.S.A.* **71,** 3607–3610.

Price, J., and Mudge, A. W. (1983). *Nature (London)* **304,** 241–243.

Reichardt, L. F., and Patterson, P. H. (1977). *Nature (London)* **270,** 147–151.

Shiosaka, S., Takatsuki, K., Sakanaka, M., Inagaki, S., Takagi, H., Senba, E., Kawai, Y., and Tohyama, M. (1981). *J. Comp. Neurol.* **203,** 173–188.

Smith, J., Fauquet, M., Ziller, C., and Le Douarin, N. (1980). *Nature (London)* **282,** 853–855.

Teitelman, G., Baker, H., Joh, T. H., and Reis, D. J. (1979). *Proc. Natl. Acad. Sci. U.S.A.* **76,** 509–513.

Teitelman, G., Gershon, M. D., Rothman, T. P., Joh, T. H., and Reis, D. J. (1981a). *Dev. Biol.* **86,** 348–355.

Teitelman, G., Joh, T. H., and Reis, D. J. (1981b). *Proc. Natl. Acad. Sci. U.S.A.* **78,** 5225–5229.

Teitelman, G., Joh, T. H., Grayson, L., Reis, D. J., and Iacovitti, L. (1985). *J. Neurosci.* **5,** 29–39.

Tischler, A. S., and Greene, L. A. (1975). *Nature (London)* **258,** 341–342.

Wakshull, E., Johnson, M. I., and Burton, H. (1979). *J. Neurophysiol.* **42,** 1410–1425.

CHAPTER 13

DEVELOPMENT OF NEURONAL PROPERTIES IN NEURAL CREST CELLS CULTURED *IN VITRO*

Catherine Ziller

INSTITUT D'EMBRYOLOGIE
ANNEXE DU COLLÈGE DE FRANCE
CENTRE NATIONAL DE LA RECHERCHE SCIENTIFIQUE
NOGENT-SUR-MARNE, FRANCE

I. Introduction

The early events leading in embryonic life to the great diversity of cell types encountered in the peripheral nervous system (PNS) of vertebrates have been the subject of active investigations in recent years. It has been well established that nearly all the constituents of the PNS derive from the neural crest (see Weston, 1970; Le Douarin, 1982, for reviews). A number of investigations are related to the understanding of the mechanisms which govern the transformation of the apparently identical and undifferentiated neural crest cells into the differentiated ganglion cells of the PNS.

The precursors of the elements constituting the PNS have to make several choices between different phenotypic expressions. The fact that

177

a neural crest cell becomes a neuron or a Schwann cell, a sensory or an autonomic neuron, a catecholamine (CA)- or an acetylcholine (ACh)-synthesizing cell may depend on several factors, for instance on the state of determination of premigratory neural crest cells. Recent studies (Ziller *et al.,* 1983; Weston, 1984; Sieber-Blum and Sieber, 1984) have shown that some heterogeneity exists within the apparently homogeneous cell population of the neural crest in the quail embryo. However, this does not exclude the essential role played by the environment in the differentiation of neural crest cells. It is well known from a great number of *in vivo* studies on avian embryos that the phenotype expressed by the neural crest-derived cells depends largely on the various tissues in which they have settled (Le Douarin, 1982). Thus, although the neural crest is regionalized with respect to its differentiation into neural derivatives (Le Douarin and Teillet, 1973, 1974), plasticity appears as a prominent trait of neural crest cells and of the developing PNS (Black and Patterson, 1980; Smith *et al.,* 1981; Le Douarin, 1981).

The analysis of the state of determination of neural crest cells and of the environmental factors which are involved in the differentiation of PNS precursor cells may be greatly facilitated by an *in vitro* approach. Therefore, many workers in recent years have studied the behavior of neural crest cells in culture. Most of these investigations deal with the appearance of the autonomic properties, with particular emphasis on neuronal development and neurotransmitter choice. The main results of these *in vitro* experiments will be reviewed here.

II. Emergence of Cholinergic and Adrenergic Phenotypes in Explanted Neural Crest Cells

Relatively pure crest cell populations can be isolated from 2-day-old chick or quail embryos. The method devised by Cohen and Konigsberg (1975) has been used by several authors. A piece of neural primordium (neural tube plus crest) is explanted from the trunk region of the embryo into a culture medium. As soon as it attaches to the dish, neural crest cells begin to migrate, forming a monolayer after 24 hours. The neural tube can then be removed, leaving a more or less pure population of neural crest cells behind. However, the presence of the neural tube during the first hours of culture may influence the development of the neural crest cells. Therefore, we have mostly used another technique which consists in excising microsurgically small fragments of the tip of the neural folds just before closure of the neural tube. This operation can be done in the head region of quail embryos at stage 7 (Zacchei, 1961) and in the trunk region at stages 11–12. Cephalic neural crest can also be obtained from stage 9–10 embryos: here, the neural crest cells are migrating laterally away from the mesencephalic vesicle and can be dissected out, together with the overlying ectoderm but entirely free of cells of the neural tube (Ziller *et al.*, 1979, 1983; Fauquet *et al.*, 1981). If required, the ectoderm can be removed either mechanically or enzymatically. However, since we have established that it does not influence the differentiation of neural crest cells, as far as our criteria are concerned, we usually leave it in place.

Using cytological, biochemical, immunocytochemical, and ultra-structural procedures, we studied the appearance of neuronal traits and especially autonomic phenotypes in cultures of neural crest re-moved from different axial levels of the embryo and grown under a variety of culture conditions.

The fact that autonomic properties can arise in neural crest cul-tures was shown in 1977 by Cohen, who observed differentiation of adrenergic cells using crest derived from the trunk neural primor-dium, and by Greenberg and Schrier (1977) with cephalic crest cells, which expressed cholinergic properties. The autonomic traits were evi-denced in the case of CA by formaldehyde-induced fluorescence (FIF) and by the observation of dense core vesicles at the electron micro-scopic level (Cohen, 1977); cholinergic differentiation was revealed by demonstrating the enzymatic activity of choline acetyltransferase (CAT), the key enzyme for ACh synthesis (Greenberg and Schrier, 1977). The logical conclusion from these experiments was that neural crest cells express *in vitro* the autonomic phenotype which is charac-teristic of their level of origin: trunk crest, the normal precursor of sympathetic ganglia and adrenomedullary cells (Le Douarin and Teillet, 1973), gives rise to adrenergic cells, whereas the cephalic crest, from which the parasympathetic ciliary ganglion derives (Le Lièvre, 1976; Narayanan and Narayanan, 1978), produces cholinergic cells in culture. However, in these cultures, the neuronal differentiation re-mained incomplete, in particular as far as morphological and cytologi-cal traits were concerned.

The emergence of the neuronal phenotype from cultured neural crest cells has been extensively studied in our laboratory since 1979. We have mainly used neural crest explants directly excised from the neural fold of the head and the trunk and migrating crest from the mesencephalic region. The neural crest fragments were placed without dissociation into 35-mm-diameter plastic culture dishes containing a film of culture medium (600 μl) which allows attachment of the tissues to the dish. After a few hours, 1 ml of culture medium, consisting of Dulbecco's minimal essential medium (DMEM), supplemented with 15% horse or fetal calf serum (HS or FCS) and 2% chick embryo extract

(CEE), was added. The histological, cytochemical, biochemical, and immunocytochemical assays were usually performed after a period of 7–10 days, during which the medium was not changed. In this system we compared the developmental potentialities of truncal and cephalic crest cells (Ziller *et al.*, 1979; Fauquet *et al.*, 1981). The cultures grew actively in the presence of FCS and HS. Most cells in the cultures were flat, polygonal, or stellate and no cells with neuronal morphology were obvious. The only differentiation typical for neural crest visible with phase-contrast microscopy in the living cultures was the appearance of melanocytes, mainly in the trunk crest cultures.

Using biochemical methods, we investigated the ability of these cultures to synthesize one or both of the major neurotransmitters of the autonomic nervous system, acetylcholine and catecholamines. The tests were performed at the time of explantation and after 7 days in culture. CAT activity was measured and ACh and CA synthesis determined with the radiochemical assay described by Hildebrand *et al.* (1971) and Mains and Patterson (1973), in which conversion by living cells of tritiated choline and tritiated tyrosine to ACh and CA, respectively, is quantitated after electrophoretic analysis. The results of these experiments were the following:

1. Excised noncultured migrating or premigratory mesencephalic neural crest cells were able to synthesize ACh before culture (Smith *et al.*, 1979). It is noteworthy that neural crest cells both in the neural fold at the time of closure of the neural tube and in the migratory state possess another marker of the cholinergic system, the enzyme for ACh degradation acetylcholinesterase (AChE) (Cochard and Coltey, 1982).

2. No CA synthesis was detected in noncultured neural crest cells.

3. After 7 days in culture in the presence of FCS, low quantities of CA were produced by both types of cells, cephalic as well as truncal, which continued to synthesize low amounts of ACh.

4. In the presence of HS, no CA synthesis was detected in trunk and cephalic crest cultures. ACh synthesis was higher in mesencephalic crest cultures with HS than with FCS. This observation is consistent with the results obtained by Greenberg and Schrier on mesencephalic neural crest (1977).

5. Irrespective of the age of the culture, CA synthesis remained lower than ACh synthesis.

In summary, cholinergic properties, demonstrated from the outset in neural crest cells, preceded adrenergic expression. The two phenotypes could be displayed together in cultures of both trunk and cranial neural crest, but our culture conditions did not allow the exclusive expression of either phenotype at will. Moreover, adrenergic expression was always weak. In particular, CA histofluorescence was never found in the cultured neural crest cells.

III. Influence of Environment on Autonomic Differentiation in Cultured Neural Crest Cells

A. COCULTURES OF NEURAL CREST CELLS WITH TISSUES OF MESODERMAL ORIGIN

In an attempt to provide the cultured neural crest cells with conditions that might induce them to reach a higher differentiated status, we associated them with various embryonic tissues that form the environment in which autonomic ganglia differentiate *in vivo* and which are known, or suspected, to have an inductive effect on neural crest-derived cells: somitic mesenchyme, notochord (Cohen, 1972; Norr, 1973; Teillet *et al.*, 1978), heart, or hindgut. These cocultures also synthesized both neurotransmitters, ACh and CA, producing considerably higher amounts of these neurotransmitters than neural crest cells alone. However, as in isolated neural crest cultures, no morphologically differentiated neurons developed in the cocultures of 2-day neural crest plus mesodermal tissues.

Another type of coculture was realized by explanting trunk crest cells taken from quail embryos at 3 days of incubation, together with the sclerotomal part of the somite, in which the neural crest cells are migrating but have not begun to aggregate into dorsal root ganglia and sympathetic chains. Therefore, sclerotome cultures are in fact a mixed population of a mesodermal tissue with neural crest-derived cells. Freshly removed sclerotome explants produced ACh but no CA, and accordingly no FIF was observed *in situ* at 3 days. Sclerotomes were cultured in DMEM + 15% HS or FCS + 2% CEE as explants, using the same technique as for excised neural crest.

A striking difference between sclerotome and pure neural crest cultures was observed by phase-contrast microscopy: after 1–2 days in the former cultures, cells with typical neuronal morphology, round cell bodies, and long processes were seen. Their neuronal phenotype was confirmed by silver staining, histochemical revelation of AChE activity (Fauquet *et al.*, 1981), and the immunocytochemical demonstration of the presence of neurofilament proteins in the cytoskeleton (Fig. 1) and of tetanus toxin binding sites on the cell surface (Ziller and Le Douarin, 1983).

The appearance of autonomic neuronal traits was routinely moni-

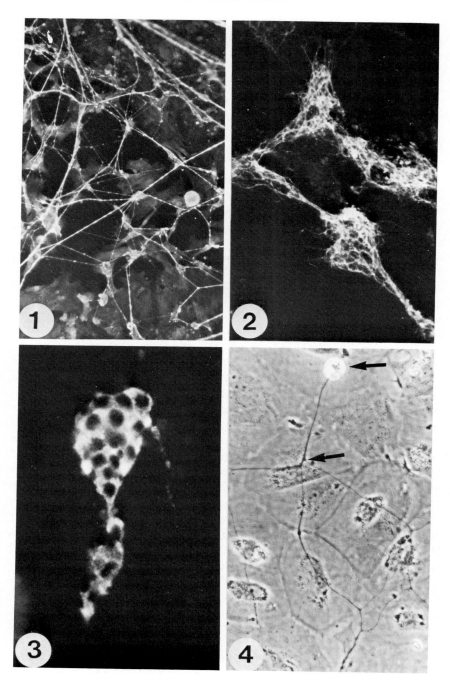

tored biochemically and histochemically from day 1 to day 9 *in vitro.* The principal result was that, in contrast to cultures of pure pre-migratory trunk crest cells, the cultures of sclerotomes became adrenergic as early as the second day of culture. ACh was produced already at the time of explantation and throughout the culture period after 7 days in the presence of FCS. In culture the CA/ACh ratio was well in favor of CA. The adrenergic differentiation in sclerotome cultures was confirmed by histochemical observations after glyoxylic acid fixation (Furness and Costa, 1975; König, 1979), which revealed CA-specific fluorescence (GIF) in a number of cells (Fig. 2). Moreover, electron microscope examination of the explants showed the presence of numerous dense core vesicles in cells and processes.

Another substance thought to have a neurotransmitter role, the neuropeptide somatostatin, was studied recently in the developing PNS in our laboratory (Garcia-Arraras *et al.,* 1984) and was shown to be expressed *in vitro* by the neural crest-derived cells in cultured sclerotome (Fig. 3). Interestingly, the cells with somatostatin-like immunoreactivity also contained CA fluorescence and tyrosine hydroxylase (T-OH), the key enzyme for CA synthesis (Garcia-Arraras *et al.,* 1984a,b).

B. Influence of Culture Medium Composition on Autonomic Expression in Neural Crest Cultures

In contrast to our own results with isolated neural crest cells, which in our *in vitro* conditions displayed low synthetic activity and never produced fully differentiated neurons, the neural crest cultures described by other authors (Cohen, 1977; Kahn *et al.,* 1980; Sieber-Blum and Cohen, 1978, 1980; Loring *et al.,* 1982; Kahn and Sieber-Blum, 1983), and which were derived from total neural primordia, reached a much higher state of differentiation. In particular, cells with neuronal morphology and CA fluorescence were regularly obtained after a few

---

FIG. 1. Neurofilament protein (150 kDa) immunoreactivity in neuronal cell bodies and processes of a 7-day neural crest/sclerotome culture. Indirect immunofluorescence staining. ×160.

FIG. 2. Fluorescence due to CA in cells of a 7-day neural crest/sclerotome culture. The culture was fixed with glyoxylic acid according to the technique of König (1979). ×130.

FIG. 3. Somatostatin-like immunoreactivity in a neural crest/sclerotome culture after 6 days *in vitro.* The immunoreactive cells are grouped in a ganglion-like structure. Indirect immunofluorescence staining. ×490.

FIG. 4. Differentiated neuron (arrows) and polygonal flat cells in a culture of mesencephalic neural crest after 8 days in the presence of serum-free medium. Phase-contrast micrograph. ×260.

days in culture. This discrepancy could be explained by differences in the composition of the culture medium. We noticed, for instance, that the levels of ACh and CA synthesis depended significantly on the nature of the serum used. However, the critical point seemed to be the amount of CEE added to the medium. When the medium contained only 2% CEE (Fauquet *et al.*, 1981), CA synthesis was low and no morphologically differentiated neurons appeared. A concentration of 10–15% CEE is necessary to induce terminal differentiation of adrenergic neurons, as stressed by Howard *et al.* (1982). In a recent series of experiments (Smith, Fauquet, Ziller, and Le Douarin, unpublished results), we obtained a number of adrenergic neurons exhibiting CA histofluorescence and T-OH immunoreactivity in mesencephalic and truncal neural crest cells after 6–10 days in culture in the presence of 15% FCS and 15% CEE. The chick embryo seems therefore to contain factor(s) with the capacity of inducing adrenergic differentiation in autonomic neuronal precursors.

C. Stimulation of Adrenergic Differentiation in Neural Crest Cultures by Glucocorticoids

Among the factors extrinsic to the neural crest which may play a crucial role in its development are certain hormones, such as glucocorticoids. The latter have been shown to control (although indirectly through their action on nonneuronal cells) neurotransmitter choice in sympathetic neurons from the superior cervical ganglion (SCG) of newborn rat cultured *in vitro* (Fukada, 1980; McLennan *et al.*, 1980). Their influence on cultured neural crest cells has been studied recently by Smith and Fauquet (1984). Glucocorticoid treatment strikingly enhanced adrenergic differentiation in cultures of sclerotomes, CA synthesis being considerably increased in the presence of the hormones. The effect was specific for the adrenergic phenotype in that ACh synthesis was unaffected by the treatment.

An interesting finding was that glucocorticoids had no effect on isolated premigratory neural crest cells, but stimulated significantly the development of CA-producing cells from premigratory neural crest grown in the presence of somitic and notochordal mesoderm. A possible inference from this is that in this system as well as in the neuronal cultures of newborn rat SCG neurons the glucocorticoids act indirectly on the neural crest cells via the mesodermal environment. Alternatively, it may be assumed that glucocorticoids do not trigger adrenergic differentiation, but that they selectively enhance catecholaminergic properties in neural crest cells that have already been exposed to a signal coming from the neighboring mesodermal tissues.

D. Influence of Substrate and Culture Conditions on Expression of the Adrenergic Phenotype

Culture conditions have been described in which quail neural crest cells, instead of spreading on the culture dish, form clusters on the sides of the explanted 2-day neural tube (Loring *et al.*, 1981; Glimelius and Weston, 1981). These clusters of pure neural crest cells can be harvested from the neural tube after 1 or 2 days and subcultured on substrata favoring their attachment, e.g., plastic, collagen, glycosaminoglycans (Loring *et al.*, 1982). It was shown that the proportion of melanocytes and of cells displaying FIF depended on the substrate. On plastic, approximately 100% of the neural crest cells became pigmented. On a somitic mesoderm-conditioned substrate or on fibronectin, melanogenesis was nearly entirely suppressed and differentiation of CA-synthesizing cells was increased (see also Sieber-Blum *et al.*, 1981). Therefore, the choice of phenotype by cultured neural crest cells can be affected by environmental stimuli arising from the substrate. The authors suggested that some extracellular matrix components present in embryonic tissue where the neural crest cells normally migrate may induce the adrenergic phenotype in cells that would otherwise be able to choose a different pathway of differentiation (Sieber-Blum *et al.*, 1981).

E. Neural Crest Cells in Serum-Free Medium: Segregation of a Neuronal Precursor Cell Line

Reduced serum concentration or removal of the serum from the culture medium elicits neuronal differentiation in neuroblastoma cell lines (Seeds *et al.*, 1970; Schubert *et al.*, 1971; Bottenstein and Sato, 1979; Solomon, 1979) and in teratocarcinoma lines (Pfeiffer *et al.*, 1981; Darmon *et al.*, 1981, 1982). These observations, as well as the frequently outlined similarities between embryonic and tumoral cells, prompted us to perform neural crest cell cultures in serum-free medium (Ziller *et al.*, 1981, 1983; Ziller and Le Douarin, 1983). The defined medium we used, described by Brazeau *et al.* (1981), contains several hormones and growth factors. Premigratory neural crest cells of the cephalic and trunk regions and migrating mesencephalic neural crest cells were cultured in this medium. As early as 15–20 hours after explantation, when migrating mesencephalic neural crest cells were used, about 5–10% of the cells present in the culture differentiated, acquiring a typical neuronal morphology with round, refringent perikarya and long, thin, branched processes which made contacts with other cells. The majority of the neural crest cells developed into flat, polygonal cells, forming a layer on which the network of neuron-like cells extended. After 1–3 days in culture in the defined medium, the neurite-bearing cells exhibited cytological and immunological traits characteristic for neurons: silver staining, AChE activity, tetanus toxin binding, neurofilament protein in the cytoskeleton. Electrical membrane excitability was observed in these neurons (Bader *et al.*, 1983). The differentiated neurons could survive for several weeks in the serum-free medium, together with the flat cells of the underlayer. Little or no growth of the cell population took place, in contrast to the noticeable cell multiplication occurring in neural crest cultured in serum-supplemented medium. Migrating mesencephalic neural crest cells adhered within 1–2 hours to uncoated plastic dishes although the defined medium is devoid of specific attachment factors. It has been shown that migrating mesencephalic crest cells produce fibronectin (Newgreen and Thiery, 1980) and thus presumably have the ability to condition the culture substrate and to favor cell adhesion. In contrast, the culture of premigratory neural crest cells in the defined medium was possible only if the dish was previously treated with a culture of somitic mesoderm, which had deposited a fibronectin-rich substrate on the plastic (Newgreen and Thiery, 1980).

An important observation was that the neurons in mesencephalic neural crest cultures in defined medium developed from precursor cells which did not undergo cell division *in vitro* prior to differentiation. After tritiated thymidine incorporation during the first 2 days of culture, less than 10% of the differentiated neurons were labeled, whereas more than 90% of the flat cells had incorporated the radioactive precursor. This finding demonstrates that the defined medium allows a certain category of determined neuronal precursors to differentiate without further maturation.

When younger neural crest cells, taken from the neural fold, were cultured in the serum-free medium, a smaller proportion of the cell population differentiated into neurons, and their differentiation occurred 24 hours later than when migrating cephalic neural crest cells were taken. This suggests that the neuronal precursors present in the neural crest at the neural fold stage undergo a maturation before reaching the "migrating" stage. Indeed, when "migrating" cephalic neural crest, excised from an embryo which had been injected with radioactive thymidine at the neural fold stage (4 hours earlier), was cultured in nonradioactive, serum-free medium, a significant proportion of the neurons that developed was labeled. This demonstrated that the neuronal precursors had divided during the transition from the premigratory to the "migrating" stage. It should be noted that although "migrating" neural crest cells are capable of virtually immediate neuronal differentiation in culture their terminal differentiation *in vivo* into trigeminal or ciliary neurons will occur only 2 days later.

What is the nature of these neurons which develop so rapidly in neural crest cultures in serum-free medium? The expression of neurotransmitter-related phenotypes in the cultures was investigated. The differentiated neurons were not adrenergic: no GIF was demonstrable and, at the level of the population, biochemical assays showed that no tritiated tyrosine was converted to CA. Nor could we demonstrate that the neurons are cholinergic. The cultures did in fact synthesize some ACh after 3–7 days *in vitro,* but there was no increase of the synthetic activity in comparison with excised cephalic crest before culture. A possibility is that these cells (or some of them) might be sensory neurons. A number of them have a unipolar morphology, with a cell body bearing a single, branched, or T-shaped neurite (Fig. 4). However, no cytological evidence characterizing specifically these neurons as sensory is available.

When the defined medium was supplemented with serum (FCS, HS, chick or quail serum), no neurons appeared, and if serum was added after 1–3 days of culture in defined medium, when the neurons were already differentiated, they disappeared within 24–48 hours. The neurites retracted and disintegrated and neurofilament immunoreactivity was lost. In contrast, the flat, nonneuronal cells of the culture, which had a very low proliferation rate after a few days in defined medium, resumed multiplication as soon as the serum was added.

If the cultures initiated in defined medium were supplemented with serum and 15% CEE, a few cells exhibited a neuronal morphology as well as CA histofluorescence (GIF) 6–10 days later. These adrenergic cells were derived from precursors which had divided during the culture period, as was shown by histoautoradiography after tritiated thymidine incorporation (Ziller, Fauquet, and Le Douarin, unpublished results). These adrenergic neurons may correspond to a second population of neuronal precursors which require factors present in the CEE to undergo terminal differentiation, whereas the neurons of the first population developing in the defined medium disappear when the serum and the CEE are added.

An explanation for this finding is that the two kinds of neurons obtained in the cultures reflect a heterogeneity among the precursors present in the crest. The fact that the adrenergic phenotype was acquired in our cultures when the precursor cells were allowed to proliferate in the presence of CEE is consistent with the recent observations of Kahn and Sieber-Blum (1983), who showed that appearance of adrenergic cells was inhibited by blocking cell proliferation with cytosine arabinoside.

In all the neural crest cultures in which neurons develop, both in defined and in serum- and CEE-supplemented medium, a high proportion of the cell population is represented by flat, polygonal, or stellate cells whose nature remains unknown. At explantation, the neural crest comprises not only neuronal precursors, but also progenitors of supporting cells, Schwann cells, melanocytes, and in the case of cephalic neural crest, mesectodermal cells. Except for the melanocytes, which in the presence of serum and CEE appear abundantly as early as the third or fourth day of culture, the other cell types cannot be characterized with the cytological or immunological techniques available to date.

IV. Plasticity of Neurotransmitter Expression in Developing Peripheral Ganglia

A classical example of plasticity in autonomic neurons is the case of the superior cervical ganglion (SCG) of the newborn rat (see Patterson, 1978, for a review). Dissociated SCG neurons cultured in the absence of nonneuronal cells develop adrenergic functions, whereas they synthesize ACh when cultured in the presence of nonneuronal cells (muscle or cardiac cells) or in medium conditioned by such cells. The rate of CA synthesis decreases, while the ACh synthetic activity increases proportionally to the number of nonneuronal cells or the amount of conditioned medium added to the cultures. Single SCG neurons grown *in vitro* provide the same results, some of them exhibiting both adrenergic and cholinergic functions at the same time (Landis, 1976; Furshpan *et al.*, 1976). Thus, external cues can induce a phenotypic change in a postmitotic neuron. The cholinergic factor present in medium conditioned by heart or muscle cells is also produced by C6 glioma cells and has been partially purified (Weber, 1981).

Transplantation experiments between quail and chick performed by our group have demonstrated a remarkable plasticity of neural crest-derived cells even when they have stopped migrating and have aggregated to form ganglia. Developing autonomic or spinal ganglion cells from quail embryos, back-transplanted into a 2-day chick embryo, migrated and colonized the normal sites of arrest of neural crest in the host. When they were located in the sympathetic ganglia or in the adrenal gland, they displayed adrenergic histofluorescence (Le Douarin *et al.*, 1978; Le Lièvre *et al.*, 1980). Thus, cells originating from the parasympathetic cholinergic ciliary ganglion have the capacity to differentiate into CA-producing cells, when they are subjected to an appropriate environment in the host. An outstanding question is whether the cells which express a novel phenotype in these conditions are postmitotic neurons of the grafted ganglion or cells of the non-neuronal population. Experiments with the nodose ganglion (Le Lièvre and Le Douarin, 1982) and ciliary ganglion (Dupin, 1984) answered this question. When chimeric nodose ganglia, in which either the non-neuronal or the neuronal cells had the quail nuclear marker were grafted into a 2-day chick, nonneuronal quail cells became adrenergic when they colonized the sympathetic ganglia of the host; in contrast, when the neurons of the grafted ganglion were of quail origin, no adrenergic cells were seen in the crest-derived structures of the chick.

Tritiated thymidine incorporation experiments carried out in our laboratory (Dupin, 1984) on chick embryos which had received grafts of quail ciliary ganglia of 4- to 15-day embryos also showed that postmitotic neurons did not survive in the host. It was thus demonstrated that in the new environment created by the graft conditions the nonneuronal cells of the ganglion and not the neurons can modify their developmental program.

It can therefore be concluded that, even relatively late in embryonic life, the developing peripheral ganglia contain cells which retain multiple differentiation potentialities (Le Douarin, 1984). An *in vitro* study which has recently been undertaken in our laboratory leads to the same conclusion (Xue *et al.*, 1985). When dissociated DRG of 7- to 9-day quail embryos are grown in culture in the presence of medium supplemented with FCS and 5–10% CEE, some cells exhibit adrenergic features after 4 days *in vitro*. GIF and T-OH immunoreactivity are seen, and CA synthesis is detected biochemically. *In situ,* no FIF has ever been found in the quail DRG at any developmental stage up to hatching (the latest time point examined). A number of the adrenergic cells in the DRG cultures are labeled after tritiated thymidine incorporation, thus demonstrating that cell divisions occur in the differentiating subpopulation. It is noteworthy that a low concentration of CEE (2%) does not allow adrenergic differentiation in DRG cultures. Similar results are obtained with nodose ganglion cell cultures.

V. Concluding Remarks

As far as neurotransmitter choice and neuronal expression are concerned, the *in vitro* experiments briefly described here have enlightened significantly the problem of neural crest diversification. At the time of its formation and before starting its migration, the neural crest already displays cholinergic traits. Whether ACh is synthesized only by certain cells or by the whole neural crest cell population is still unknown.

When they are isolated from their normal embryonic environment, premigratory neural crest cells exhibit specific requirements for further development, especially for adrenergic and neuronal phenotype expression. CA histofluorescence is obtained only when the culture medium is enriched with some compounds derived from the developing embryo. The secondarily acquired adrenergic phenotype is expressed, together with cholinergic features, in cultured neural crest cells. At the level of the cell population, the premigratory neural crest appears therefore to possess potentialities for both major phenotypes of the autonomic embryonic nervous system.

Once they have started to migrate, the neural crest cells have reached a more advanced developmental status which allows them to differentiate more rapidly and more completely when they are explanted in culture. Migrating trunk crest cells produce fully developed adrenergic neurons when they are taken from the embryos together with the sclerotome moiety of the somite (Fauquet *et al.,* 1981) or when they are allowed to undergo maturation in clusters on the neural tube for 1 or 2 days prior to seeding in culture on an appropriate substrate (Loring *et al.,* 1982).

Another neural crest cell population which is able to give rise to morphologically differentiated neurons *in vitro* is the migrating mesencephalic neural crest, provided that it is cultured in a serum-free medium (Ziller *et al.,* 1983).

Most experiments, *in vivo* and *in vitro,* stress the crucial role played by external cues, not only in the development of young neural crest cells but also in the phenotypic expression of embryonic PNS ganglion cells already engaged in given differentiation pathways. This means that at least a proportion of neural crest cells and of neural crest-derived cells in developing ganglia are probably pluripotent, and surely bipotent as far as the cholinergic and adrenergic phenotypes are concerned. The environment may act as an inducer on pluripotent cells, as is the case, for instance, in cultures of developing SCG neurons in which a CA-producing cell can become cholinergic in a conditioned medium (Landis, 1976; Furshpan *et al.,* 1976). The role of environmental factors could also be to select already committed precursors from a heterogeneous population. That the neural crest is not a homogeneous population of uncommitted cells has indeed been demonstrated by several *in vitro* approaches. Early segregation of a neuronal precursor line has been revealed by culture in a serum-free medium (Ziller *et al.,* 1983); it has also been shown that the neural crest contains melanogenic progenitors that are competent for terminal differentiation at the onset of its migration, whereas progenitors of adrenergic neurons have to go through several additional cell divisions before they reach the same stage (Kahn and Sieber-Blum, 1983; Sieber-Blum and Sieber, 1984). Cell line divergence has also been evidenced in the developing PNS by *in vivo* investigations with quail–chick chimeras. In particular, it seems that separate pools of precursor cells for the sensory and the autonomic derivatives exist in the neural crest (Le Douarin, 1984).

Investigations on the early emergence of cell diversity during embryonic life will be greatly facilitated by the development of new marking techniques. Cell-specific monoclonal antibodies, for instance, should be a useful tool to demonstrate antigenic diversity in an apparently homogeneous population of embryonic cells. As far as the development of the PNS is concerned, several monoclonal antibodies directed against derivatives of the neural crest have been described recently (Ciment and Weston, 1982; Barald, 1982; Vincent and Thiery, 1984). The production of monoclonal antibodies that recognize cell surface antigens specific for several different embryonic cell types including neural crest derivatives is now in progress in our laboratory.

ACKNOWLEDGMENTS

The author thanks Professor N. Le Douarin and Dr. J. Smith for helpful comments on this chapter. This work was supported by the CNRS, the MRI, and Basic Research Grant No. 1-866 from the March of Dimes Birth Defects Foundation.

REFERENCES

Bader, C. R., Bertrand, D., Dupin, E., and Kato, A. C. (1983). *Nature (London)* **305**, 808–810.
Barald, K. F. (1982). *In* "Neuronal Development" (N. Spitzer, ed.), pp. 110–119. Plenum, New York.
Black, I. B., and Patterson, P. H. (1980). *Curr. Top. Dev. Biol.* **15**, 27–40.
Bottenstein, J. E., and Sato, G. H. (1979). *Proc. Natl. Acad. Sci. U.S.A.* **76**, 514–517.
Brazeau, P., Ling, N., Esch, F., Bohlen, P., Benoit, R., and Guillemin, R. (1981). *Regul. Peptide* **1**, 255–264.
Ciment, G., and Weston, J. A. (1982). *Dev. Biol.* **93**, 355–367.
Cochard, P., and Coltey, P. (1982). *Dev. Biol.* **98**, 221–238.
Cohen, A. M. (1972). *J. Exp. Zool.* **179**, 167–182.
Cohen, A. M. (1977). *Proc. Natl. Acad. Sci. U.S.A.* **74**, 2899–2903.
Cohen, A. M., and Konigsberg, I. R. (1975). *Dev. Biol.* **46**, 262–280.
Darmon, M., Bottenstein, J., and Sato, G. (1981). *Dev. Biol.* **85**, 463–473.
Darmon, M., Stallcup, W. B., and Pittman, Q. J. (1982). *Exp. Cell Res.* **138**, 73–78.
Dupin, E. (1984). *Dev. Biol.* **105**, 288–299.
Fauquet, M., Smith, J., Ziller, C., and Le Douarin, N. M. (1981). *J. Neurosci.* **1**, 478–492.
Fukada, K. (1980). *Nature (London)* **287**, 553–555.
Furness, I. B., and Costa, M. (1975). *Histochemistry* **41**, 335–352.
Furshpan, E. J., McLeish, P. R., O'Lague, P. H., and Potter, D. D. (1976). *Proc. Natl. Acad. Sci. U.S.A.* **73**, 4225–4229.

Garcia-Arraras, J. E., Chanconie, M., and Fontaine-Pérus, J. (1984a). *J. Neurosci.* **4**, 1549–1558.

Garcia-Arraras, J. E., Fauquet, M., Chanconie, M., and Smith, J. (1984b). In press.

Glimelius, B., and Weston, J. A. (1981). *Cell Diff.* **10**, 57–67.

Greenberg, J. H., and Schrier, B. K. (1977). *Dev. Biol.* **61**, 86–93.

Hildebrand, J. G., Barker, D. L., Herbert, E., and Kravitz, E. A. (1971). *J. Neurobiol.* **2**, 231–246.

Howard, M. J., Bronner-Fraser, M., and Tomosky-Sykes, A. (1982). *Soc. Neurosci. Abstr.* **8**, 257.

Kahn, C. R., and Sieber-Blum, M. (1883). *Dev. Biol.* **95**, 232–238.

Kahn, C. R., Coyle, J. T., and Cohen, A. M. (1980). *Dev. Biol.* **77**, 340–348.

König, R. (1979). *Histochemistry* **61**, 301–305.

Landis, S. C. (1976). *Proc. Natl. Acad. Sci. U.S.A.* **73**, 4220–4224.

Le Douarin, N. M. (1981). *Ciba Found. Symp.* **83**, 19–50.

Le Douarin, N. M. (1982). "The Neural Crest." Cambridge Univ. Press, London and New York.

Le Douarin, N. M. (1984). *In* "Cellular and Molecular Biology of Neuronal Development" (I. B. Black, ed.), pp. 3–28. Plenum, New York.

Le Douarin, N. M., and Teillet, M.-A. (1973). *J. Embryol. Exp. Morphol.* **30**, 31–48.

Le Douarin, N. M., and Teillet, M.-A. (1974). *Dev. Biol.* **41**, 162–184.

Le Douarin, N. M., Teillet, M.-A., Ziller, C., and Smith, J. (1978). *Proc. Natl. Acad. Sci. U.S.A.* **75**, 2030–2034.

Le Lièvre, C. (1976). Thèse d'Etat, Université de Nantes.

Le Lièvre, C., and Le Douarin, N. M. (1982). *Dev. Biol.* **94**, 291–310.

Le Lièvre, C., Schweizer, G. G., Ziller, C., and Le Douarin, N. M. (1980). *Dev. Biol.* **77**, 362–378.

Loring, J., Glimelius, B., and Weston, J. A. (1981). *Dev. Biol.* **82**, 86–94.

Loring, J., Glimelius, B., and Weston, J. A. (1982). *Dev. Biol.* **90**, 165–174.

MacLennan, I. S., Hill, C. E., and Hendry, I. A. (1980). *Nature (London)* **283**, 206–207.

Mains, R. E., and Patterson, P. H. (1973). *J. Cell Biol.* **59**, 329–345.

Narayanan, C. H., and Narayanan, Y. (1978). *J. Embryol. Exp. Morphol.* **47**, 137–148.

Newgreen, D. F., and Thiery, J. P. (1980). *Cell Tissue Res.* **211**, 269–292.

Norr, S. C. (1973). *Dev. Biol.* **34**, 16–38.

Patterson, P. (1978). *Annu. Rev. Neurosci.* **1**, 1–17.

Pfeiffer, S. E., Jakob, H., Mikoshiba, K., Dubois, P., Guénet, J. L., Nicholas, J. F., Gaillard, J., Chevance, G., and Jacob, F. (1981). *J. Cell Biol.* **88**, 57–66.

Schubert, D., Humphreys, S., de Vitry, F., and Jacob, F. (1971). *Dev. Biol.* **25**, 514–547.

Seeds, N. W., Gilman, A. G., Amano, T., and Nirenberg, M. W. (1970). *Proc. Natl. Acad. Sci. U.S.A.* **66**, 160–167.

Sieber-Blum, M., and Cohen, A. M. (1978). *J. Cell Biol.* **76**, 628–638.

Sieber-Blum, M., and Cohen, A. M. (1980). *Dev. Biol.* **80**, 96–106.

Sieber-Blum, M., and Sieber, F. (1984). *Dev. Brain Res.* **14**, 241–246.

Sieber-Blum, M., Sieber, F., and Yamada, K. M. (1981). *Exp. Cell Res.* **133**, 285–295.

Smith, J., and Fauquet, M. (1984). *J. Neurosci.* **4**, 2160–2172.

Smith, J., Fauquet, M., Ziller, C., and Le Douarin, N. M. (1979). *Nature (London)* **282**, 852–855.

Smith, J., Le Douarin, N. M., Fauquet, M., and Ziller, C. (1981). *Br. Soc. Dev. Biol. Symp.* **5**, 129–145.

Solomon, F. (1979). *Cell* **16**, 165–169.

Teillet, M.-A., Cochard, P., and Le Douarin, N. M. (1978). *Zoon* **6**, 115–122.

Vincent, M., and Thiery, J. P. (1984). *Dev. Biol.* **103,** 468–481.

Weber, M. J. (1981). *J. Biol. Chem.* **256,** 3447–3453.

Weston, J. A. (1970). *Adv. Morphogen.* **8,** 41–114.

Weston, J. A. (1984). *Int. Cell Biol. Congr. Tokyo* Abstr. SS 8-4, 179.

Xue, Smith, J., and Le Douarin, N. M. (1985). In press.

Zacchei, A. M. (1961). *Arch. Ital. Anat. Embriol.* **66,** 36–62.

Ziller, C., and Le Douarin, N. M. (1983). *March Dimes Birth Defects Found.* **19,** 251–261.

Ziller, C., Smith, J., Fauquet, M., and Le Douarin, N. M. (1979). *Prog. Brain Res.* **51,** 59–74.

Ziller, C., Le Douarin, N. M., and Brazeau, P. (1981). *C.R. Acad. Sci. Paris* **292,** 1215–1219.

Ziller, C., Dupin, E., Brazeau, P., Paulin, D., and Le Douarin, N. M. (1983). *Cell* **32,** 627–638.

CHAPTER 14

PHENOTYPIC DIVERSIFICATION IN NEURAL CREST-DERIVED CELLS: THE TIME AND STABILITY OF COMMITMENT DURING EARLY DEVELOPMENT

James A. Weston

DEPARTMENT OF BIOLOGY
UNIVERSITY OF OREGON
EUGENE, OREGON

I. Introduction: Cellular Metaplasia Provides an Opportunity to Analyze Phenotypic Stability and Commitment

One of the most important issues in developmental biology is how phenotypic differences arise and become stabilized in cell lineages of multicellular embryos. To address this problem, we must be able to

195

CURRENT TOPICS IN
DEVELOPMENTAL BIOLOGY, VOL. 20

study the developmental abilities of individual cells and, at the same time, determine with certainty the extent of stable, propagable restrictions in developmental ability. In addition, for reasons that are discussed below, it is important to establish when and under what conditions such restrictions are imposed on a cell or its progeny.

Because of its possible implications concerning the nature of stable (heritable) restrictions that occur during development, and the possibility that some apparently stable determinative restrictions may be reversible, the phenomenon of metaplasia has attracted considerable attention from developmental biologists. Metaplasia can be defined as the adventitious expression of phenotypic traits characteristic of one tissue by cells of another differentiated tissue type. One explanation of metaplasia might be that some stable restrictions in a developmental lineage are reversible under certain conditions. The appearance of a new cellular phenotype suggests the possibility of qualitative changes in genomic expression, perhaps due to the activation of previously inactive regulatory genes (see Jones, 1985). However, for reasons to be considered below, it seems more likely that adventitious phenotypes become detectable in cells during metaplasia not by the activation of new or previously inactive genetic determinants, but by the enhancement of expression of portions of the genome already activated in a differentiated cell population (cf. Agata *et al.,* 1983; Clayton *et al.,* this volume).

Alternatively, since metaplasia is often characterized at the level of tissues rather than individual cells, the phenomenon might be caused by the proliferation of a minority population within the tissue. Such a population may be restricted to a particular developmental lineage, but remain undifferentiated and able to proliferate rapidly when provided with appropriate environmental cues. The present lack of evidence for the existence of such subpopulations in tissues undergoing metaplasia reduces the likelihood, but does not eliminate this alternative. It is premature to do so, moreover, because our failure to detect the presence of minor subpopulations may be primarily due to the lack of appropriate, sensitive, cell type-specific markers to look for them.

In order to determine whether cellular metaplasia ("transdifferentiation"; see Okada, 1980, and Kondoh and Okada, this volume) accounts for the observed changes in tissue phenotype, and to be able to examine the possibility that the program of use of genetic determinants in individual cells has actually been altered, one or more of the following criteria must first be satisfied: the adventitious phenotype must be expressed either (1) in postmitotic cells, (2) in cells clonally derived from a progenitor cell expressing another phenotype, or (3) in cells that simultaneously express traits characteristic of a different cell type (as revealed by suitable dual-marking procedures).

In general, the phenomenon of metaplasia has led to two important generalizations. First, metaplasia appears under conditions that involve the destabilization of tissues that accompanies cell proliferation and migration, i.e., the conditions that exist in embryonic, regenerating, and neoplastic tissues (see Okada, 1980; Lopashov, 1983). For example, metaplasia occurs when cell dedifferentiation and proliferation precede the formation of a regeneration blastema (see Brockes, 1984) or the appearance of lentoids from cultured preneuronal (Kondoh et al., 1983) or glial cells (Moscona et al., 1983) of the avian embryonic retina. Similarly, metaplasia is observed in a variety of tumors (e.g., teratocarcinomas: see Sherman, this volume; neuroblastomas: see Sueoka and Droms, this volume; neurofibromas: see Weston, 1981).

Second, the diversity of cellular phenotypes arising by metaplasia often appears to be limited to those characteristic of the "parental" tissue lineage (e.g., for glial cell progenitors, see Temple and Raff, 1985; for muscle/connective tissues, see Nathanson, this volume; for neural crest-derived cells, see below). Apparent exceptions to this last generalization, for example the case in which δ-crystallin immunoreactivity is detected in aggregates of dissociated limb bud mesenchyme (Kodama and Eguchi, 1982), might result from cross-reactivity of the antibody used in the phenotypic assay with intermediate filament components in the mesodermally derived cells (see Kodama and Eguchi, 1983).

Therefore, a systematic analysis of metaplasia, using experimental material in which the three criteria listed above can be satisfied, might ultimately provide useful insights into the nature and extent of genomic restrictions that normally lead progressively from pluripotent embryonic cell populations to intermediate cells in a developmental lineage and finally to apparently terminally differentiated cells.

II. The Neural Crest Gives Rise to Diverse Cellular Phenotypes during Embryonic Development

Successful analysis of the process of sequential developmental restrictions during normal development requires that several experimental criteria be satisfied. First, sensitive cell type-specific markers must be available by which the expression of normal or adventitious phenotypes can be observed in individual cells at early developmental stages. Second, the developmental precursors of such cells should be known and accessible to experimental manipulation. Finally, we must be able to perturb the environments and follow the developmental fates of individual identified cells and their progeny.

The neural crest, a *transient* stem cell population (cf. Till, 1982) of vertebrate embryos, is well suited to such an analysis since it contains a population of cells, recognizable in the early embryo, that gives rise to a large diversity of cellular phenotypes. The precursor population can be isolated and characterized, *in vivo* and *in vitro,* with useful phenotypic markers. The differentiation of crest-derived cells, moreover, can be perturbed by manipulating environmental cues *in vivo* or *in vitro,* so that the time and stability of commitment to specific developmental pathways can be assessed.

A. DIVERSITY OF NEURAL CREST-DERIVED PHENOTYPES

The neural crest is known to give rise to cells that disperse in the developing embryo, localize in characteristic sites, and ultimately produce a remarkable diversity of cellular phenotypes. These cell types include pigment cells of the integument and iris, neurons and glial cells of the peripheral (sensory, autonomic, and enteric) nervous systems, and neurosecretory tissues in the adrenal, heart, lungs, and pharyngeal structures (e.g., carotid body and the C cells of the thyroid). Finally, ectomesenchyme from the cranial neural folds ultimately differentiate as skeletal and connective tissue of the head and face (see Le Douarin, 1982; Noden, 1983, 1984; Weston, 1982; Weston *et al.,* 1984).

B. THE EMBRYONIC ENVIRONMENT INFLUENCES PHENOTYPIC DIVERSITY AMONG CREST-DERIVED CELLS

There is a variety of experimental evidence to support the notion that the embryonic environment affects the pathway and extent of migration of crest cells (see Weston, 1982; Weston *et al.,* 1984; Erickson, 1984, for reviews), and that crest cells respond to local environmental cues as they differentiate (for reviews, see Le Douarin, 1982; Noden, 1980). This evidence has been derived from a number of heterotopic and heterochronic grafting experiments, as well as from culture experiments in which metaplasia occurs in crest cell populations in response to altered environmental cues that they encounter. The general result from such experiments has been that local environmental cues, rather than the source of the crest-derived cells, determine their phenotypic expression (see Le Douarin, 1982; Weston, 1982; and below).

C. The Role of Environmental Cues in Phenotypic Expression of Crest-Derived Cells Is Presently Unknown

It should be emphasized that the heterotopic grafting experiments, and most of the culture experiments, have been performed with cell *populations* rather than single cells. For this reason, conclusions about the developmental abilities of individual cells, and their responses to environmental cues, must be made with extreme caution. In fact, as is the case with cellular metaplasia discussed above, the results might be explained in two distinct ways. The first alternative, induction or instruction, stipulates that environmental cues act on developmentally labile cells to change the use of genomic determinants by the responding cells. The second alternative, selection, suggests that local environmental factors promote survival and proliferation of developmentally distinct crest cell *subpopulations*. In contrast to the conventional hypothesis invoking crest cell pluripotentiality (see Weston, 1970), the latter interpretation predicts that neural crest-derived subpopulations are already present among migrating crest cells, that the developmental repertoire of these subpopulations is at least partially restricted, and that they respond differentially to different environmental cues. The implicit, but crucial, distinction between the two alternatives is the time during embryonic development that the responding cells undergo restrictions of their developmental abilities. It is essential, therefore, to ascertain when and under what conditions developmental restrictions occur in identified crest-derived cells.

III. Neural Crest Cells Undergo a Progressive Series of Developmental Restrictions

In general, the results of experiments involving cultures of crest derivatives, and a variety of heterochronic grafting operations in which crest-derived tissues are transplanted back into the migratory spaces of younger host embryos, suggest that the developmental repertoire of crest cells becomes progressively restricted during development. For example, the existence of heterogeneity in early cultured crest populations was suggested by the cloning experiments of Sieber-Blum and Cohen (1980) (see also Kahn and Sieber-Blum, 1983). Their results showed that, whereas some of the crest-derived progenitor cells were at least bipotent, others produce clones that contain only one cell type. Moreover, in heterochronic grafts of crest-derived structures (e.g., sensory and autonomic ganglia; branchial arches) into the crest migratory spaces of younger host embryos, it seems clear that although some cells are able to migrate, localize, and differentiate appropriately, other crest-derived cells in the grafted tissues are unable to do so (Erickson *et al.*, 1980; Le Lievre *et al.*, 1980; Ciment and Weston, 1983).

The presence of neural crest subpopulations at early migratory stages has now been directly demonstrated by the use of monoclonal antibodies (Barald, 1982; Ciment and Weston, 1982; Vincent and Thiery, 1984; Vulliamy *et al.*, 1981; Wood *et al.*, 1982). The mechanisms responsible for the generation of these subpopulations still remain to be elucidated, but they are likely to function at very early stages of crest cell dispersal, possibly even before differentiation antigens are detectable. Although local environmental cues encountered at the earliest stages of crest cell dispersal may influence developmental fates, the possibility remains that intrinsic mechanisms can generate subpopulations within the premigratory crest upon which environmental factors act (see Girdlestone and Weston, 1985).

A. SEGREGATION OF DEVELOPMENTALLY RESTRICTED CELL POPULATIONS PROBABLY OCCURS IN A PRECISE SEQUENCE

It has been postulated, but not yet substantiated, that neural crest cells undergo phenotypic diversification by a *sequence* of developmental restrictions in the crest lineage (Weston, 1981, 1982). Although it is not yet known what the normal order of restrictions is, or when they occur, they would be expected to produce subpopulations of cells with different degrees of developmental restriction. Within each of these developmentally intermediate subpopulations, local environmental stimuli might be expected to elicit expression of one or more of the phenotypes that remain in its repertoire. In the absence of appropriate local cues, some intermediate cell types might fail to thrive. If such cells survive, parts of their repertoire might remain latent, or be segregated by subsequent developmental events into more highly restricted intermediate populations. The presence and properties of such intermediate cell types require further experimental characterization.

Such crest-derived populations have been demonstrated to exist with partial restrictions on the range of developmental abilities remaining in their repertoire. For example, the cranial crest can produce every known neural crest derivative when grafted into suitable embryonic locations, whereas early restrictions seem to have been imposed on the trunk neural crest that prevent it from producing skeletal and connective tissue derivatives in otherwise appropriate embryonic environments (see Noden, 1980; evidence summarized by Weston, 1982). Similar restrictions seem to be imposed prior to or immediately after the onset of crest cell migration to limit the ability of crest cells to produce sensory neurons (Erickson *et al.*, 1980; Le Lièvre *et al.*, 1980; Ayer-Le Lièvre and Le Douarin, 1982; see also Girdlestone and Weston, 1985). In addition, nonneuronal cells of sensory and sympathetic ganglia can transiently give rise to melanocytes when their association with neurons is disrupted, but this ability is stably (but reversibly) segregated in older embryos (Nichols and Weston, 1977; Ciment *et al.*, 1986). Finally, results of heterochronic and heterotopic grafting of crest-derived branchial arch mesenchyme suggest that these cells undergo progressive partial restrictions during development (Ciment and Weston, 1983; see following section).

B. Cells with Partial Developmental Restrictions Are Present in Migrating Crest Cell Populations

The extent of the developmental restrictions or the range of capabilities remaining in identified crest-derived subpopulations can, in principle, be used to define various developmentally intermediate cell classes. In recent years, this has been done by identifying populations on the basis of the expression of particular phenotypic traits (e.g., by using cell type-specific monoclonal antibodies or morphological criteria). Then these populations were shown to produce a number of crest cell derivatives under different environmental conditions, but, in addition, to have some limitations in their developmental repertoire (Ayer-Le Lièvre and Le Douarin, 1982; Ciment and Weston, 1982, 1983, 1985; Le Lièvre *et al.*, 1980).

Recently, for example, we have been able to test the developmental abilities of an early neural crest cell subpopulation by isolating an apparently homogeneous population of neural crest-derived cells that accumulates in the posterior branchial arches during normal development of avian embryos. The crest-derived cells that populate these structures contain a neurofilament-associated protein (Ciment *et al.*, 1985), recognized by the monoclonal antibody E/C8 (Ciment and Weston, 1982). These E/C8-immunoreactive mesenchymal cells are contiguous with the mesodermal mesenchyme of the primordial gut at the time when crest-derived cells are first known to enter it (Allan and Newgreen, 1980). Since they arise from hindbrain crest that normally gives rise to enteric neurons (Le Lièvre and Le Douarin, 1975), it seemed likely that at least some of the E/C8-positive branchial arch cells are enteric neuron precursors.

Using this apparently homogeneous E/C8-positive cell population, we tested the hypothesis that they were a partially restricted cell type in the crest lineage leading to neurogenesis [i.e., able to give rise to neurons, but not to some other crest derivative(s), under appropriate conditions]. In culture, branchial arch cells exhibit neuronal traits, as judged by their neuron-specific immunoreactivities (Weston *et al.*, 1984). Moreover, when cocultured with aneural gut from 4-day chicken embryos, quail-derived branchial arch cells entered the mesenchyme of the gut and formed E/C8-positive enteric ganglia and fibers (Ciment and Weston, 1983). Heterospecific, heterotopic grafts of branchial arch cells indicate that they can give rise to neuronal as well as other neural crest derivatives (e.g., glandular and connective tissues) *in vivo* (Ciment and Weston, 1985). In contrast, unlike their rhombencephalic neural crest cell antecedents, 4-day branchial arch cells lack melanogenic ability under comparable environmental conditions *in vivo* and *in vitro* (Ciment and Weston, 1983). By the fifth day of development, crest-derived posterior branchial arch mesenchyme also appears to have lost the ability to give rise to neurons. These results suggest that, at least as populations, the branchial arch cells and the neural crest from which they originate progressively change their responsiveness to some presently undefined environmental cues.

IV. Environmental Modulation of Phenotype in Cultured Crest-Derived Cells Probably Does Not Cause Qualitative Changes in Gene Regulation

Results from a number of laboratories seem to support two generalizations about the role of environmental cues in neural crest development: first, local environmental cues encountered at the earliest stages of crest cell dispersal influence their developmental fates; and second, intrinsic mechanisms are likely to generate subpopulations within the premigratory crest upon which environmental factors can act. As mentioned above (Section II,C), however, it is often difficult to determine with certainty whether altered environments induce homogeneous, pluripotent neural crest cell populations to express distinct phenotypes, or if the environment differentially exerts its effects on developmentally distinct subpopulations.

A. CREST-DERIVED NEURONS AND MELANOCYTES ALTER THEIR PHENOTYPE IN RESPONSE TO ENVIRONMENTAL CUES

There are a number of reports that the transmitter phenotype of crest-derived neuronal precursors of the autonomic nervous system depends on cues encountered in the environment during early crest cell development in the embryo or in culture (see Black *et al.*, 1984; Patterson, 1978). In one particular case, there seems to be clear evidence that individual, identified, postmitotic adrenergic neurons become cholinergic under certain culture conditions (Furshpan *et al.*, 1976; Patterson, 1978). This result clearly eliminates explanations based on proliferation of neuronal precursor subpopulations in nervous tissue, and demonstrates that dual expression and release of transmitters by single cells occur. Moreover, there is good evidence that identified neurons contain a variety of different neuroactive substances (Black *et al.*, 1984; Hokfelt *et al.*, 1984), and that the proportion of these substances varies with developmental stage and activity. Whatever the nature of the change in transmitter phenotype, however, it is clear that this apparent metaplasia represents a relatively minor modulation of phenotypic expression in a restricted population of crest-derived neurons. Therefore, the usefulness of this phenomenon for understanding the mechanisms of restriction of developmental potential may be rather limited.

Similarly, there are progressive changes in pigment phenotype in cultured amphibian pigment cells from xanthophores to melanophores (Ide, this volume). The phenotype of pigment cells is clearly complex, involving not only the enzymes that produce the pigment molecules, but also the "scaffolding" within the membranous organelles (e.g., pterinosomes or melanosomes) that contain both the enzymes and pigments. However, pigment cells of one type sometimes contain pigmentary organelles characteristic of other types (Bagnara, 1972). In addition, since these vesicular organelles often contain enzymes characteristic of both cell types (see Matsumoto and Obika, 1968), it seems likely that changes in the pigments within them could result from relatively minor changes in enzymatic or cofactor activity (see Weston, 1970; Bagnara and Hadley, 1973).

B. CREST-DERIVED GLIA CELL PROGENITORS OF THE PERIPHERAL NERVOUS SYSTEM CAN TRANSIENTLY GIVE RISE TO MELANOCYTES

Some of the nonneuronal cells of nascent avian sensory ganglia are transiently able to undergo melanogenesis when interactions with neurons are disrupted or prevented. Thus, cultured ganglia from 4-day avian embryos produce populations of melanocytes, which are identified not only by the presence of melanosomes, but also by cell morphology and their ability to invade feather germs when cultured embryo spinal ganglia (DRG) are grafted into the wing buds of unpigmented host embryos. Based on these observations, we have suggested that glia and melanocytes share a common, partially restricted, progenitor cell type (Nichols and Weston, 1977; Nichols *et al.*, 1977; Weston, 1982). Recent immunocytochemical results are consistent with this suggestion. For example, the monoclonal antibody R24 (Dippold *et al.*,

1980), made against human melanoma cells, also binds to cultured
crest-derived melanocytes and nonneuronal cells from avian embryo
spinal ganglia (Girdlestone and Weston, 1985). This antibody recognizes the ganglioside GD3 (Pukel *et al.*, 1982), which, as expected, is
produced both by cultures of nonneuronal cells from avian embryonic
DRG and by crest-derived melanocytes.

The ability of avian DRG to give rise to adventitious melanocytes
in culture (see above) is normally lost after the fifth day of development. Ciment *et al.* (1986) have recently shown, however, that older
DRG regain their melanogenic abilities when treated in culture with
the tumor promoting drug 12-*O*-tetradecanoylphorbol-13-acetate
(TPA). In contrast to controls, DRG from embryos as old as 9 days can
produce pigment cells when they are cultured in the presence of TPA
(Fig. 1). Unlike the situation in crest cell cultures in which TPA-
treatment appears to delay melanogenesis and, presumably, the commitment to this cellular phenotype (Glimelius and Weston, 1981a),
TPA seems to reverse a restriction that is imposed on cells of the
ganglion after the fourth day of incubation. It remains to be determined what effect TPA has on the expression of other crest phenotypes, and particularly whether the TPA affects glial cell differentiation in a complementary way.

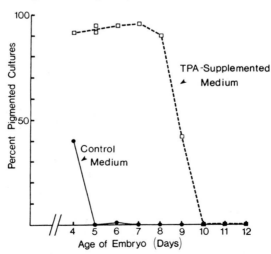

FIG. 1. Melanogenic ability of embryonic quail DRG in culture. Filled points indicate
that ganglia from embryos older than 5 days lose the ability to produce pigment cells in
control cultures. Open points reveal that ganglia cultured in the presence of 10^{-7} M
TPA regain their ability to produce adventitious melanocytes. (From Ciment *et al.*,
1985b.)

C. Culture Conditions Can Alter the Proportions of Alternative Crest-Derived Phenotypes

When clusters of crest cells (Loring *et al.*, 1981) are cultured on plastic substrata, most of the cells undergo melanogenesis. In contrast, when cluster-derived crest cells were cultured on somite fibroblast-derived microexudate substrata (Culp, 1974; Schwartz *et al.*, 1979; Loring *et al.*, 1982), the cultures exhibited a dramatic reduction in the proportion of melanocytes. At the same time, the proportion of cells exhibiting catecholamine fluorescence (FIF), characteristic of adrenergic neurons, was increased on somite microexudate-conditioned substrata. Some of these cells transiently expressed both FIF and melanosomes (Loring *et al.*, 1982). Although the significance of the dual expression of crest phenotypes is unclear, these results support the idea that individual crest-derived cells can undergo metaplasia in response to developmental cues associated with nondiffusible matrix components.

Recently, we have used the neuron-specific antibody, A2B5 (Eisenbarth *et al.*, 1979), to investigate the time course of appearance of cells with neuronal phenotype in cultures of neural crest cells. When neural crest cell clusters were isolated from 20- to 24-hour explants of neural tube, about 1% of the crest cells were already A2B5 positive. When these cluster-derived cells were subcultured for 1 or more additional days, additional A2B5-positive cells appeared (Fig. 2). The dramatic increase in the proportion of A2B5-immunoreactive cells in secondary cultures of 1-day crest clusters was also observed in secondary cultures of neural crest cell outgrowth from 2-day neural tube explants. In contrast to these results, secondary cultures of crest cell clusters that remained associated with the neural tube explants for 2 days ("2-day clusters") never exhibited the increase of A2B5-immunoreactive cells seen in cultures of 1-day clusters. In such cultures of 2-day crest cell clusters, only the initial population of A2B5-positive cells with uni- or bipolar morphology could be detected, and during the culture period, the proportion of these A2B5-immunoreactive cells declined, probably because of proliferation of A2B5-negative cells.

From [³H]TdR incorporation studies, it seems likely that the cultured crest-derived cells are postmitotic at the time that they express their A2B5 immunoreactivity (Girdlestone and Weston, 1985). The absolute increase in the number of A2B5-positive cells must, therefore, be due either to recruitment from the A2B5-negative population, or to proliferation of an A2B5-negative precursor. The data summarized in Fig. 3 show that the latter alternative is probably cor-

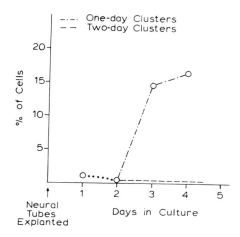

FIG. 2. Appearance of A2B5-immunoreactive cells in cultures of 1- and 2-day neural crest cell clusters. Decline in the proportion of A2B5-positive cells in cultures of cells from 2-day clusters is due to proliferation of A2B5-negative cells (see Fig. 3). (From Girdlestone and Weston, 1985.)

rect, and that virtually all cells that were A2B5 positive at the end of a 4-day culture period incorporated label on the second day. During the next 2 days of culture the labeling index for A2B5-immunoreactive cells dropped progressively, whereas most of the cell population as a whole still incorporated label on the fourth day of culture. We conclude that the second group of A2B5-positive cells arose from A2B5-negative cells that divided at least once in culture (Girdlestone and Weston, 1985). It is not yet clear, however, whether the increase in the absolute number of postmitotic immunoreactive cells can be accounted for solely by the proliferation of the progeny of a developmentally restricted lineage antecedent, or if all crest cells are able to express A2B5 (neuronal) immunoreactivity.

As mentioned above, the appearance of the second wave of A2B5-immunoreactive cells in crest cell cultures is markedly dependent on early culture conditions. New A2B5-positive cells fail to appear in cultures of cluster-derived cells that remain associated with each other on the explanted neural tube for 2 days. In this context, it is of interest that virtually all crest cells become pigmented in secondary cultures of crest clusters obtained from 2-day-old neural tube explants, whereas only about 10–20% of the cells of the crest outgrowth and cells from 1-day clusters undergo melanogenesis (Glimelius and Weston, 1981b). Thus, melanocytes and A2B5-immunoreactive cells seem to appear reciprocally in crest cell cultures. This result is also reminiscent of the

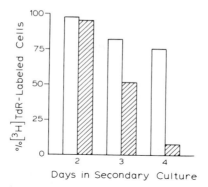

FIG. 3. Proportion of cultured crest cells incorporating [³H]thymidine. Secondary cultures of 1-day crest clusters (see Fig. 2) were labeled for 24 hours on the second, third, and fourth day of culture. All cultures were fixed at the end of the fourth day of culture, and processed for A2B5 immunoreactivity and autoradiography. Open bars represent the proportion of [³H]TdR-labeled cells in the total population. Hatched bars represent the proportion of [³H]TdR-labeled A2B5-positive cells. (From Girdlestone and Weston, 1985.)

inverse relationship between the proportions of melanocytes and FIF-positive cells that appear in cultures of cluster-derived crest cells on different substrata (see above; Loring *et al.*, 1982).

The reciprocal expression of neuronal and pigment cells can be reconciled by what has been observed in cultured sensory ganglia (see Section IV,B above; Nichols and Weston, 1977; Nichols *et al.*, 1977), in which melanogenesis occurs when interaction between neurons and glial cell precursors is disrupted. It is also consistent with the observation that increased expression of S100 protein, characteristic of glial cells, requires close association between differentiated neurons and glial progenitor cells (Holton and Weston, 1982; see Weston, 1982). In all of these situations, melanogenesis seems to be favored when environmental conditions in crest cell cultures do not permit survival or differentiation of neurons, and in consequence, the interaction between nascent neurons and the intermediate glial/melanocyte progenitors cannot occur.

V. Metaplasia by Neural Crest-Derived Cells May Be Useful for Understanding Progressive Developmental Restrictions

In the preceding recitation of evidence about metaplasia of crest-derived cells, one generalization is inescapable: we do not yet have enough data to make firm conclusions about the differentiative potential of individual cells! Unfortunately, analysis of this problem is complicated by a "biological uncertainty principle." Thus, if a cell were apparently undifferentiated, it is difficult to predict what pathway of differentiation it is determined to follow (that is, what phenotypes it can express). On the other hand, if the cell in question already expresses differentiated traits, one cannot reconstruct what its developmental antecedent could have become under other circumstances.

Cell differentiation obviously involves the coordinated expression of many phenotypic traits that are manifest at molecular, structural, and functional levels of organization. It is also clear, of course, that detection of expression of individual phenotypic traits is not sufficient to define a cell's phenotype. Nevertheless, in order to establish when and under what conditions developmental restrictions occur in the neural crest lineage, it remains essential that specific markers for a large number of crest phenotypes be systematically applied on individual crest-derived cells or on homogeneous populations of cells from identified precursors. Then, we can begin an effective analysis of the mechanisms of commitment that precede phenotypic diversification in embryonic cell populations.

ACKNOWLEDGMENTS

I am very grateful to my colleagues, Gary Ciment, Michael Marusich, Kathleen Morrison-Graham, and Kris Vogel, for reading and criticizing drafts of this chapter, and for helping me to try to understand the problem of phenotypic diversification and stability in embryonic cell populations. Work from my laboratory, which was discussed here, has been supported by grants from the NIH (DE-04316) and the NSF (PCM-8218899).

REFERENCES

Agata, K., Yasuda, K., and Okada, T. B. (1983). *Dev. Biol.* **100**, 222–226.
Allan, I. J., and Newgreen, D. F. (1980). *Am. J. Anat.* **157**, 137–154.
Ayer-Le Lièvre, C. S., and Le Douarin, N. M. (1982). *Dev. Biol.* **94**, 291–310.
Bagnara, J. (1972). *In* "Pigmentation: Its Genesis and Control" (V. Riley, ed.), pp. 171–180. Appleton, New York.
Bagnara, J., and Hadley, M. E. (1973). "Chromatophores and Color Change." Prentice-Hall, New York.
Barald, K. F. (1982). *In* "Neuronal Development" (N. C. Spitzer, ed.), pp. 101–119. Plenum, New York.
Black, I. R., *et al.* (1984). *Science* **225**, 1266–1270.
Brockes, J. P. (1984). *Science* **255**, 1280–1287.
Ciment, G., and Weston, J. A. (1982). *Dev. Biol.* **93**, 355–367.
Ciment, G., and Weston, J. A. (1983). *Nature (London)* **305**, 424–427.
Ciment, G., and Weston, J. A. (1985). *Dev. Biol.* **111**, 73–83.
Ciment, G., Glimelius, B., Nelson, D. M., and Weston, J. A. (1986). Submitted.
Ciment, G., Ressler, A., Letourneau, P., and Weston, J. A. (1985). *J. Cell Biol.,* in press.

Culp, L. A. (1974). *J. Cell Biol.* **63,** 71–83.
Dippold *et al.* (1980). *Proc. Natl. Acad. Sci. U.S.A.* **77,** 6114–6118.
Eisenbarth, G. S., Walsh, F. S., and Nirenberg, M. (1979). *Proc. Natl. Acad. Sci. U.S.A.* **76,** 4913–4917.
Erickson, C. (1986). *In* "Cellular Basis of Morphogenesis. Developmental Biology: A Comprehensive Synthesis" (L. Browder, ed.). Plenum, New York, in press.
Erickson, C., Tosney, K., and Weston, J. A. (1980). *Dev. Biol.* **77,** 142–156.
Furshpan, E. J., MacLeish, P. R., O'Lague, P. H., and Potter, D. D. (1976). *Proc. Natl. Acad. Sci. U.S.A.* **73,** 4225–4229.
Girdlestone, J., and Weston, J. A. (1985). *Dev. Biol.* **109,** 274–287.
Glimelius, B., and Weston, J. A. (1981a). *Dev. Biol.* **82,** 95–101.
Glimelius, B., and Weston, J. A. (1981b). *Cell Differ.* **10,** 57–67.
Hokfelt, T., Johansson, O., and Goldstein, M. (1984). *Science* **225,** 1326–1334.
Holton, B., and Weston, J. A. (1982). *Dev. Biol.* **89,** 64–81.
Jones, P. A. (1985). *Cell* **40,** 485–486.
Kahn, C. R., and Sieber-Blum, M. (1983). *Dev. Biol.* **95,** 232–238.
Kodama, R., and Eguchi, G. (1982). *Dev. Biol.* **91,** 221–226.
Kodama, R., and Eguchi, G. (1983). *Dev. Growth Differ.* **25,** 261–270.
Kondoh, H., Takagi, S., Nomura, K., and Okada, T. S. (1983). *Wilhelm Roux's Arch. Dev. Biol.* **192,** 256–261.
Le Douarin, N. M. (1982). "The Neural Crest." Cambridge Univ. Press, London and New York.
Le Lièvre, C. S., and Le Douarin, N. M. (1975). *J. Embryol. Exp. Morphol.* **34,** 125–154.
Le Lièvre, C. S., Schweitzer, G. G., Ziller, C. M., and Le Douarin, N. M. (1980). *Dev. Biol.* **77,** 362–378.
Lopashov, G. V. (1983). *In* "Developmental Biology. An Afro-Asian Perspective" (S. C. Goel and R. Bellairs, eds.), pp. 87–96. Indian Soc. Devel. Biol., Poona.
Loring, J., Glimelius, B., Erickson, C., and Weston, J. A. (1981). *Dev. Biol.* **82,** 86–94.
Loring, J., Glimelius, B., and Weston J. A. (1982). *Dev. Biol.* **90,** 165–174.
Matsumoto, J., and Obika, M. (1968). *J. Cell Biol.* **39,** 233–250.
Moscona, A. A., Brown, M., Degenstein, L., Fox, L., and Soh, B. M. (1983). *Proc. Natl. Acad. Sci. U.S.A.* **80,** 7239–7243.
Nichols, D. H., and Weston, J. A. (1977). *Dev. Biol.* **60,** 217–225.
Nichols, D. H., Kaplan, R., and Weston, J. A. (1977). *Dev. Biol.* **60,** 226–237.
Noden, D. M. (1980). *In* "Current Research Trends in Prenatal Cranio-Facial Development" (R. M. Pratt and R. Christiansen, eds.), pp. 3–25. Elsevier, Amsterdam.
Noden, D. M. (1983). *Dev. Biol.* **96,** 144–165.
Noden, D. M. (1984). *Anat. Rec.* **206,** 1–13.
Okada, T. S. (1980). *Curr. Top. Dev. Biol.* **16,** 349–380.
Patterson, P. H. (1978). *Annu. Rev. Neurosci.* **1,** 1–17.
Pukel, C. S., Lloyd, K. O., Travassos, L. R., Dippold, W. G., Oettgen, H. F., and Old, L. J. (1982). *J. Exp. Med.* **155,** 1133–1147.
Schwartz, C. E., Hoffman, L., Hellerqvist, C. G., and Cunningham, L. W. (1979). *Exp. Cell Res.* **118,** 427–430.
Sieber-Blum, M., and Cohen, A. M. (1980). *Dev. Biol.* **79,** 170–180.
Temple, S., and Raff, M. C. (1985). *Nature (London)* **313,** 223–225.
Till, J. E. (1982). *J. Cell. Physiol. Suppl.* **1,** 3–11.
Vincent, M., and Thiery, J.-P. (1984). *Dev. Biol.* **103,** 468–481.
Vulliamy, T., Rattray, S., and Mirsky, R. (1981). *Nature (London)* **291,** 418–420.
Weston, J. A. (1970). *Adv. Morphog.* **8,** 41–114.

210   JAMES A. WESTON

Weston, J. A. (1982). *In* "Cell Behaviour" (R. Bellairs, A. Curtis, and B. Dunn, eds.), pp. 429–470. Cambridge Univ. Press, London and New York.
Weston, J. A., Ciment, G., and Girdlestone, J. (1984). *In* "The Role of Extracellular Matrix in Development" (R. Trelstad, ed.), pp. 433–460. Liss, New York.
Wood, J. N., Hudson, L., Jessell, T. M., and Yamamoto, M. (1982). *Nature (London)* **296,** 34–38.
Ziller, C. E., Dupin, E., Brazeau, P., Paulin, D., and LeDouarin, N. M. (1983). *Cell* **32,** 627–638.

CHAPTER 15

ON NEURONAL AND GLIAL DIFFERENTIATION OF A PLURIPOTENT STEM CELL LINE, RT4-AC: A BRANCH DETERMINATION

Noboru Sueoka and Kurt Droms*

DEPARTMENT OF MOLECULAR, CELLULAR AND DEVELOPMENTAL BIOLOGY
UNIVERSITY OF COLORADO
BOULDER, COLORADO

I. Introduction

Development of eukaryotes is the result of a series of hierarchial or nonhierarchial branchings of precursor cells (or pluripotent stem cells). Here, hierarchial branching implies that a predetermined order of segregation of cell types exists; nonhierarchial branching means that the order of segregation of cell types from stem cells is not fixed, but is more or less random (stochastic). Nonhierarchial branching has been observed in the branching of certain pluripotent stem cell systems (e.g., hematopoietic stem cells; Suda *et al.*, 1983, 1984). Numerous examples of hierarchial branching exist in lower eukaryotes. In general, the derivative cells from the last branching point undergo maturation processes to become terminally differentiated cell types (Fig. 1). In this chapter, the term *maturation* is used to describe the cases in which terminal cells express overt differentiation properties in response to external agents, often without cell division. Under matura-

---

* Present address: School of Pharmacy, University of Colorado, Boulder, Colorado.

*CURRENT TOPICS IN
DEVELOPMENTAL BIOLOGY, VOL. 20*

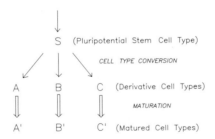

FIG. 1. Determinational branching and maturation. In this schema, branching does not necessarily imply that one stem cell gives rise to two different terminal cell types by one division; one daughter cell may remain as a stem cell similar to the parent and the other daughter cell may become a different cell type, either a terminal cell type, or a precursor of yet another branching. Branching, then, is effected in the population. In a few cases it is known that branching can take place as different paths of maturation, as seen in astrocyte–oligodendrocyte (Raff *et al.,* 1983) and adrenergic–cholinergic neurons (Potter *et al.,* 1983).

tion conditions, the majority of a homogeneous population of precursor cells responds to an agent and expresses terminal differentiation properties. Whereas gene expression during maturation processes has been studied by molecular biological means in a number of cases (reticulocytes, adipocytes, liver cells, oviduct, mammary epithelium, myocells, etc.), branching at the determinational level has not been studied in multicellular eukaryotes except in the case of immunoglobulin biosynthesis.

We have been studying neuronal and glial differentiation of a cell line, RT4-AC, which was clonally isolated from a rat peripheral neurotumor RT4 (review Sueoka *et al.,* 1982). The tumor was induced by an injection of ethylnitrosourea (ENU) into a newborn BDIX rat, as described by Druckrey *et al.* (1969). The primary culture of RT4 consisted of four morphologically distinct cell types (RT4-AC, RT4-B, RT4-D, and RT4-E) (Imada and Sueoka, 1978). Upon clonal isolation of these cell types, we have established that RT4-AC gives rise to the other three cell types with a frequency of approximately 10^{-5} (cell type conversion; Imada and Sueoka, 1978), whereas RT4-D and RT4-E are stable. RT4-B in turn gives rise to yet another cell type, RT4-F. The latter are large cells that are difficult to culture and have not been studied further.

Detailed studies of neural as well as other characteristics revealed a clear-cut segregation of neuronal and glial properties and tumorigenicity among different derivative cell types upon cell type conversion (Fig. 2). In this system, the stem cell type (RT4-AC) expresses some of both neuronal properties (veratridine-stimulated Na^+ influx and K^+ efflux; Tomozawa and Sueoka, 1978; Tomozawa et al., 1984) and glial properties [S100 protein (S100P) and glial fibrillary acidic protein (GFAP); Imada and Sueoka, 1978; Tomozawa and Sueoka, 1978]. The systematic and consistent segregation of these properties is a most remarkable and clear phenomenon that alludes to a fundamental aspect of cell differentiation (branch determination) and, therefore, is worthy of detailed studies on its mechanism.

We have also observed differential maturational responses of the derivative cell types (RT4-B, RT4-D, and RT4-E) to dibutyryl-cAMP (dbcAMP) (Droms et al., 1985). In this system, there are no detectable, systematic abberations in the karyotype, at the G banding level, that accompany each cell type conversion (Haag et al., 1984). Based on the study of this system, we will discuss several points about determination and maturation of neuronal and glial properties and suggest some points more general to the understanding of differentiation and development.

II. Dual or Precocious Expression of Neuronal and Glial Properties in the Stem Cell Type, RT4-AC

At first glance the expression of some of both neuronal and glial properties found in RT4-AC (Fig. 2) may seem peculiar, i.e., against a general observation that the precursor cells usually do not express terminal differentiation characteristics. Upon closer examination, however, there are several cases in which stem cells in normal development express more than one of the terminal differentiation properties which are separately expressed in the different derivative cell types. In the nervous system, ependymal cells in the brain express both neuronal [14-3-2 protein or neuron-specific enolase (NSE)] and glial (S100) proteins in embryonic day 17 mice, and their descendant cells in the hypothalamus segregate these properties into neurons and glial cells (DeVitry et al., 1980). Another example of dual expression of neuronal and glial properties has been reported for primary cultures of Schwann cells from sciatic nerves of newborn rabbits (Chiu et al., 1984). In this case, morphologically identifiable Schwann cells exhibited electrophysiological excitability (Chiu et al., 1984). Incidentally, the RT4 tumor was originally isolated from the sciatic nerve region of the rat. Recently, an additional example has been reported; in this case chick ciliary ganglia grown in cell culture for 1 day exhibited both neuronal [high-affinity uptake for epinephrine and specific receptors for nerve growth factor (NGF)] and a glial marker cell-surface antigen (0.4 antigen, an oligodendrocyte cell-specific antigen; Rohrer and Sommer,

1983). In tumor-derived neural cell lines, dual expression is also not unusual. More than half of the neural cell lines isolated from ENU-induced rat brain tumors expressed both neuronal and glial properties (Schubert *et al.*, 1974). In addition, a human retinoblastoma cell line, Y-79, shows precursor cells expressing both a neuronal protein (NSE) and a glial marker protein (GFAP) and these properties segregate into neuronal- and glial-type cells (Kyritsis *et al.*, 1984). In nonnervous tissue, stem cells of mouse intestinal epithelium exhibit more than one terminal property, which are segregated later (Leblond and Cheng, 1976).

In contrast, our study on the development of rat superior cervical ganglia shows that whereas catecholamine synthesis (a neuronal property) precedes morphological distinction of neurons and glial cells (embryonic day 13) (DeChamplain *et al.*, 1968), S100P is not produced until the postnatal stage, when neurons and glial cells are morphologically clearly distinguishable (Droms and Sueoka, 1985). These results and the results of DeVitry *et al.* (1980) indicate that the actual synthesis of a specialized gene product does not necessarily indicate the "opened" or "closed" state of the corresponding gene at the determination level [or at the epigenotypic level, according to the terminology of Coon and Cahn (1966)].

III. Cell Type Conversion and Coordinate Gene Expression

We have repeatedly observed three types of cell type conversion (RT4-AC to RT4-B, to RT4-D, and to RT4-E; Fig. 2). In order to examine the pluripotential nature of the stem cell type, we have established 12 clonal RT4-AC cell lines by isolating and growing single cells from a population of RT4-AC that itself started from a single cell isolate. Ten out of 12 cell lines showed one or more cell type conversions, as shown in Table I (Imada and Sueoka, 1978).

These results on the RT4 system establish the following points. (1) An individual RT4-AC cell has the potential to generate all three derivative cell types, and there is no fixed order among the three types of cell type conversion (nonhierarchial branching). (2) The cell division leading to the cell type conversion event does not generate two different derivative cell types; instead either one daughter cell remains as RT4-AC and the other converts to a derivative cell type or both daughter cells become the same derivative cell type. This point is clear from the segregation patterns shown in Table I. Moreover, when more than one cell type is generated within a plate, the derivative cell patches are not close together. (3) The rate of cell type conversion is on the order of 10^{-5}, which, so far, we have not been able to change by

TABLE I

CELL TYPE CONVERSIONS OF CLONALLY
ISOLATED RT4-AC SUBLINES[a]

Cell lines[c]	Differentiated morphological cell types[b]		
	RT4-B	RT4-D	RT4-E
RT4-51-AC14-1			+
RT4-51-AC14-2	+		
RT4-51-AC14-3	+	+	
RT4-51-AC20-1		+	
RT4-51-AC23-1			+
RT4-51-AC23-3			+
RT4-51-AC23-4			+
RT4-51-AC24-2	(+)[d]	+	+
RT4-51-AC24-3			+
RT4-51-AC24-4			+

[a] From Imada and Sueoka (1978). The variant cell types (derivative cell types) come as patches which are well delineated from surrounding RT4-AC cells; when more than two cell types are observed in a plate, the two patches are not side by side, indicating their independent origins. These results show that a clonal RT4-AC cell line can potentially generate all three derivative cell types and that each cell type conversion event does not give rise to two different derivative cell types.

[b] Single RT4-AC cells were propagated for 40 days until plates were confluent. In 10 of 12 RT4-AC clonal cultures, areas containing cells with RT4-B, RT4-D, or RT4-E morphologies were observed. The presence of such converted cells on day 40 is indicated by a +.

[c] Single RT4-AC cells were isolated from RT4-AC51-2, a subclone of RT4-51, by two sequential single-cell isolation procedures with glass cloning cylinders.

[d] RT4-51-AC24-2 gave rise to RT4-D and RT4-E by day 40 and to RT4-B by day 55.

altering culture conditions. The low frequency of conversion may result either from inappropriate culture conditions to stimulate conversion or from the possibility that the conversions we observe in culture are due to mutations in the regulatory genes. (4) The number of coordinately affected genes by a cell type conversion is large and consistent.

In addition to these properties presented in Table I, a number of cell surface proteins are different among the three derivative cell types (Imada and Sueoka, 1980). These results indicate that the number of genes affected by a regulator gene involved in cell type conversion is large.

The segregation of various properties in the RT4 cell types is tightly coupled, forming a consistent pattern (Fig. 2). This highly consistent pattern of marker segregation accompanying cell type conversions has been shown in the cases of S100P synthesis, tumorigenicity, veratridine-stimulated Na^+ influx, and cell morphology (Imada and Sueoka, 1978; Imada et al., 1978; Tomozawa and Sueoka, 1978). These results suggest that a large number of genes are coordinately regulated by a few regulator genes. For the sake of simplicity, we consider a hypothesis that three major regulator genes (R_B, R_D, and R_E) are involved and these regulator genes produce specific regulator molecules (repressors or activators). According to this hypothesis, cell type conversion is the result of the production of regulator molecules (repressor) or a loss of them (activator).

The functional difference between R_B and R_E in the above hypothesis can not currently be defined; both of them turn off the synthesis of

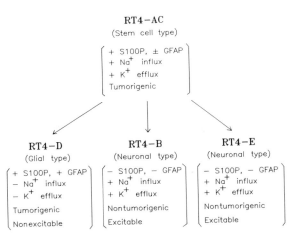

FIG. 2. Conversion coupling in the RT4 family. When cell type conversion occurs, expression or repression of several properties is coupled in the derivative cell type, e.g., expression of S100P and GFAP is coupled; these properties are either expressed (in RT4-D) or repressed (in RT4-B and RT4-E) together. The coupling feature is observed consistently with respect to morphology, tumorigenicity, veratridine-stimulated Na^+ influx and K^+ efflux, and electrical excitability of the membrane (Imada and Sueoka, 1978; Imada et al., 1978; Tomozawa and Sueoka, 1978; Tomozawa et al., 1984).

S100P and GFAP without turning off the voltage-dependent Na^+ and K^+ channels, and yet they should be responsible for making RT4-B and RT4-E different in morphology, in the pattern of cell-surface proteins, and in the response to 5-azacytidine. Treatment with 5-azacytidine generates myoblasts in the population of RT4-B cells, but not in RT4-E (Tomozawa, 1985).

IV. Cell Type Conversion of RT4-AC and Enhancement of Gene Expression

All of the properties we have examined (veratridine-stimulated Na^+ influx and K^+ efflux, presence of S100P and GFAP, and tumorigenicity) under the standard conditions so far are markedly enhanced in the derivative cell lines which express the property in question, compared to the expression of these properties in the stem cell line. This general phenomenon is termed *conversion enhancement*. An extreme case of conversion enhancement is seen in the cyclic-AMP effect on derivative cell types that is not detectable in the stem cell type, as described in the following section. Some enhancement may be seen as qualitative as well as quantitative differences. Thus, in the case of veratridine-stimulated Na^+ influx and K^+ efflux, the additional stimulation to veratridine by scorpion toxin is more pronounced in the stem cell type than in derivative neuronal types, RT4-B and RT4-E (Tomozawa *et al.*, 1984). The conversion enhancement is also seen in the tumorigenicity in RT4-D (Imada *et al.*, 1978), where in addition to faster growth of tumors of RT4-D than RT4-AC, RT4-D produces more plasminogen activator than RT4-AC. These cases of enhancement indicate that there are some additional (or secondary) regulatory mechanisms which are necessary for the fine tuning of differentiation.

V. Maturational Expression of Genes That Are Not Closed by Cell Type Conversion (or Branch Determination)

All the properties listed in Fig. 2 are observed under the standard culture conditions [Dulbecco's modified Eagle's medium (DMEM) with 12.5% horse serum and 2.5% fetal calf serum with 5% CO_2]. When these cells are cultured with 1 mM dibutyryl cAMP (db cAMP) or 1 mM db cAMP plus testololactone (10 μg/ml), the neuronal cell types, RT4-B and RT4-E, extend long axon-like processes within 3–4 days, whereas RT4-AC and RT4-D do not (Droms *et al.*, 1985). This seemingly neuronal response to db cAMP is thus confined to the neuronal cell types. Moreover, the fact that RT4-AC, which expresses some neuronal properties (veratridine-stimulated Na^+ influx and K^+ efflux), does not respond indicates that there is a class of genes whose expression can be induced by maturation factors *after* cell type conversion but not before. This is an extreme case of conversion enhancement.

Although db cAMP and testololactone do not induce the extension of processes in RT4-D, RT4-D cells respond in a different way; with the same concentrations of db cAMP and testololactone as used for RT4-B and RT4-E, the cellular proliferation of RT4-D is inhibited and the cell size increases severalfold (Fig. 3). We have found at least one bio-

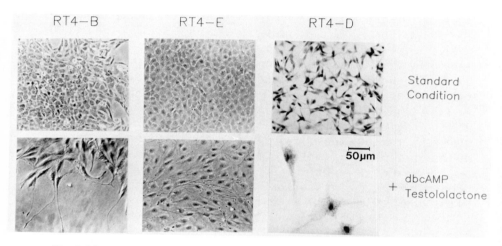

Fig. 3. Maturation response of RT4 cell lines to db cAMP and testololactone. The RT4 cell lines were cultured in 60-mm plastic tissue culture dishes or 60-mm petri dishes with microscope cover glasses (for autoradiographic experiments) in the standard medium (DMEM plus 12.5% horse serum and 2.5% fetal calf serum) or in standard medium with 1 mM db cAMP and 10 μg/ml testololactone (Squibb). The responses of all three cell lines to the inducers are different. RT4-B and RT4-E cells respond by extending long cellular processes at between 2 and 3 days induction. The growth of RT4-B cells is inhibited by the treatment to a greater extent than that of RT4-E cells, and RT4-E cells need to be confluent to maximally express the extension of processes. The RT4-D line does not express the extension of processes, but does exhibit a characteristic morphological change, increasing cell size. A characteristic biochemical differentiation has also been observed in RT4-D cells in response to the inducing conditions, but is not seen in the other RT4 cell lines (data not shown). This change is the induction of a high-affinity uptake system for GABA. After 4 days in the inducing conditions, [³H]GABA (34.9 Ci/mmol, final concentration 0.03 μM) was added to the incubation medium (Hepes buffer, Krebs–Ringer solution, KRH; Greene and Rein, 1977) and the cells were incubated for 4 hours at 37°C. Cells were rinsed three times with KRH, fixed with 3% paraformaldehyde, and coated with NTB-2 emulsion (Eastman Kodak). After 4 days exposure at 4°C the cover glasses were developed for autoradiography and stained with hematoxylin and eosin. This uptake has also been quantitated by direct measurement of accumulated radioactivity (Droms et al., 1985).

chemical property that is induced by the treatment: high-affinity up-take of γ-aminobutyric acid (GABA) is induced in RT4-D cells by the treatment. In contrast, the other cell types including RT4-AC do not drastically change the rate of uptake of low concentrations of GABA (Droms *et al.*, 1985).

Identical agents (maturation agents), therefore, can induce entirely different responses in different RT4 cell types that are derived by cell type conversion from a common stem cell type, RT4-AC. These results show that cell type conversion causes such fundamental changes in the stem cell type that expression of a number of genes, either precociously expressed or not, is coordinately affected. In this regard, at least two classes of genes can be recognized which are controlled by cell type conversion; one class is those which are expressed in the stem cell type and turned off by cell type conversion, and the other is those which can not be expressed in the stem cell type and yet have the potential to be expressed in the derivative cells under the right maturation conditions. Even in the latter cases, the difference in response to an external agent among derivative cell types is determined by the process of cell type conversion, as seen in the effect of db cAMP and testololactone. We do not know what other maturation factors are involved in the expression of the so-far unexpressed neuronal and glial properties in RT4, but it is safe to say that there should exist other classes of genes for which maturation conditions have not yet been found.

VI. The RT4 System Supports the Notion That the Tumor Has Originated from a Single Neural Stem Cell

The neuronal–glial dual expression of the RT4 stem cell type (RT4-AC) suggests that the RT4 tumor originated from a single neural stem cell. This point is supported by the fact that RT4-AC gives rise to neuronal and glial cell types and also by the fact that a single chromosomal abberation (an extra G band in the long arm of one of the two chromosomes, No. 4) appears stably in all the RT4 cell lines. Druckrey *et al.* (1969) showed that a single injection of ENU into newborn rats produces predominantly peripheral nerve tumors after about 200 days, whereas a single injection of ENU into 17-day-gestation pregnant rats causes predominantly central nervous system tumors in the offspring 200 days later. As the original investigators (Druckrey *et al.*, 1969) suggest, these facts support the notion that around these critical periods of injection, a group of cells are particularly susceptible to the primary effect of the carcinogen on the genome.

It is interesting to note that the dual expression seen in RT4-AC is also true in a number of brain tumors induced by ENU (Schubert *et al.*, 1974), indicating that the stem cell origin of neurotumors is a general phenomenon rather than peculiar to the peripheral nervous system. The fact that RT4-AC segregates into neuronal and glial cell types further suggests the possibility that, in the majority of cases, neuronal or glial tumors and their cell lines may have derived from single neoplastic cells of the stem cell type which subsequently segregate lineages before cell lines are established. The stem cell origin of tumorigenesis advocated by Cairns (1975) is supported in a number of cases, at least, by the nervous system, and in addition, the nervous system can produce tumors originating from pluripotent stem cells, at least in response to ENU.

VII. Concluding Remarks

We have preliminary results that methylation is not a likely candidate for cell type conversion. We have examined the effect of 5-azacytidine on RT4-B and RT4-E for its possible effect on the reexpression of S100. In our efforts so far, we have failed to observe a single cell of RT4-B and RT4-E that produces S100P detectable by immunostaining after 24-hour exposure in 20–100 μg/ml of 5-azacytidine and subsequent growth for several days in the standard medium without the drug (Y. Tomozawa and N. Sueoka, unpublished observations).

The mechanism of cell type conversion and the coupling features of various properties are the major concern of our current research. The presence of coordinate regulator genes seems to be essential. The structural genes of cell type-specific marker proteins should also have common signal sequences that are recognized by the regulator molecules. Since gene clones of these proteins are becoming available, the mechanism of cell type conversion is amenable to study at the molecular level.

ACKNOWLEDGMENT

I acknowledge the contributions of those who participated in this work and am grateful to Mary Haag for improvement of the chapter. This work was supported by NIH (NS15304) and by American Cancer Society (CD-1).

REFERENCES

Cairns, J. (1975). *Nature (London)* **225,** 197–200.
Chiu, S. Y., Schrager, P., and Ritchie, J. M. (1984). *Nature (London)* **311,** 156–157.
Coon, H. G., and Cahn, R. D. (1966). *Science* **153,** 1116–1119.
DeChamplain, J., Malforms, T., Olsen, L., and Sachs, C. (1968). *Acta Physiol. Scand.* **80,** 276–288.
DeVitry, F., Picart, R., Jacque, C., Legault, L., Deponey, P., and Tixier-Vidal, A. (1980). *Proc. Natl. Acad. Sci. U.S.A.* **77,** 4165–4169.

Droms, K., and Sueoka, N. (1985). In preparation.

Droms, K., Storfer, F., and Sueoka, N. (1985). In preparation.

Druckrey, H., Preussmann, R., and Ivankovic, S. (1969). *Ann. N.Y. Acad. Sci.* **163,** 676.

Greene, L. A., and Rein, G. (1977). *Brain Res.* **129,** 247–263.

Haag, M. M., Soukup, S. W., and Sueoka, N. (1984). *Dev. Biol.* **104,** 240–246.

Imada, M., and Sueoka, N. (1978). *Dev. Biol.* **66,** 97–108.

Imada, M., and Sueoka, N. (1980). *Dev. Biol.* **79,** 199–207.

Imada, M., Sueoka, N., and Rifkin, D. (1978). *Dev. Biol.* **66,** 109–116.

Kyritsis, A. P., Tsokos, M., Triche, T. J., and Chader, G. J. (1984). *Nature (London)* **307,** 471–473.

Leblond, C. P., and Cheng, H. (1976). *In* "Stem Cells of Renewing Cell Population" (A. B. Cairnie, P. K. Lala, and D. G. Osmond, eds.), pp. 7–31. Academic Press, New York.

Potter, D. D., Furshpan, E. J., and Landis, S. C. (1983). *Fed. Proc. Fed. Am. Soc. Exp. Biol.* **42,** 1626–1632.

Raff, M. C., Miller, R. H., and Noble, M. (1983). *Nature (London)* **303,** 390–396.

Rohrer, H., and Sommer, I. (1983). *J. Neurosci.* **3,** 1683–1693.

Schubert, D., Heinemann, S., Carlisle, W., Tarikas, H., Kimes, B., Patrick, J., Steinbach, J. H., Culp, W., and Brandt, B. L. (1974). *Nature (London)* **249,** 224–227.

Suda, T., Suda, J., and Ogawa, M. (1983). *Proc. Natl. Acad. Sci. U.S.A.* **80,** 6689–6693.

Suda, T., Suda, J., and Ogawa, M. (1984). *Proc. Natl. Acad. Sci. U.S.A.* **81,** 2520–2524.

Sueoka, N., Imada, M., Tomozawa, Y., Droms, K., Chow, T. P., and Leighton, T. (1982). *In* "Stability and Switching in Cellular Differentiation" (R. M. Clayton and D. E. S. Truman, eds.), pp. 165–176. Plenum, New York.

Tomozawa, Y. (1985). In preparation.

Tomozawa, Y., and Sueoka, N. (1978). *Proc. Natl. Acad. Sci. U.S.A.* **75,** 6305–6309.

Tomozawa, Y., Sueoka, N., and Miyake, M. (1984). *Dev. Biol.* **108,** 503–512.

CHAPTER 16

TRANSITORY DIFFERENTIATION OF MATRIX CELLS AND ITS FUNCTIONAL ROLE IN THE MORPHOGENESIS OF THE DEVELOPING VERTEBRATE CNS

Setsuya Fujita

DEPARTMENT OF PATHOLOGY
KYOTO PREFECTURAL UNIVERSITY OF MEDICINE
KAWARAMACHI, KYOTO, JAPAN

I. Introduction

The period of cytogenesis of the CNS in birds and mammals can be divided into three stages. During stage I of cytogenesis, the wall of the neural tube is composed of a pseudostratified columnar epithelium that proliferates steadily to increase in area and thickness. Cells constituting the epithelium are homogeneous in morphology and behavior, and are called matrix cells (Fujita, 1963a).

In stage II of cytogenesis, i.e., the stage of neuron production, some of the matrix cells become determined at the early G_1 phase and differentiate into neuroblasts (Fujita, 1964). Immediately after the determination, they lose the potency for DNA replication (Fujita, 1964, 1975a) and detach from the ventricular surface to migrate into the outermost part of the neural tube to form the mantle layer. It is now widely accepted that, at least in the CNS, specificity of the individual neurons is already determined when the neuroblasts are differentiated from the matrix cells as a function of locus and time of their production

223

(Jacobson, 1970). The same individual matrix cell gives birth to a series of progressively different neurons as stage II proceeds. What is the mechanism involved in this progressively changing differentiation of neuroblasts? Delineation of this mechanism will contribute to the proper understanding of the phenomenon of transdifferentiation (Okada, 1980).

Analysis of neuron production has made it clear that there is always a fixed end to neuron production by matrix cells in every part of the CNS, i.e., the end of stage II. Thereafter, only neuroglial cells are produced. The matrix cells change, via transitory intermediate forms known as ependymoglioblasts, into ependymal cells and glioblasts (Fujita, 1963a, 1965a; Vaughn, 1969). This is stage III, or the stage of neuroglia production.

To understand this sequential nature of the differentiative behavior of matrix cells, we have proposed a hypothesis of major differentiation of matrix cells (Fujita, 1975b) which postulates that during neurogenesis the matrix cells undergo progressive gene inactivation (Fujita, 1965b). According to this hypothesis, matrix cells at different loci and times have different potencies of differentiation, and the potency becomes progressively more narrow as the progeny proceeds through the mitotic cycle. Within the frame of the major differentiation (Fujita, 1965a), matrix cells located at various parts of the CNS can express reversible phenotype as a kind of transitory differentiation. Fujita (1965b) called this type of reversible expression of phenotype "minor differentiation." However, although it is called "minor," this transitory differentiation of matrix cells is extremely important in the proper development of the CNS. This importance will be discussed in the following sections.

II. Matrix Cells and Their Proliferation Kinetics

The wall of the neural tube is composed solely of matrix cells at the beginning of neurogenesis. When they divide, they assume a rounded form just beneath the ventricular surface. Tritiated thymidine autoradiographs taken shortly after administration of the label reveal that incorporation of the thymidine is restricted to the nuclei of matrix cells located in the outermost half of the neural tube wall (Sidman et al., 1959; Fujita, 1962). This zone is called the S zone, or the zone of DNA synthesis. In embryos killed several hours after injection of the label, many labeled nuclei are found moving upward and reaching the ventricular surface, where they divide. Thus, tritiated thymidine autoradiography reveals the dynamic aspects of the elevator movement of matrix cells (Fig. 1) (Fujita, 1963a) that takes place in accordance with

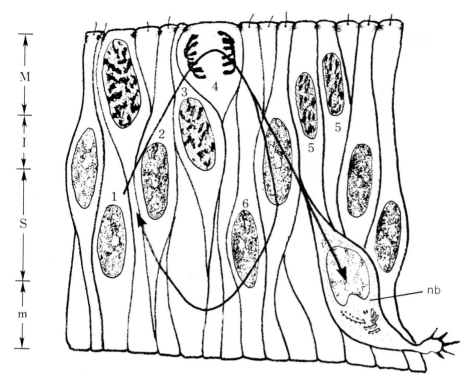

Fig. 1. Diagrammatic representation showing elevator movement of matrix cells and differentiation of a neuroblast. Numbers indicate order of progression in the cell cycle. M, M zone; I, intermediate zone or I zone; S, DNA synthesis zone or S zone; m, mantle layer; nb, neuroblast. M, I, and S zones comprise the matrix layer. (Adapted from Fujita, 1963a.)

the mitotic cycle. Their nuclei synthesize DNA in the depth of the matrix layer, ascend toward the ventricular surface during the G_2 period, and reach the surface to perform mitosis during the M phase. After division, both nuclei of the daughter cells descend toward the S zone during the G_2 period and, when they enter the S zone again, resume DNA synthesis to repeat mitotic cycles.

In analyzing the elevator movement of the matrix cells, cell cycle parameters including the length of the S phase were first measured in the neural tube of the chick embryo (Fujita, 1962). Since then, many reports have appeared on the kinetic behavior of matrix cells. Table I depicts some of those involving chicken, mouse, and human matrix cells. In chicken embryos as well as mouse and human fetuses, the cell

TABLE I

CHANGES IN KINETIC PARAMETERS OF MATRIX CELLS DURING STAGE II OF CYTOGENESIS

Species	Developmental age[a] (days)	Matrix cell	Cell cycle time (hours)	DNA synthesis time (hours)	Reference
Chick	I 2	Neural tube	5	—	Fujita (1962)
	I 2.3	Mesencephalon	6.5	—	Fujita et al. (1982)
	I 4.5	Optic tectum	10.0	4.1	Fujita (1962)
	I 6	Optic tectum	16	5.6	Fujita (1962)
Mouse	E 7	Ectoderm	5	—	Snow (1976)
	E 9	Neural tube	6.5	4.0	Hara (1975)
	E 10	Spinal cord	8.5	4	Kauffman (1966)
	E 10	Neural tube	8.0	4.5	Hara (1975)
	E 10	Neural tube	8.5	4.6	Langman and Welch (1967)
	E 11	Spinal cord	10.5	5.4	Kauffman (1968)
	E 11	Neural tube	11	5.5	Atlas and Bond (1965)
	E 13	Cerebral hemisphere	15.5	6.9	Hoshino et al. (1973)
	E 15	Cerebral hemisphere	11	7.5	Langman and Welch (1967)
	E 17	Cerebral hemisphere	26	10.4	Hoshino et al. (1973)
	P 2	Retina	28	12.5	Denham (1967)
	P 10	External granular layer	21	8.1	Fujita (1967)
Human	E 8 weeks	Cerebral hemisphere	3.5 day	11	Fujita (1963b)
	E 12 weeks	Cerebral hemisphere	11.8 day	42	Fujita (1963b)

[a] I, Incubation; E, embryonic; P, postpartum.

cycle parameters steadily increase in length as the development proceeds. The data shown in Table I were collected from reports of experiments carried out independently in different laboratories. It is remarkable that all the data fit in so well with the notion of a steady increase in the length of the cell cycle and the time of DNA synthesis in parallel with the development of the embryo, irrespective of the species of animal.

III. Major vs Minor Differentiation of Matrix Cells

It is now generally believed that the progression of cell differentiation during ontogeny is achieved by the progressive restriction of differentiative potencies of the cell. It is possible that this limitation of the potencies is realized by progressive and irreversible inactivations accumulating in functional subunits of chromosomal DNA (Fujita, 1965b, 1975b; Monesi, 1969). Chromosomes of multicellular organisms are composed of a number of replicons. This hypothesis of progressive gene inactivation on cell differentiation suggests that, as a cell is differentiated, some of the replicons of the chromosomes are irreversibly inactivated to limit future potency of the cell. It was pointed out (Fujita, 1965a) that the DNA portions that are irreversibly inactivated become incapable of transcription, are shortened and condensed in the interphase, and replicate late in the S phase. It is important to note that these acquired features of the inactivated DNA subunits are unchanged even after the cells pass through their subsequent mitoses.

According to this hypothesis, the length of the S phase, or the DNA synthesis phase, is expected to become longer in differentiated cells in comparison with that of their undifferentiated precursors. This relationship is illustrated schematically in Fig. 2.

At the beginning of ontogenesis, there is an undifferentiated cell in which none of the replicons are irreversibly inactivated. DNA replicons in such a cell are thought to replicate their DNA rather syn-

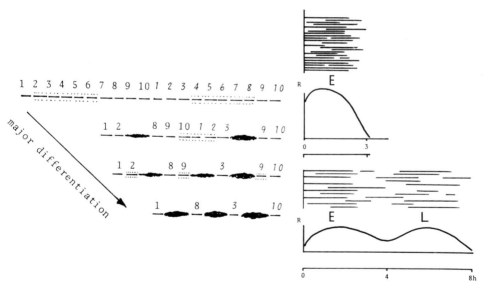

FIG. 2. Schematic representation of changes in the rate of DNA synthesis in relation to cytodifferentiation. On the left, hypothetical changes in the chromosome during development are illustrated. In a totally undifferentiated cell (top), the replicons composing the chromosomes are not irreversibly inactivated. All replicons, as shown in the graphs on the top right, are early replicating (E), and the curve of the rate of DNA synthesis (R) is expected to form a simple pulse shape. The absolute length of the S phase should be short. As the major differentiation proceeds, many replicons are irreversibly inactivated, as shown condensed in this diagram. They become late replicating (L) and add another peak to the curve of DNA synthesis. The length of the S phase also becomes longer. (Diagram according to Fujita, 1965b.)

chronously in the early S phase, and the overall rate of DNA synthesis would show a simple pulse-shaped curve, as shown in the upper right-hand graph of Fig. 2. The length of time for DNA synthesis is expected to be short, whereas in a more differentiated cell the irreversibly inactivated replicons increase in number and the S phase becomes longer. The curve of the overall rate of DNA synthesis of the cell is thought to show multiple peaks, as schematically shown in the lower right-hand diagrams of Fig. 2.

If we adopt this scheme of the genetic basis of cell differentiation (Fujita, 1965b), it is possible to use the length of the S phase of a cell as a measure of progression of the major differentiation in developing cell systems.

IV. Progression of the Major Differentiation in Matrix Cells

The observations shown in Table I indicate that there is an unmistakable tendency toward steady elongation of the length of the S phase in the matrix cell population during embryonic development. Can we regard this elongation of the S phase as a reflection of the state of the major differentiation of matrix cells, or is it merely a superficial coincidence?

To answer these questions, we analyzed curves of the overall rate of DNA synthesis of mouse matrix cells in embryonic metencephalon and in the external granular layer of postnatal cerebellum (Fujita, 1975b). The external granular cells are thought to be direct progeny of the metencephalic matrix cells. The results of our analysis are reproduced in Fig. 3. The matrix cells in metencephalon at embryonic day 11 have an S phase of approximately 5.5 hours, and the curve of the rate of DNA synthesis shows a simple bimodal pulse shape. In contrast, the S phase of the external matrix cells in the external granular layer of the mouse at postnatal day 10 is 8.1 hours, and the curve of the rate of DNA synthesis shows two additional peaks, suggesting the occurrence of more late-replicating loci in the chromosomes of the external matrix cells when they are derived from the metencephalic matrix cells.

We infer that the length of the S phase can be regarded as a measure of the major differentiation of matrix cells and that there is a steady progression of the major differentiation of matrix cells along the time axis of embryonic development. This measure is of value in estimating the degree of cell differentiation of rapidly proliferating embryonic stem cells that rarely express their differentiation as the phenotype, such as matrix cells of the CNS.

When genes essential for neuronal differentiation, such as one or several structural genes for proteins of excitable membranes, trans-

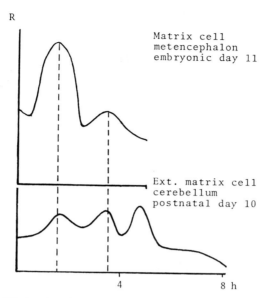

FIG. 3. Curves (R) showing the rate of DNA synthesis during the S phase of matrix cells in developing mouse brains. (From Fujita, 1975.) Upper graph is the rate of DNA synthesis in metencephalic matrix cells at embryonic day 11, and the lower is that of external matrix cells of postnatal cerebellum. Since the external matrix cells in the external granular layer of cerebellum are direct descendants of metencephalic matrix cells, the curve of DNA synthesis in the former (lower picture) can be regarded as a development from that of the latter (upper curve). The shapes of these two curves seem to conform to the hypothesis that matrix cells accumulate progressive gene inactivation in their genome as they proceed through mitotic cycles, thereby realizing diversified major differentiation.

mitters, or neurofilaments, are inactivated in a cell, it can no longer differentiate neuroblasts, but necessarily produces only nonneuronal elements, neuroglia, and ependyma. The sequential production of neurons and neuroglia, i.e., the consecutive occurrence of stage I, II, and III, can thus be understood.

V. Intermediate Filaments as Markers of Cell Differentiation in the Developing CNS

Three kinds of intermediate filaments are recognized in cells of the CNS as characteristic structural elements of neurons and neuroglia. Neither neurofilament (cf. Calvert and Anderton, 1982) nor gliafibrillary acidic protein (GFAP) (cf. Hatfield *et al.*, 1984) is specific to the cells of the CNS. However, it is widely accepted that, within the CNS, GFAP (Eng *et al.*, 1971) is present only in neuroglial cells of the astrocytic variety. Several investigators (cf. Antanitus *et al.*, 1976; Choi and Lapham, 1978; Levitt *et al.*, 1983) have claimed that a quantity of GFAP is present in matrix cells of early human and monkey brains.

If matrix cells of stage I and II expressed GFAP in any great amount, thereby indicating their astrocytic differentiation, as these authors claimed, such matrix cells would demonstrate striking examples of transdifferentiation since they differentiate into neurons, oligodendroglia, ependymal cells, or choroid epithelium in subsequent embryonic development. Does such remarkable transdifferentiation take place in the cytogenesis of the CNS?

On the other hand, vimentin (Franke *et al.*, 1978) is not specific to nervous tissue, but is originally found in fibroblasts. Vimentin is abundant in some neuroglia of the epithelial type and in reactive astrocytes. It is interesting that vimentin is also found in immature neuroglial cells in late embryonic and early postnatal nervous tissue (Dahl, 1981; Schnitzer *et al.*, 1981). Few investigations have been made on the mode of existence of intermediate filaments in matrix cells. We addressed this problem using immunohistochemical and immunochemical techniques.

Three kinds of antisera (anti-vimentin, anti-neurofilament, and anti-GFAP) were used. Together with commercially available antibodies and homemade antisera obtained from rabbits immunized with the respective purified intermediate filament proteins, we were able to use some special anti-GFAP preparations. These were the generous gifts of Drs. Eng (anti-GFAP) and Mori (anti-astroprotein). The immunohistochemical reaction was performed by the PAP method or avidin–biotin–peroxidase complex method for maximal staining. The same materials of chicken, mouse, and human embryonic brains were cut in half. One part was used for histochemistry and the other for biochemical assay of the respective proteins by SDS–polyacrylamide gel electrophoresis (PAGE) and immunoblotting (Western blotting).

Matrix cells in the neural tube of chicken, mouse, and human embryos were stained positive with anti-vimentin (Fig. 4). As the development proceeded, vimentin immunoreactivity gradually became localized in the peripheral processes of matrix cells. In stage II, neuroblast cell bodies and early axons were stained positive, whereas axons and cell somata of mature neurons showed negative immunoreactivity, in agreement with the previous report of Schnitzer *et al.* (1981).

Neurofilament antibodies visualized early axons and axon hillocks, but the perikaryon of neuroblasts rarely showed definitive immunoreactivity. The appearance of neurofilament immunoreactivity in embryonic CNS, except for that in the mitotic cells, exactly corre-

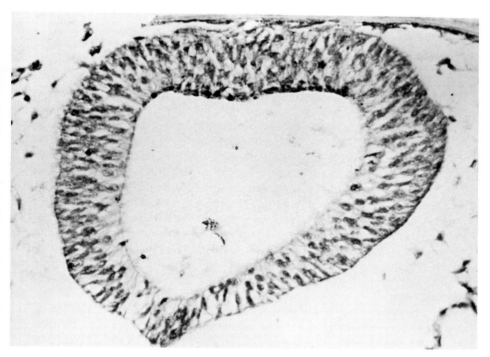

FIG. 4. Positive vimentin immunoreactivity in matrix cells of the neural tube of chicken embryo at 2-day incubation. Anti-vimentin antisera react positively with matrix cell cytoplasm of chicken, as well as in mouse and human neural tubes. ×215.

sponded to the commencement of stage II of cytogenesis in the respective regions of the neural tube. Curiously, the cytoplasm of all the mitotic cells in the early stage of embryonic development was stained strongly with anti-neurofilament antibodies irrespective of dermal origin; mitotic figures of matrix cells in the neural tube (of ectodermal origin), those of mesenchymal cells surrounding the neural tube (of mesodermal origin), as well as those in the primitive gut epithelium (of endodermal origin) were positive. However, as development progressed, the mitotic cells gradually lost reactivity to the neurofilament antibodies, and in the latter half of stage II and thereafter the immunoreactivity to anti-neurofilament antibodies became restricted to axons and distal perikarya of neurons and neuroblasts. Therefore, it is not reasonable to regard this immunoreactivity in mitotic cells as a sign of neuroblast differentiation; rather it is a transitory expression of neurofilament-like immunoreactivity in immature cells of three-dermal origin. The immunoreactivity is particularly strong in mitotic

cells. The molecular mechanism of this peculiar expression of neurofilament immunoreactivity of mitotic cells is, however, not yet
understood.

By immunohistology, GFAP-positive cells were first found in the
optic tectum of the chicken embryo on the eleventh day of incubation.
On this day, the GFAP antiserum revealed the presence of thin, short,
cellular processes in the molecular layer that run roughly perpen-
dicular to the pial surface. SDS–PAGE and the Western blotting technique detected GFAP in the optic tectum of the chicken for the first
time on the eleventh day of incubation. Thereafter, the amount of
GFAP revealed by these techniques increased gradually. In cerebral
hemispheres and spinal cord of the mouse and rat, it has been demonstrated that GFAP-positive cells are found only in late fetal or early
postnatal development and that no evidence of the presence of GFAP
in early embryonic life is found in the CNS of mice and rats (cf. Bignami and Dahl, 1974; Woodhams *et al.*, 1981). We were able to confirm
these findings with immunohistological staining and biochemical
analysis of the same brains by SDS–PAGE and Western blotting techniques (Fig. 5).

Critical examination of anti-GFAP antisera by the immunoblotting
technique is essential to avoid erroneous conclusions due to contamination and cross-reactions to other antigens such as vimentin or
other intermediate filaments. Such contaminations have frequently
been found in many samples of antisera or even in antigens with which
the antisera are raised (Yokoyama *et al.*, 1981).

We have examined the cerebral pallium of human embryos and
fetuses ranging in age from 7 to 22 weeks of pregnancy. GFAP-
positive cells were first found in the medial wall of the hemisphere at
16 weeks of pregnancy in the hippocampal area, where stage III
cytogenesis commences earlier than in the other regions. At this time,
the subependymal layer of the hippocampus is already thinned out to
assume the typical morphology of stage III ventricular cytoarchitec-
ture. Biochemical analysis of the other half of the brains used for
immunochemical examination detected the first appearance of the
GFAP molecule in the cerebral hemisphere at 15 weeks of pregnancy
in hippocampal tissue and a subsequent increase in the amount of
GFAP throughout the hemispheres.

These observations on phenotypic expression of GFAP with chick-
en, mouse, rat, and human brains conform to the previous notion based
on autoradiographic studies that stage III of cytogenesis, i.e., the stage
of neuroglia formation, follows stage II of cytogenesis, or the stage of
neuron production. Roessmann *et al.* (1980) and Takashima and Beck-

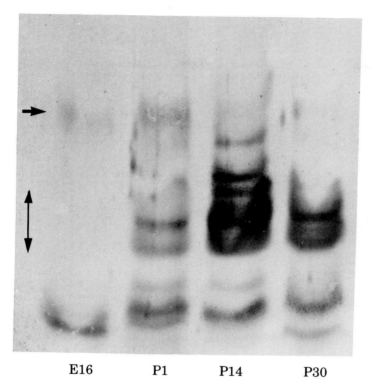

E16   P1   P14   P30

FIG. 5. Appearance of GFAP in the developing mouse brain studied by immunoblotting technique. Intermediate filament proteins extracted from the cerebral hemispheres of mice were examined by the Western blotting method. This figure depicts the results in mouse brains ranging in age from embryonic day 16 to postnatal day 30. Arrow indicates 57K vimentin and the line with two arrowheads represents 46K GFAP and its degradation products. GFAP becomes detectable in the mouse cerebral hemisphere only postnatally.

er (1983) studied human fetal and postnatal brains using GFAP antibodies and found that GFAP-positive ependymal and glial cells appear at the fifteenth week of pregnancy in the ependymal layer of the third ventricle and in the hemisphere only at the early postnatal period. These dates correspond to a period some time after the commencement of stage III cytogenesis in the respective regions of the brain.

It becomes clear that vimentin is positive in immature cells, such as matrix cells, young neuroblasts, and ependymoblasts, and in some immature astrocytes. In contrast, GFAP is not found in matrix cells, neuroblasts, or oligodendroglia. Therefore, we conclude that there is no evidence of transdifferentiation of GFAP-positive astrocytic cells into neuroblasts or oligodendroglia.

VI. Bundle Formation of Matrix Cells at Stage II and Its Function in Corticogenesis of the Vertebrate CNS

Throughout the period of neuronogenesis, i.e., stage II of cytogenesis, in the cerebral hemisphere of human fetuses and the optic tectum of the chicken, the outer cytoplasmic processes of matrix cells are seen to form bundles that run radially through the entire thickness of the brain wall (Fig. 6) (Hattori and Fujita, 1974). Rakic (1971) first described similar structures as "fascicles" of cell processes in the

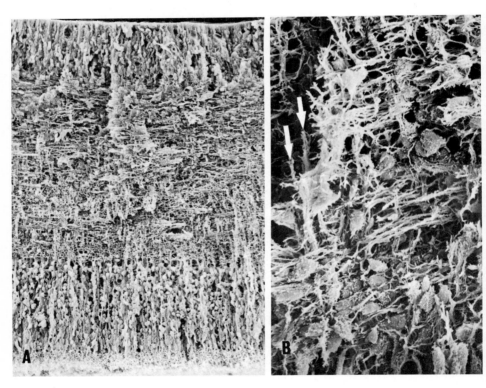

FIG. 6. Fractograph of human fetal cerebral hemisphere at 13.5 weeks of pregnancy (occipital region). (A) In the wall of the hemisphere, four layers are discernible. From top to bottom: matrix layer, intermediate layer, cortical plate, and molecular layer. Note that radial fibers composed of fasciculated processes of matrix cells span the entire thickness of the wall. Horizontal fibers are mostly axons. ×118. (B) Higher magnification. Arrows indicate some of the bundles of peripheral processes of matrix cells. ×507.

cerebral cortex of monkey fetuses. These bundles are not encountered in the regions where cortical structure is not formed, for example, in the ganglionic eminence, thalamus, or hypothalamus.

It is likely that bundle formation can occur because the lateral cell membranes of these matrix cells in the corticogenetic region can adhere to each other. The bundles, since they are made up of processes of many matrix cells, can be maintained as stable structures despite the periodic (but asynchronous) withdrawal of individual processes due to the elevator movement of matrix cells.

The functions of the bundles are twofold (Fig. 7A). First, they guide neuroblasts in their migration into the cortical anlage, which lies distant from the matrix layer. For this function, the dynamic adhesion of neuroblasts to matrix cells undoubtedly plays an important role. The second function of the bundle is to guide the processes of matrix cells themselves during their elevator movement. Without these guide rails, smooth performance of the elevator movement would be difficult, especially when the wall of the cerebral hemisphere (cf. Fig. 6) or optic tectum becomes thick. With the help of this guidance function, the matrix cells also send out their processes to the subpial layer, and their fine delicate branches proliferate to form the molecular layer.

Therefore, if the processes of matrix cells do not form bundles (Fig. 7B), neuroblasts cannot be guided properly. These neuroblasts have to migrate along the individually dispersed processes of matrix cells, which are lost periodically. When the process of a matrix cell is lost, migrating neuroblasts are forced to drop down to lie above the preceding neuroblasts, thereby forming neuronal piles one next to the other. Consequently, the arrangement of neurons in such a region would have an "outside-in" pattern in contrast to the "inside-out" configuration in corticogenetic regions. Further, matrix cells themselves could not send out their processes to the subpial region, and the molecular layer would not form. Hattori and Fujita (1974) and Fushiki (1981) investigated the distribution of matrix cell processes in developing chicken, mouse, and human brains by scanning electron microscopic fractography and found that a lack of matrix cell bundles and a lack of molecular layer formation are actually characteristic features detectable at stage II of cytogenesis in noncorticogenetic areas such as hypothalamus and thalamus (Fig. 8). When cytogenesis enters stage III and matrix cells change into ependymoglioblasts, the bundles of peripheral processes are no longer discernible, and only individual processes of the ependymoblasts and ependymal cells, with migrating neuroblasts clinging to them, are found for some time in the early period of stage III cytogenesis.

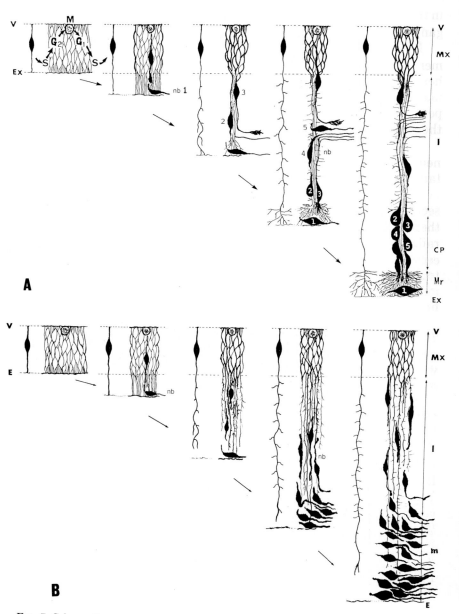

FIG. 7. Schematic representation showing developmental changes in morphology of matrix cells in corticogenetic and noncorticogenetic regions and their functional difference in guiding neuroblasts in respective regions. (A) Histogenesis in the corticogenetic region. Keeping pace with thickening of the wall of the brain (from left to

Thus the transitory formation of bundles by stage II matrix cells seems to play an essential role in the corticogenetic process in birds and mammals.

VII. Transitory Expression of the Matrix Cell Abnormality in the Reeler Mutation

Reeler is a recessive single gene mutation of the mouse that results in abnormal migration of neuroblasts in the corticogenetic regions. Since the earliest study of Hamburgh (1963), it has been pointed out that the arrangement of cortical neurons is reversed, large neurons lying immediately beneath the pia mater without a molecular layer (Fig. 9). Birth date analysis with tritiated thymidine autoradiography has revealed (Cavines, 1982) that the actual "inside-out" pattern of cortical neurons is reversed to an "outside-in" pattern. The molecular layer is virtually lost in the reeler cortex.

We have paid particular attention to the fact that the reeler abnormality of corticogenesis bears a striking similarity to the histogenetic pattern of noncorticogenetic regions in normal animals. Fushiki (1981) and Fujita *et al.* (1982) investigated the embryonic development of reeler mutants as well as their littermates and found that, in mouse embryos homozygous with reeler genes, bundles of matrix cells are virtually lacking in the cerebral hemispheres in stage II cytogenesis (Fig. 10), whereas in control animals of the same litter these structures are clearly discernible with scanning electron microscopic fractography. These findings were soon confirmed by Mikoshiba *et al.* (1984).

The abnormality of the reeler mutation that shows the disorganization of cortical structures is not due to any defect in the migratory

---

right), matrix cells elongate and the processes of tall matrix cells tend to adhere to each other and form bundles. The first neuroblast produced before the bundle is formed (i.e., of Cajal–Retzius neuron) migrates, according to the "outside-in" principle, to the subpial position and becomes embedded in the molecular layer (nb 1). Subsequent neuroblasts produced in the group of the matrix cells that form a bundle migrate along the bundle to the cortical plate and reach the deepest position adjacent to the molecular layer and are arranged in an "inside-out" pattern (nb 2, 3, 4, 5). (B) Histogenesis in the noncorticogenetic region. Matrix cells in this region do not form bundles. As the wall thickens, many matrix cells do not reach the pial surface and the molecular layer cannot be formed. Neuroblasts migrating along the individual processes of matrix cells are shaken off as the latter perform the elevator movement. The neuroblasts are piled up in the mantle layer (m) in order of their time of birth so that an "outside-in" pattern is created. The older neurons are schematically illustrated as larger in size. V, Ventricular surface; Mx, matrix layer; I, intermediate layer; CP, cortical plate or anlage; Mr, molecular layer; Ex or E, external or pial surface; nb, neuroblast.

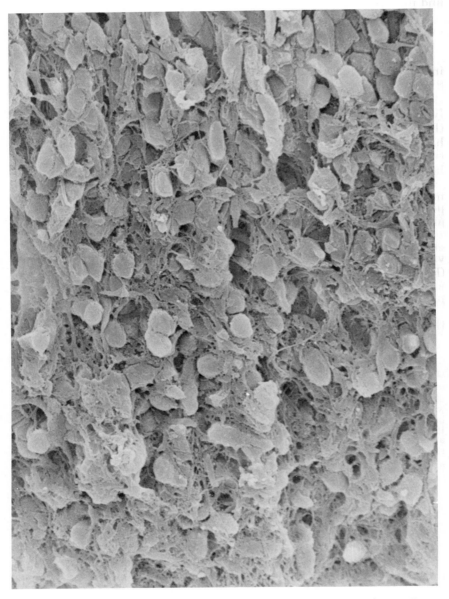

FIG. 8. Fractograph of diencephalic region of a mouse at embryonic day 14. Intermediate and mantle layers are depicted. Note the absence of bundles of matrix cell processes. Although there are some foci of loose aggregation of matrix cell processes, no organized structures of bundles are detectable. Clusters of neuroblasts are scattered, but no columnar arrangement of neuroblasts is discernible. Compare with Fig. 7B. ×850.

activity of neuroblasts, but is thought to be due to a deficiency in the specific factor(s) governing the adhesion of the matrix cells during stage II cytogenesis in the corticogenetic regions, so that bundle formation becomes impossible. Recent studies of Mikoshiba (1985) using chimera of reeler and wild-type mice have definitively demonstrated that the abnormality of reeler corticogenesis is not intrinsic to neuroblasts, but is attributable to the environment of neuroblast migration.

Recently, Shur (1982) reported that, in the cell membrane fraction of immature brain tissue of the reeler mutant, galactosyltransferase activity is specifically deficient. This enzyme has attracted the attention of cell biologists as a candidate for the cell-to-cell adhesion molecule since Roseman (1970) proposed a hypothesis of cell adhesion by means of glycosylation enzyme–substrate complex.

It is possible that galactosyltransferase or some other adhesion molecule (Takeichi et al., 1981; Edelman, 1983) may play an important role in the mutual adhesion of matrix cells expressed only transitorily in the histogenetic processes of corticogenetic regions. Nevertheless, it may play a crucial role in the morphogenesis and construction of functional integrity in the corticogenesis of the vertebrate CNS.

VIII. Concluding Remarks

When vertebrate neurogenesis begins, the wall of the neural tube is composed solely of matrix cells. They are apparently homogeneous in morphology and kinetics for some time during development, but actually change their major differentiation while they pass through stages I, II, and III of cytogenesis.

At stages I and II, matrix cells are vimentin positive but GFAP negative. Neurofilament protein is produced in neuroblasts immediately after their production, whereas GFAP (gliafibrillary acidic protein) becomes detectable only when ependymal and astroglial cells are differentiated at stage III.

During stage II cytogenesis, peripheral processes of matrix cells in the corticogenetic regions, such as the cerebral hemisphere or optic tectum, form bundles so that neuroblasts are transferred along these structures to the cortical anlage to be arranged in an "inside-out" pattern. Matrix cells also send out their processes along these bundles to form the molecular layer. In contrast, in the noncorticogenetic regions such as the diencephalon, the processes exist singly so that neuroblasts are shaken off, as the matrix cell rounds up in each mitotic phase, and pile up in an "outside-in" fashion. Since matrix cells are not able to send out their processes to the pial surface, the molecular layer is not formed.

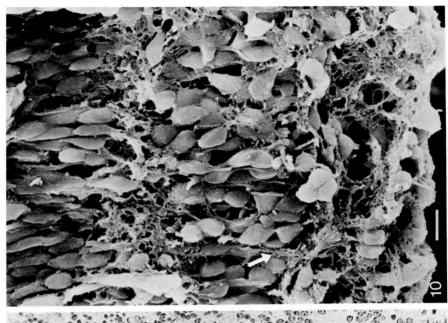

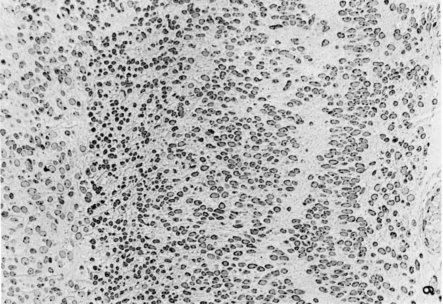

Examining reeler embryos at stage II of cytogenesis, we found a lack of bundle formation of mutant matrix cells in the corticogenetic regions. Anomalies in corticogenesis in the reeler mutation seem to be derived from the lack of bundle formation. The mutual adhesion of matrix cells that appears transiently during stage II of normal cytogenesis is lacking in the mutant. The nature of the adhesion molecule(s) involved in the corticogenesis of normal matrix cells that are lost in the reeler mutant awaits future investigation.

REFERENCES

Antanitus, D. S., Choi, B. H., and Lapham, L. W. (1976). *Brain Res.* **103**, 613–616.
Atlas, M., and Bond, V. P. (1965). *J. Cell Biol.* **26**, 19–24.
Bignami, A., and Dahl, D. (1974). *Nature (London)* **252**, 55–56.
Calvert, R., and Anderton, B. H. (1982). *FEBS Lett.* **145**, 171–175.
Cavines, V. S., Jr. (1982). *Dev. Brain Res.* **4**, 293–302.
Choi, B. H., and Lapham, L. W. (1978). *Brain Res.* **148**, 295–311.
Dahl, D. (1981). *J. Neurosci. Res.* **6**, 741–748.
Denham, S. (1967). *J. Embryol. Exp. Morphol.* **18**, 53–66.
Edelman, G. M. (1983). *Science* **219**, 450–457.
Eng, L. F., Vanderhaegen, J. J., Bignami, A., and Gerstl, B. (1971). *Brain Res.* **28**, 351–358.
Franke, W. W., Schmid, E., Osborn, M., and Weber, K. (1978). *Proc. Natl. Acad. Sci. U.S.A.* **75**, 5034–5038.
Fujita, S. (1962). *Exp. Cell Res.* **28**, 52–60.
Fujita, S. (1963a). *J. Comp. Neurol.* **120**, 37–42.
Fujita, S. (1963b). *Recent Adv. Res. Nerv. System (Tokyo)* **7**, 117–143.
Fujita, S. (1964). *J. Comp. Neurol.* **122**, 311–328.
Fujita, S. (1965a). *J. Comp. Neurol.* **124**, 51–59.
Fujita, S. (1965b). *Nature (London)* **206**, 742–744.
Fujita, S. (1967). *J. Cell Biol.* **32**, 277–288.
Fujita, S. (1973). *J. Comp. Neurol.* **151**, 25–34.
Fujita, S. (1975a). *J. Comp. Neurol.* **155**, 195–202.
Fujita, S. (1975b). *Symp. Cell. Chem.* **27**, 97–105.
Fujita, S., Hattori, T., Mikoshiba, K., and Tsukada, Y. (1982). *Recent Adv. Res. Nerv. System (Tokyo)* **26**, 433–445.

FIG. 9. Semithin section of cerebral hemisphere of a homozygous reeler mutant at embryonic day 16. Contrary to the normal "inside-out" arrangement, larger neurons tend to be located nearer to the pial surface (bottom), showing an "outside-in" pattern. Notice the absence of a molecular layer. At the top of the picture is part of a matrix layer. ×195.

FIG. 10. Fractograph of the cerebral hemisphere of a reeler embryo at embryonic day 16. The photograph shows the outer half of the cortical anlage and pial surface (bottom) of a homozygous reeler fetus at embryonic day 16. No typical molecular layer is present. Peripheral processes of matrix cells do not form bundles, but run singly in diverse directions. Scrutiny of fractographs reveals only rarely a partial and irregular aggregation of matrix cell processes, as indicated by the arrow. ×975.

Fushiki, S. (1981). *J. Kyoto Pref. Univ. Med.* **90,** 645–667.

Hamburgh, M. (1963). *Dev. Biol.* **8,** 165–185.

Hara, H. (1975). *Beitr. Pathol.* **154,** 293–307.

Hatfield, J. S., Skoff, R. P., Maisel, H., and Eng, L. (1984). *J. Cell Biol.* **98,** 1895–1898.

Hattori, T., and Fujita, S. (1974). *J. Electronmicrosc.* **23,** 269–276.

Hoshino, K., Matsuzawa, T., and Murakami, U. (1973). *Exp. Cell Res.* **77,** 89–94.

Jacobson, M. (1970). "Developmental Neurobiology." Holt, New York.

Kauffman, S. L. (1966). *Exp. Cell Res.* **42,** 67–73.

Kauffman, S. L. (1968). *Exp. Cell Res.* **49,** 420–424.

Langman, J., and Welch, G. W. (1967). *J. Comp. Neurol.* **131,** 15–26.

Levitt, P., Cooper, M. L., and Rakic, P. (1983). *Dev. Biol.* **96,** 472–484.

Mikoshiba, K. (1985). *Acta Histochem. Cytochem.* **18,** 113–124.

Mikoshiba, K., Nishimura, Y., and Tsukada, Y. (1984). *Dev. Neurosci.* **6,** 18–25.

Monesi, V. (1969). *Handb. Mol. Cytol.* pp. 472–499.

Okada, T. S. (1980). *Curr. Top. Dev. Biol.* **6,** 349–380.

Rakic, P. (1971). *Brain Res.* **33,** 471–476.

Roessmann, U., Velasco, M. E., Sindely, S. D., and Gambetti, P. (1980). *Brain Res.* **200,** 13–21.

Roseman, S. (1970). *Chem. Phys. Lipid* **5,** 270–297.

Schnitzer, J., Franke, W. W., and Schachner, M. (1981). *J. Cell Biol.* **435,** 447.

Shur, B. D. (1982). *J. Neurochem.* **39,** 201–209.

Sidman, R. L., Miale, I. L., and Feder, N. (1959). *Exp. Neurol.* **1,** 322–333.

Snow, W. H. L. (1976). *Ciba Found. Symp.* **40,** 53–70.

Takashima, S., and Becker, L. E. (1983). *Arch. Neurol.* **40,** 14–18.

Takeichi, M., Atsumi, T., Yoshida, C., Uno, K., and Okada, T. S. (1981). *Dev. Biol.* **87,** 340–350.

Vaughn, J. E. (1969). *Z. Zellforsch.* **94,** 293–324.

Woodhams, P. L., Basco, E., Hajos, F., Csillag, A., and Balazs, R. (1981). *Anat. Embryol.* **163,** 331–343.

Yokoyama, K., Mori, H., and Kurokawa, M. (1981). *FEBS Lett.* **135,** 25–30.

CHAPTER 17

PRESTALK AND PRESPORE DIFFERENTIATION DURING DEVELOPMENT OF *DICTYOSTELIUM DISCOIDEUM*

Ikuo Takeuchi, Toshiaki Noce, and Masao Tasaka*

DEPARTMENT OF BOTANY
FACULTY OF SCIENCE
KYOTO UNIVERSITY
KYOTO, JAPAN

I. Introduction

Although the development of the cellular slime molds is among the simplest, it nevertheless contains major features observable in more complex higher organisms. A homogeneous cell population produced during the growth phase aggregates and forms a tissue in which two types of cells differentiate in a linear pattern. They appear in a certain proportion irrespective of the tissue size and are interconvertible until they terminally differentiate (Bonner, 1967; Loomis, 1975, 1982; Raper, 1984).

These features, together with the ease with which the organisms can be handled experimentally, have attracted many researchers who, although they belong to different disciplines, share a common interest in the analysis of development. The development of the cellular slime molds is thus regarded as a model system, to which different approaches—from the molecular to the mathematical—are now applicable. We shall concentrate our discussion on the differentiation of the two cell types (prestalk/prespore and stalk/spore) within the developing tissue of *Dictyostelium discoideum,* the most widely studied species

* Present address: Division of Developmental Biology, The National Institute for Basic Biology, Okazaki, Japan.

of the cellular slime molds. The convertibility of cell types will be discussed in relation to the reproductive strategy of differentiation.

II. Prestalk and Prespore Cells

The development of D. discoideum leads to formation of a fruiting body that consists of three parts: a spore head, a supporting cellular stalk, and a basal disk surrounding the base of the stalk. Spores are ellipsoidal cells enveloped by a spore coat. Both the stalk and basal disk are composed of vacuolate dead cells encased by a cellulosic cell wall. This is the only stage in the life cycle at which the cells are enclosed by the cell wall and are immobile. The basal disk is missing from most of the related species. In most of the following discussion, we shall regard stalk and disk cells as a single cell type, since the latter are in the minority and are apparently much like the former.

It was shown by the pioneering work of Raper (1940) that during the formation of a fruiting body the anterior one-third (in length) of a slug differentiates into stalk cells, while the majority of the rest becomes spores. Basal disk cells are thought to be derived from the rearmost region.

Differences between the anterior prestalk and the posterior prespore cells of a slug have long been noted by their staining characteristics. Various methods of histochemical (Bonner et al., 1955; Krivanek and Krivanek, 1958; Mine and Takeuchi, 1967) and vital (Bonner, 1952) staining have revealed differences in intensity between the two regions of slugs. That the substances specifically present in spores are localized in prespore cells was first shown by immunocytochemical staining using anti-spore serum (Takeuchi, 1963, 1972). Anti-D. mucoroides-spore serum gave specific immunofluorescence staining to cytoplasmic granules of prespore cells as well as to the surface of spores. Major antigens for the antiserum were recently identified spore coat proteins (Devine et al., 1983) that are specifically synthesized in prespore cells (Devine et al., 1982). The antigens are contained in prespore vacuoles (PSV) (Ikeda and Takeuchi, 1971) that are specifically present in prespore cells (Hohl and Hamamoto, 1969; Maeda and Takeuchi, 1969; Gregg and Badman, 1970). During sporulation, the PSVs fuse with the plasma membrane, and the antigens are exocytosized. Cellulose is secreted later, and the antigens constitute the outermost layer of the spore coat (Hohl and Hamamoto, 1969; Maeda, 1971). In contrast to PSVs, autophagic vacuoles are almost exclusively located in prestalk cells (Maeda and Takeuchi, 1969); they first appear in every cell soon after the vegetative stage (Chastellier and Ryter, 1977), but they then disappear in prespores at the late aggregation stage while they develop further in prestalks (Yamamoto *et al.*, 1981). The vacuoles are responsible for the strong vital staining of prestalks with neutral red or Nile blue sulfate (Yamamoto and Takeuchi, 1983).

By the use of two-dimensional gel analyses, prestalk and prespore cells were found to differ in the kind of proteins they synthesize (Alton and Brenner, 1979). The difference is mostly due to transcriptional control as shown by the fact that the *in vitro* translational products of mRNAs extracted from both cell types represent the majority of the proteins synthesized *in vivo*. Comparison of prestalk- and prespore-specific proteins with stalk- and spore-specific proteins revealed that the majority of spore-specific proteins are synthesized in prespore cells, whereas very few stalk-specific proteins are present in prestalk cells (Borth and Ratner, 1983; Morrissey *et al.*, 1984). The pattern of protein synthesis in prestalk cells rather resembles that of preaggregating cells. Most stalk-specific proteins are synthesized during culmination.

Certain enzymes are specific for either cell type. UDP-galactose: polysaccharide transferase is localized in prespore cells (Newell *et al.*, 1969) and an isozyme of acid phosphatase (acid phosphatase 2) (Oohata, 1983) is localized in prestalk cells. The latter appears to be derived from a single gene product common to both acid phosphatase 2 and 1 (cell type nonspecific) through prestalk-specific modification (Loomis and Kuspa, 1984). Some enzymes are known to be predominantly present in either prestalk or prespore cells. Enzymes such as trehalase (Jefferson and Rutherford, 1976), cyclic AMP phosphodiesterase (Brown and Rutherford, 1980), *N*-acetylglucosaminidase (Hamilton and Chia, 1975), β-glucosidase, β-galactosidase (Oohata and Takeuchi, 1977), alkaline phosphatase (Hamilton and Chia, 1975), and cathepsin B-like protease (Fong and Rutherford, 1978) showed higher levels of activity in prestalk cells. In contrast, adenylate cyclase (Merkle and Rutherford, 1984), cyclic AMP-dependent protein kinase (Schaller *et al.*, 1984), trehalose-6-phosphate synthetase, and UDP-galactose epimerase (Hamilton and Chia, 1975) were shown to be more active in prespore cells. Some of these enzymes do not become predominant in either cell type until the culmination stage.

III. Regulation of Prestalk/Prespore Differentiation

Despite the fact that prestalk and prespore cells are differentiated in slugs, their developmental fate is by no means determined: when a slug is dissected into fragments, each fragment gives rise to a normal fruiting body if it is allowed to regulate itself before terminal differentiation (Raper, 1940).

One of the characteristics of prestalk/prespore differentiation is that the proportion of the two cell types is roughly constant irrespective of the size of a slug (Bonner, 1957; Takeuchi *et al.*, 1977; Stenhouse and Williams, 1977). When the normal proportion is disturbed by dissection, prestalk and prespore cells are converted to each other in order to restore the proportion (Bonner *et al.*, 1955; Sakai, 1973). It takes more time for prestalks to become prespores than for prespores to become prestalks. It was argued by Borth and Ratner (1983) that this is because prespore differentiation is accompanied by synthesis of many new proteins, whereas prestalk differentiation is not. Such conversion of the cell types in slug fragments is not accompanied by cell division (Bonner and Frascella, 1952).

Some mutants are known that produce terminal structures consisting of only spores or stalk cells but that nevertheless show the normal prestalk/prespore pattern at earlier stages (Morrissey *et al.*, 1981). It was shown in the case of a *spory* mutant that after the prespore cells differentiate into spores, the remaining prestalk cells that are unable to become mature stalk cells are now converted to prespores and then to spores (Amagai *et al.*, 1983). A similar mechanism was found to be responsible for producing a *stalky* mutant as well (Morrissey *et al.*, 1984). These indicate that prestalk and prespore cells are interconvertible up to nearly the last moment before terminal differentiation.

IV. Differentiation Patterns of Prestalk/Prespore Cells

Taking advantage of the fact that prespore cells of *D. discoideum* specifically react with anti-*D. mucoroides*-spore serum (Takeuchi, 1963, 1972), the process of prespore differentiation was examined both with cells disaggregated from tissues (Hayashi and Takeuchi, 1976; Forman and Garrod, 1977) and with tissue sections (Takeuchi *et al.*, 1978). Cells reactive with the antiserum first appear in a cell aggregate that is about to form a tip (Figs. 1 and 2). As the tip elongates, prespore cells increase in number and accumulate underneath the tip. Formation of prestalk/prespore pattern is thus completed within a standing slug. Thereafter, the proportion of prespore cells remains about 80% throughout the development. The timing of prespore differentiation as described here agrees well with the studies using other prespore markers such as UDP-galactose transferase (Sussman and Osborn, 1964), PSVs (Müller and Hohl, 1973), prespore-specific mRNAs (Barklis and Lodish, 1983; Mehdy *et al.*, 1983), and monoclonal antibodies, which will be described shortly. The appearance of prespore cells also coincides with the crucial time in the developmental cycle when a large number of developmentally regulated proteins and

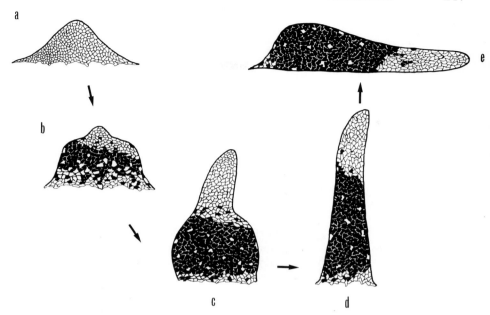

FIG. 1. Differentiation of prespore cells within a developing cell aggregate of the strain NC4. Sections were made at various stages of aggregation and slug formation and stained with fluorescein isothiocyanate (FITC)-conjugated anti-*D. mucoroides*-spore serum. Black cells: stained prespore cells; white cells: unstained cells. (From Takeuchi *et al.*, 1978.)

mRNAs begin to be synthesized (Alton and Lodish, 1977; Blumberg and Lodish, 1980).

In contrast to prespore differentiation, the process of prestalk differentiation was not clear until recently because no specific markers for prestalk cells were known. It was vaguely thought that prestalks arose concurrently with prespores through bifurcation of undifferentiated cells. As it was difficult to obtain prestalk-specific polyclonal antibodies, we tried to produce monoclonal antibodies against prestalk or prespore cells and were successful in obtaining several of them (Tasaka *et al.*, 1983).

Both prestalk (C1)- and prespore (B6)-specific monoclonal antibodies were used to analyze the processes of differentiation, and the results are summarized in Fig. 2. B6 and other prespore-specific monoclonal antibodies gave essentially the same results as the antispore serum described above. However, C1 (prestalk- and stalk-specific) antibody yielded an entirely different result for prestalk differentiation. Cells reactive with C1 first appear soon after the vegetative stage and

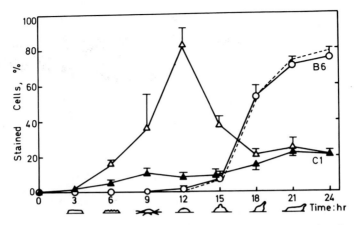

F<small>IG</small>. 2. Developmental changes of NC4 cells stained with monoclonal and polyclonal antibodies. Cells developing on Millipore filters were at times disaggregated, fixed, and stained with B6 (○) or C1 (△) antibodies by the use of FITC-conjugated anti-mouse IgG, and the percentages of stained cells were determined. (▲) Percentages of cells strongly stained with C1. The broken line indicates the changes of cells stained with FITC-conjugated anti-*D. mucoroides*-spore serum. The abscissa indicates the time after starvation. The developmental stages attained are illustrated below in diagrams. Bars indicate standard deviations. (From Tasaka *et al.*, 1983.)

then increase in number, so that about 80% of cells contain the C1 antigen at the early aggregation stage. As prespore differentiation begins at the late aggregation stage, the number of cells containing the antigen decreases to approximately 20% by the slug stage.

These results showed that prestalk cells begin to differentiate prior to aggregation, much earlier than prespore differentiation. This agrees with other studies using different prestalk markers. By the use of cDNA clones specific for prestalk differentiation, Mehdy *et al.* (1983) showed that prestalk mRNAs appear after 7.5 hours of development, before the completion of cell aggregation, in contrast to prespore mRNAs which are undetectable until about 15 hours. Similar results were also obtained by Barklis and Lodish (1983), who found that a class of prestalk mRNAs is present at a low but detectable level in vegetative cells and increases after 4 hours of development. Some wheat germ agglutinin (WGA)-binding proteins are specific for prestalk or prespore cells (West and McMahon, 1979). Okamoto and Kumagai recently found that prestalk proteins first appear in early aggregative cells, whereas prespore proteins are not detectable until the late aggregation stage (unpublished).

The above results indicate that after the vegetative stage all the cells first proceed to the pathway of prestalk differentiation, but that after aggregation some cells lose the stalk antigen and instead begin to synthesize spore antigens. The latter process appears analogous to the cell type conversion occurring within the prestalk isolate of a slug, as described in the preceding section.

Another feature of the synthesis of the stalk antigen by preaggregating cells is their heterogeneity: there is a marked variation among the cells in the initiation and the extent of the synthesis. As these cells are collected randomly to an aggregation center, an early aggregate consists of cells considerably variable in the cellular content of the antigen. This suggests the possibility that when prespore cells begin to differentiate, cells containing less stalk antigen may be converted to prespore cells, whereas those containing more remain prestalk cells.

V. Prestalk/Prespore Differentiation Tendencies

The above possibility was attested to by examining the expression of the stalk antigen within cell populations that showed differentiation tendencies toward either prestalk or prespore cells. It was shown by Watts and Ashworth (1970) that an axenic strain, AX2, is able to grow in either the presence or the absence of glucose in a growth medium. Leach et al. (1973) further showed that when glucose-grown G^+ cells are mixed with non-glucose-grown G^- cells and allowed to develop together the former tend to occupy the posterior part of the slug, whereas the latter tend to occupy the anterior part. Recently, we have shown that this is because when they are mixed G^- cells have a tendency to differentiate into prestalks, whereas G^+ cells differentiate into prespores (Tasaka and Tekeuchi, 1981). Accordingly, we examined the production of the stalk antigen in G^- and G^+ cells (Noce and Takeuchi, 1985).

Unlike the wild-type cells that begin to form the C1 antigen after the vegetative stage, some of the axenically growing AX2 cells prematurely synthesize the antigen. A similar premature appearance during the growth phase of some developmentally regulated enzymes has been noted (Ashworth and Quance, 1972). The extent of the expression of the C1 antigen, however, differs considerably between G^- and G^+ cells: a much higher percentage of the former, in contrast to the latter, contain the antigen during the growth phase (Fig. 3). During starvation, both G^- and G^+ cells increase the C1 content, but the difference between the two remains about the same. That the two populations also differ in the cellular content of C1 is shown by flow cytometry.

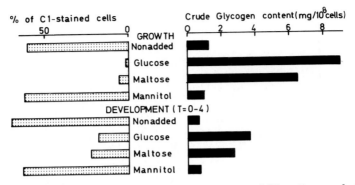

FIG. 3. The effects of sugars on the cellular contents of C1 antigen and glycogen during the growth and development (preaggregation) phases. For the growth phase effects, AX2 cells were grown in an axenic medium in the presence or absence of sugars (86 mM for glucose and mannitol, 43 mM for maltose) and harvested at the exponential phase ($1–5 \times 10^6$ cells/ml). For the development phase effects, washed, non-glucose-grown cells were shaken in 20 mM phosphate buffer, pH 6.4 (1×10^7 cells/ml) in the presence or absence of the sugars and collected after 4 hours (for C1 determination) and 5 hours (for glycogen determination). The cells were stained with C1 antibody by the use of FITC-conjugated anti-mouse IgG and the percentages of stained cells were determined. For determination of crude glycogen, the cells were digested in boiling 30% KOH for 20 minutes, and glycogen was precipitated with 60% ethanol. After washing, total carbohydrate of the fraction was determined by anthrone assay, using purified glycogen as a standard.

In addition to glucose, metabolizable sugars such as mannose, galactose, and maltose show similar (though weaker) inhibitory effects on the production of the C1 antigen, but nonmetabolizable sugars do not (Fig. 3). The addition of metabolizable sugars to starving cells as well as to growing cells is effective in reducing the C1 content. The presence of these (but not nonmetabolizable) sugars during both the growth phase and the early development gives rise to an increase in the cellular content of glycogen (Fig. 3). A similar effect of glucose on the glycogen content of vegetative cells was previously noted by Hames and Ashworth (1974).

We investigated the developmental behavior of G^- and G^+ cells that contained more and less C1 antigen, respectively. G^- cells were vitally stained with fluorescein isothiocyanate (FITC), mixed with unstained G^+ cells, and allowed to develop together. Both types of cell aggregated randomly to collection points. Immunofluorescence staining of these cells with the C1 antibody revealed that the majority of FITC-labeled G^- cells contained more C1 antigen than unlabeled G^+ cells (Fig. 4). G^- cells are randomly distributed within an early cell

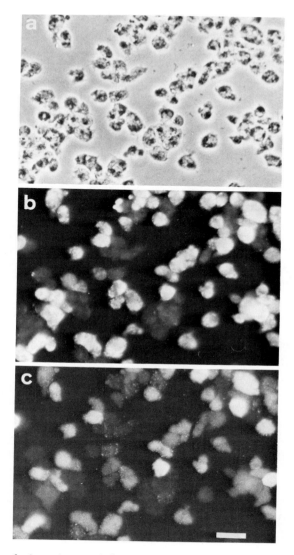

FIG. 4. A developing mixture of glucose- and non-glucose-grown cells of AX2. Non-glucose-grown cells vitally stained with FITC (dissolved in DMSO) were mixed with glucose-grown cells (DMSO treated) at a ratio of 1:3 and allowed to develop on cover glasses thinly coated with agar. After 6 hours of development, cells were fixed *in situ* and stained with C1 antibody by the use of tetramethyl-rhodamine isothiocyanate (TRITC)-conjugated anti-mouse IgG. (a) Phase-contrast micrograph; (b) fluorescent micrograph for FITC; (c) fluorescent micrograph for TRITC. Bar, 20 μm.

aggregate, but when a tip is formed, they are largely localized in the tip region and finally occupy the prestalk region of a slug. Immunostaining showed that these cells have a high C1 antigen content.

These findings indicate that cells that have produced more C1 antigen prior to aggregation are sorted out to the tip region of an aggregate and remain prestalk cells. In contrast, those having produced less are sorted out to the basal portion (underneath the tip) of the aggregate, where cells are known to initiate synthesis of the spore antigen (cf. Fig. 1). It is thus concluded that the C1 antigen content of preaggregating cells reflects the tendency of the cells toward either prestalk or prespore differentiation after aggregation. The fact that the wild type and AX2 show the same staining pattern with the C1 antibody both before and after aggregation indicates that they undergo the same processes of prestalk/prespore differentiation. It has recently been shown that prestalk/prespore differentiation tendencies are related to the cell cycle phase of vegetative cells (Weijer et al., 1984; McDonald and Durston, 1984). It seems interesting to study the expression of the C1 antigen during the cell cycle.

VI. Reproductive Strategy of Cell Differentiation

The development of the cellular slime molds results in the formation of fruiting bodies consisting of spore and stalk cells. Spores are reproductive cells that give rise to the next generation upon germination, whereas stalk cells are vacuolate dead cells (Whittingham and Raper, 1960) whose function is only to support spores aloft. Possible selective advantages of this function for better spore dispersal were discussed by Bonner (1982), who pointed out that stalk formation is considered to be a case of altruism since stalk cells sacrifice themselves for effective gene preservation.

It is interesting to note that this division of labor between stalk and spore cells is reflected in prestalk/prespore differentiation. For slug migration, it was shown that prestalk cells put forth a motive force that is three times as great as that of prespore cells (Inouye and Takeuchi, 1980). In short, the former drag the latter. This is consistent with the fact that prestalks are more active in the transcription of actin genes than prespores (Tsang et al., 1982; Mehdy et al., 1983). It is also known that prestalks are more active in chemotaxis toward cyclic AMP (Matsukuma and Durston, 1979; Sternfeld and David, 1981a) or oxygen (Sternfeld and David, 1981b).

These facts indicate that prestalks consume more energy than prespores during development. This is consistent with the fact that the former develop many more autophagic vacuoles to digest the cytoplasm than the latter (Yamamoto *et al.*, 1981), for neither cell takes up any food during development. This is also reflected by the fact that prestalk cells decrease in volume at a faster rate than prespore cells (Voet *et al.*, 1984). Eventually, prestalks exhaust all the stored energy to become skeletal components of a stalk. In contrast, prespores appear to save as much of their energy as possible and survive for the next generation as spores. It is therefore not surprising to find that prespores are more active in glucose utilization and protein synthesis than prestalks (Bonner *et al.*, 1984).

The fact that the addition of metabolizable sugars to AX2 cells inhibits the production of the C1 antigen and hence increases the tendency toward prespore differentiation on the one hand and stimulates the synthesis of glycogen on the other hand suggests an interesting reproductive strategy for this organism: cells that have more glycogen tend to become spores, whereas those that have less become stalk cells. Although the glycogen content was not always shown to be correlated with the spore/stalk differentiation tendency (Weeks and Ashworth, 1972; MacWilliams, 1982), it is possible that some sugars or sugar derivatives may be related to this tendency. This suggests that cells containing more stored material are conserved for the next generation by the sacrifice of those containing less.

VII. Concluding Remarks

The development of *D. discoideum* leads to the formation of a fruiting body consisting of two cell types, stalk and spore cells (basal disk cells are much like stalk cells). They are derived from the anterior and posterior parts of a slug, respectively. Prestalk and prespore cells of the slug have many specific characteristics, at least some of which are known to be correlated with stalk/spore differentiation. The prestalk/prespore proportion is roughly constant irrespective of slug size. When the normal proportion is disturbed, the two cell types are converted into each other to restore the proportion.

Immunofluorescent staining of developing cells with prestalk/ prespore specific monoclonal antibodies indicates that after the vegetative stage all the cells first proceed to the pathway of prestalk differentiation and begin to synthesize a stalk (C1) antigen. However, the cells vary considerably in the extent of antigen synthesis, giving rise to much variation in the C1 content of preaggregating cells. After these cells are collected randomly to an aggregation center, some of them lose the C1 antigen and instead begin to synthesize spore antigens to become prespores, whereas the others retain the C1 antigen and remain prestalks. In consequence, the normal prestalk/prespore pattern becomes established within a slug.

The expression of the C1 antigen in glucose- and non-glucose-grown cells indicated that the presence of glucose during either the growth or preaggregation period suppresses the production of the C1 antigen and stimulates the synthesis of glycogen. Examination of developmental behavior of both cells revealed that cells that have produced more C1 antigen during the preaggregation period remain prestalk cells after aggregation and occupy the anterior part of a slug. In contrast, those that have produced less are converted to prespore cells and occupy the posterior part. It is thus concluded that the C1 antigen content of preaggregating cells reflects the tendency of the cells toward either prestalk or prespore differentiation after aggregation.

The reproductive strategy of prestalk/prespore differentiation was discussed with special reference to the fact that the inhibitory effect of sugars on C1 production parallels the stimulating effect on glycogen accumulation.

ACKNOWLEDGMENTS

This work was supported in part by Grants-in-Aid (Nos. 57480011 and 58119005 for Special Project Research, "Multicellular Organization") from the Ministry of Education, Science and Culture of Japan.

REFERENCES

Alton, T. H., and Brenner, M. (1979). *Dev. Biol.* **71,** 1–7.
Alton, T. H., and Lodish, H. F. (1977). *Dev. Biol.* **60,** 180–206.
Amagai, A., Ishida, S., and Takeuchi, I. (1983). *J. Embryol. Exp. Morphol.* **74,** 235–243.
Ashworth, J. M., and Quance, J. (1972). *Biochem. J.* **126,** 601–608.
Barklis, E., and Lodish, H. F. (1983). *Cell* **32,** 1139–1148.
Blumberg, D. D., and Lodish, H. F. (1980). *Dev. Biol.* **78,** 285–300.
Bonner, J. T. (1952). *Am. Nat.* **86,** 79–89.
Bonner, J. T. (1957). *Q. Rev. Biol.* **32,** 232–246.
Bonner, J. T. (1967). "The Cellular Slime Molds." Princeton Univ. Press, Princeton, New Jersey.
Bonner, J. T. (1982). *Am. Nat.* **119,** 530–552.
Bonner, J. T., and Frascella, E. B. (1952). *J. Exp. Zool.* **121,** 561–572.
Bonner, J. T., Chiquoine, A. D., and Kolderie, M. Q. (1955). *J. Exp. Zool.* **130,** 133–158.
Bonner, J. T., Sundeen, C. J., and Suthers, H. B. (1984). *Differentiation* **26,** 103–106.
Borth, W., and Ratner, D. (1983). *Differentiation* **24,** 213–219.
Brown, S. S., and Rutherford, C. L. (1980). *Differentiation* **16,** 173–183.
Chastellier, C., and Ryter, A. (1977). *J. Cell Biol.* **75,** 218–236.
Devine, K. M., Morrissey, J. H., and Loomis, W. F. (1982). *Proc. Natl. Acad. Sci. U.S.A.* **79,** 7361–7365.
Devine, K. M., Bergmann, J. E., and Loomis, W. F. (1983). *Dev. Biol.* **99,** 437–446.

Fong, D., and Rutherford, C. L. (1978). *J. Bacteriol.* **134,** 521–527.
Forman, D., and Garrod, D. R. (1977). *J. Embryol. Exp. Morphol.* **40,** 215–228.
Gregg, J. H., and Badman, W. S. (1970). *Dev. Biol.* **22,** 96–111.
Hames, B. D., and Ashworth, J. M. (1974). *Biochem. J.* **142,** 301–315.
Hamilton, I. D., and Chia, W. K. (1975). *J. Gen. Microbiol.* **91,** 295–306.
Hayashi, M., and Takeuchi, I. (1976). *Dev. Biol.* **50,** 302–309.
Hohl, H. R., and Hamamoto, S. T. (1969). *J. Ultrastruct. Res.* **26,** 442–453.
Ikeda, T., and Takeuchi, I. (1971). *Dev. Growth Differ.* **13,** 221–229.
Inouye, K., and Takeuchi, I. (1980). *J. Cell Sci.* **41,** 53–64.
Jefferson, B. L., and Rutherford, C. L. (1976). *Exp. Cell Res.* **103,** 127–134.
Krivanek, J. O., and Krivanek, R. C. (1958). *J. Exp. Zool.* **137,** 89–116.
Leach, C. K., Ashworth, J. M., and Garrod, D. R. (1973). *J. Embryol. Exp. Morphol.* **29,** 647–661.
Loomis, W. F. (1975). *"Dictyostelium discoideum:* A Developmental System." Academic Press, New York.
Loomis, W. F., ed. (1982). "The Development of *Dictyostelium discoideum.* Academic Press, New York.
Loomis, W. F., and Kuspa, A. (1984). *Dev. Biol.* **102,** 498–503.
McDonald, S. A., and Durston, A. J. (1984). *J. Cell Sci.* **66,** 195–204.
MacWilliams, H. K. (1982). *Symp. Soc. Dev. Biol.* pp. 463–483.
Maeda, Y. (1971). *Mem. Faculty Sci., Kyoto Univ. Ser. Biol.* **4,** 97–107.
Maeda, Y., and Takeuchi, I. (1969). *Dev. Growth Differ.* **11,** 232–245.
Matsukuma, S., and Durston, A. J. (1979). *J. Embryol. Exp. Morphol.* **50,** 243–251.
Mehdy, M. C., Ratner, D., and Firtel, R. A. (1983). *Cell* **32,** 763–771.
Merkle, R. K., and Rutherford, C. L. (1984). *Differentiation* **26,** 23–29.
Mine, H., and Takeuchi, I. (1967). *Annu. Rep. Biol. Works, Fac. Sci., Osaka Univ.* **15,** 97–111.
Morrissey, J. H., Farnsworth, P. A., and Loomis, W. F. (1981). *Dev. Biol.* **83,** 1–8.
Morrissey, J. H., Devine, K. M., and Loomis, W. F. (1984). *Dev. Biol.* **103,** 414–424.
Müller, U., and Hohl, H. R. (1973). *Differentiation* **1,** 267–276.
Newell, P. C., Ellingson, J. S., and Sussman, M. (1969). *Biochim. Biophys. Acta* **177,** 610–614.
Noce, T., and Takeuchi, I. (1985). *Dev. Biol.* **109,** 157–164.
Oohata, A. A. (1983). *J. Embryol. Exp. Morphol.* **74,** 311–319.
Oohata, A., and Takeuchi, I. (1977). *J. Cell Sci.* **24,** 1–9.
Raper, K. B. (1940). *J. Elisha Mitchell Sci. Soc.* **56,** 241–282.
Raper, K. B. (1984). "The Dictyostelids." Princeton Univ. Press, Princeton, New Jersey.
Sakai, Y. (1973). *Dev. Growth Differ.* **15,** 11–19.
Schaller, K. L., Leichtling, B. H., Majerfeld, I. H., Woffendin, C., Spitz, E., Kakinuma, S., and Rickenberg, H. V. (1984). *Proc. Natl. Acad. Sci. U.S.A.* **81,** 2127–2131.
Stenhouse, F. O., and Williams, K. L. (1977). *Dev. Biol.* **59,** 140–152.
Sternfeld, J., and David, C. N. (1981a). *Differentiation* **20,** 10–21.
Sternfeld, J., and David, C. N. (1981b). *J. Cell Sci.* **50,** 9–17.
Sussman, M., and Osborn, M. J. (1964). *Proc. Natl. Acad. Sci. U.S.A.* **52,** 81–87.
Takeuchi, I. (1963). *Dev. Biol.* **8,** 1–26.
Takeuchi, I. (1972). *In* "Aspects of Cellular and Molecular Physiology" (K. Hamaguchi, ed.), pp. 217–236. Univ. of Tokyo Press, Tokyo.
Takeuchi, I., Hayashi, M., and Tasaka, M. (1977). *In* "Development and Differentiation in the Cellular Slime Moulds" (P. Cappuccinelli and J. M. Ashworth, eds.), pp. 1–16. Elsevier, Amsterdam.

Takeuchi, I., Okamoto, K., Tasaka, M., and Takemoto, S. (1978). *Bot. Mag. Tokyo Special Issue* **1**, 47–60.

Tasaka, M., and Takeuchi, I. (1981). *Differentiation* **18**, 191–196.

Tasaka, M., Noce, T., and Takeuchi, I. (1983). *Proc. Natl. Acad. Sci. U.S.A.* **80**, 5340–5344.

Tsang, A. S., Mahbubani, H., and Williams, J. G. (1982). *Cell* **31**, 375–382.

Voet, L., Krefft, M., Mairhofer, H., and Williams, K. L. (1984). *Cytometry* **5**, 26–33.

Watts, D. J., and Ashworth, J. M. (1970). *Biochem. J.* **119**, 171–174.

Weeks, G., and Ashworth, J. M. (1972). *Biochem. J.* **126**, 617–626.

Weijer, C. J., Duschl, G., and David, C. N. (1984). *J. Cell Sci.* **70**, 133–145.

West, C. M., and McMahon, D. (1979). *Exp. Cell Res.* **124**, 393–401.

Whittingham, W. F., and Raper, K. B. (1960). *Proc. Natl. Acad. Sci. U.S.A.* **46**, 642–649.

Yamamoto, A., and Takeuchi, I. (1983). *Differentiation* **24**, 83–87.

Yamamoto, A., Maeda, Y., and Takeuchi, I. (1981). *Protoplasma* **108**, 55–69.

CHAPTER 18

TRANSDIFFERENTIATION OCCURS CONTINUOUSLY IN ADULT HYDRA

Hans Bode, John Dunne, Shelly Heimfeld, Lydia Huang,

Lorette Javois, Osamu Koizumi, John Westerfield, and Marcia Yaross

DEVELOPMENTAL BIOLOGY CENTER AND
DEPARTMENT OF DEVELOPMENTAL AND CELL BIOLOGY
UNIVERSITY OF CALIFORNIA, IRVINE
IRVINE, CALIFORNIA

I. Introduction

In a mature animal or plant the differentiated state of a cell is generally quite stable. The cell has a distinct morphology, a characteristic set of macromolecules that are involved in a particular physiological function, and usually does not undergo cell division. This state persists throughout the lifetime of the cell, which can coincide with the life-span of the organism as with a muscle cell, or be much shorter, as in the red blood cell. This pattern would suggest that a cell once having traversed a differentiation pathway and having reached the end state is now fixed in its structure and function.

However, ablation experiments have revealed that this is not always the case. For a number of cell types the stability of the differentiated state is not fixed, but is only metastable. In some tissues, such as the liver, partial ablation results in the resumption of cell division of the remaining cells. Here, quiescent cells do not alter their cell type,

257

but undergo a behavior not usually associated with the final differentiated state. In others a change in differentiation occurs. For example,
during Wolffian lens regeneration pigmented iris epithelial cells are
converted into lens cells (Eguchi, 1964; Yamada, 1967). During limb
regeneration in newts, fibroblasts of the dermis are converted into
chondrocytes (Namenwirth, 1974; Dunis and Namenwirth, 1977). In a
few instances wholesale conversion of cell types takes place. Regeneration of an entire plant from a single cell type is a well-known example
(e.g., Steward *et al.*, 1966; Takebe *et al.*, 1971). Also, an entire manubrium, the feeding and sexual organ of the marine coelenterate,
Podocoryne carnea, containing a variety of cell types, will regenerate
from one cell type, the striated muscle cell (see Schmid and Alder, this
volume).

The existence of cell type conversion, or transdifferentiation, in
which a cell undergoes a change in its physiological function with
accompanying structural changes, clearly demonstrates that the differentiated state is not always immutable. One aspect all these examples
have in common is that experimental intervention was necessary to
expose the metastability of the differentiation of these cells. To be sure
that the observed conversions were not simply a consequence of the
manipulations, it would be useful to identify cases in which changes in
cell type occur normally in an organism. In the following we describe
such an example.

In hydra, a freshwater coelenterate, transdifferentiation occurs
continuously. At any moment a fraction of the cells are undergoing a
change in differentiation, suggesting that the differentiated state for
some cell types in hydra is largely metastable. This behavior is readily
understood in terms of the unusual growth dynamics and the simplicity of this primitive metazoan. For ease of understanding, the
growth dynamics will be described first. Then, two examples in which
the lability of the differentiated state is particularly pronounced will
be described in detail. Finally, we will consider how this phenomenon
observed in a simple metazoan is related to examples of transdifferentiation, or metastable differentiation, in more complex animals.

II. Growth Dynamics

A. BODY PLAN AND CELL COMPOSITION

The structure of a hydra is very simple. The body column is essentially a tube composed of two epithelial layers surrounding a gastric
cavity (Fig. 1). The outer layer, the ectoderm, and the inner layer, the
endoderm, are separated by a basement membrane termed the meso-

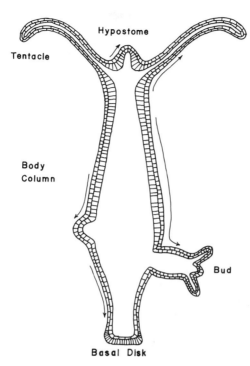

Fig. 1. Longitudinal section of an adult hydra, showing the bilayer construction. Regions of the animal are named, and the pattern of tissue movements (modified from Campbell, 1967b) is indicated by the arrows.

glea. At the apical end, the tube tapers into a dome-shaped structure called the hypostome, and below the hypostome is a whorl of tentacles. The hypostome and tentacles, which together comprise the head, have the same two-layered construction found in the body column. At the basal end of the column, the epithelial layers converge to seal off the cavity, leaving a small opening referred to as the aboral pore. This region is the basal disk or foot of the animal.

In keeping with the simple body plan, the number of cell types is few, about 15. These can be placed in three groups, each containing a stem cell, or a cell with stem-cell-like qualities, that gives rise to the other cells within the group. Two of these groups make up the cells of the epithelia, one for each layer. The ectodermal epithelium is constructed from three cell types, while there is only one well-defined cell type of the endodermal epithelium. The remaining cells reside among the epithelial cells of either layer and are all derived from the interstitial cells, a population of multipotent stem cells (e.g., Bode and David, 1978). The differentiated cells include nematocytes, nerve cells, gland cells, and mucous cells (Slautterback and Fawcett, 1958; Davis, 1972). When the animal is in a sexual state, the gametes are also derived from the interstitial cells (Brien, 1961).

B. PATTERNS OF GROWTH

Of importance for the growth dynamics is the cell division behavior of the epithelial cells of the two layers. The epithelial cells of the body column are continuously in the mitotic cycle (Campbell, 1965, 1967a). When the animals are well fed, the epithelial cells have a cell cycle time of 3–4 days (David and Campbell, 1972). The rate of cell division is lower in the hypostome, and drops to zero in the tentacles and the basal disk (Campbell, 1967a). As 75% of the epithelial cells are in the body column (Bode *et al.*, 1973), the mass of the animal doubles in less than 5 days (David and Campbell, 1972). Despite the continual production of epithelial cells, the size of the adult does not increase. This is the result of a steady state in which the production of new epithelial cells in the body column is balanced by their loss elsewhere (Brien and Reniers-Decoen, 1949; Burnett, 1961; Campbell, 1967b). Campbell (1967b) showed that about 85% of the loss is due to displacement of epithelial cells into developing buds (hydra reproduce asexually by budding). The remainder are sloughed at the extremities of the tentacles, hypostome, and basal disk. Recent evidence indicates that small numbers of epithelial cells can also be lost directly from the body column (Bosch and David, 1984).

One effect of these growth dynamics is that the axial locations of all epithelial cells are constantly changing. Cells in the upper part of the body column are continuously being displaced upward into the head, while those below are displaced downward onto a developing bud, or into the foot. This was originally demonstrated by Tripp (1928) and recently in more detail by Campbell (1967b,c, 1973). A summary of the pathway of displacement in all parts of the animal is shown in Fig. 1.

Whether an individual epithelial cell is displaced up or down the column depends on its axial location in an obvious way. As Campbell (1967b) demonstrated, the border demarking the different displacement directions is a consequence of the geometry of the situation. If tissue is lost at equal rates at the upper and lower ends, the border is in the middle of the column. If more tissue is lost at the lower end, which happens in budding animals, the border is shifted in an apical direction. This border exists only in a mathematical sense. There is nothing remarkable or unique about the epithelial cells or tissue at the border.

Being able to shift the border up or down the column by simply altering the feeding rate (Otto and Campbell, 1977) emphasizes this point.

Thus, the structure and size of a hydra are constant as in other animals. However, unlike them, the tissues are constantly growing and all the cells are continually changing their location within the animal.

III. Epitheliomuscular Cells

A. EPITHELIAL CELL TYPES OF THE ECTODERM

One example in which the constant change of axial position has an influence on the differentiated state of a cell involves the three epithelial cell types of the ectoderm. Each of the cell types is located in a particular region (tentacles, body column, or basal disk) of the animal, and the two in the extremities arise by change in differentiation from the third in the body column. To fully appreciate this example of metastability, each of the cell types and the sequence of changes that occur will be described in detail.

The ectodermal epithelial cells of the body column are termed epitheliomuscular cells. They are cuboidal to columnar in shape (Fig. 2B) and have a cytoplasmic vacuole that occupies at least 90% of the cell volume (R. D. Campbell, personal communication). At the basal end of the cell, they have muscle processes which extend away from the cell body parallel to the body axis and along the mesoglea. The epitheliomuscular cells have several functions. First, they form an epidermis, a barrier to the environment. Second, the muscle processes provide the

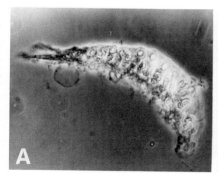
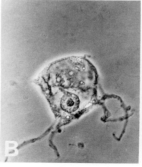
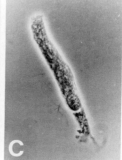

FIG. 2. Isolated ectodermal epithelial cells obtained by macerating hydra and viewed with phase microscopy: (A) battery cell; (B) epitheliomuscular cell; (C) basal disk gland cell. Each cell is oriented so that the side facing the mesoglea would be at the bottom side of the cell. (From Campbell and Bode, 1984.) ×750.

mechanism by which the body column can contract, thereby reducing its length. Third, the large cytoplasmic vacuole might function in osmoregulation (Koblick and Yu-Tu, 1965). The epithelium also conducts electrical signals (Campbell *et al.*, 1976). Finally, the epitheliomuscular cells have the quality of stem cells in that they are continually in the mitotic cycle (Campbell, 1965; David and Campbell, 1972), thereby producing the epithelial cells for the entire ectoderm.

On the tentacles, the ectodermal epithelial cells are referred to as battery cells. They are similar to, yet differ from, the ectodermal epithelial cells of the body column in several respects. They are similar having muscle processes that extend parallel to the tentacle axis, and they have a large cytoplasmic vacuole. The battery cells differ from the body epithelial cells in that they are larger and much more squamous (Fig. 2A), and in that they do not undergo cell division (Campbell, 1967a). The most striking difference is the presence of 10–20 nematocytes embedded in each battery cell. The nematocytes are completely surrounded by the battery cell, but are not within the cytoplasm of the cell. Instead, the battery cell has the shape of a multiholed doughnut, and each nematocyte is nestled in one of the holes. Every nematocyte has a hair, or cnidocil, which serves as a trigger for its discharge. The many cnidocils protrude beyond the epithelial cell into the watery environment, giving it the appearance of a battery of guns; hence, the name. These battery cells function primarily as an anchorage for the nematocytes, which are involved in the capture of prey (Ewer, 1947).

At the basal end of the animal, the cells of the basal disk are also distinctly different from those of the body column. These cells, termed the basal disk gland cells (Campbell and Bode, 1984), are smaller, more elongate or cone-shaped cells (Fig. 2C) filled with granules. The cytoplasmic vacuole is absent or greatly reduced, and they lack or have very attenuated muscle processes. These cells secrete copious amounts of mucus, which serves to attach the animal to substrates in its freshwater environment.

B. Changes in the Differentiated State of Epitheliomuscular Cells

As epitheliomuscular cells are continuously displaced from the body column onto either extremity, they must undergo a change into the cell type of that region. At the lower end the transition from epitheliomuscular cell to basal disk gland cell is abrupt and dramatic. As seen in the scanning electron micrographs (Fig. 3A and B) the border between the lower end of the body column (the peduncle) and the basal disk is the space between neighboring cells. The last cell on the body column has the characteristics of an epitheliomuscular cell, while its immediate basal neighbor is a basal disk gland cell. Thus, as body column cells are displaced across the border, the large vacuole is lost or reduced significantly, thereby reducing the size of the cell. An accompanying effect is the change from the flat, smooth surface of the cells of the peduncle to a surface covered with microvilli on the basal gland disk cells (Fig. 3B). In addition, large numbers of granules are produced, and mucus secretion begins in earnest.

These morphological changes are accompanied by equally abrupt changes in the molecular composition of the cells, which we have detected with monoclonal antibodies. The antibody, TS12, binds an antigen on the surface of the epitheliomuscular cells along the entire length of the body column, but does not bind to cells of the basal disk (Fig. 3C). Conversely, other antibodies recognize antigens on cells of the basal disk, but none on the epithelial cells of the peduncle. DB17, which recognizes a cell-surface antigen, is one of them (Fig. 3D). In both cases, the concentration change in the antigen is very large as the cells are displaced basally across the border.

At the apical end of the body column epitheliomuscular cells displaced onto the tentacles are converted into battery cells. Unlike the transitions at the basal end, not all the changes in the differentiated state occur at once. The tentacle/body column boundary is not detectable in terms of morphological differences between the cells, but is simply defined as the location (a geometrical ring) where the tentacle joins the body column. The epithelial cells on either side of the boundary are morphologically very similar. They are columnar, have a vacuole, and are devoid of embedded nematocytes. However, as an epithelial cell is progressively displaced along the tentacle toward the tip, it becomes increasingly squamous, and acquires its complement of nematocytes. The transition to a battery cell is complete by the time the epitheliomuscular cell has traversed one-quarter to one-third of the length of the tentacle (Bode and Flick, 1976).

In contrast to these gradual morphological changes, some of the antigenic changes are as abrupt as those at the peduncle/basal disk border. The same antibody, TS12, whose binding vanishes at the basal disk border also ceases to bind cells at the body column/tentacle boundary. The last cell on the body column is brightly labeled, using indirect immunofluorescence (Fig. 4C and D), while its immediate neighbor, the first cell on the tentacle, is unlabeled. Conversely, another antibody, TS19, binds an antigen that is in high concentration on the first cell of the tentacle but is not detectable on the next cell, which is the last cell still on the body column (Fig. 4A and B).

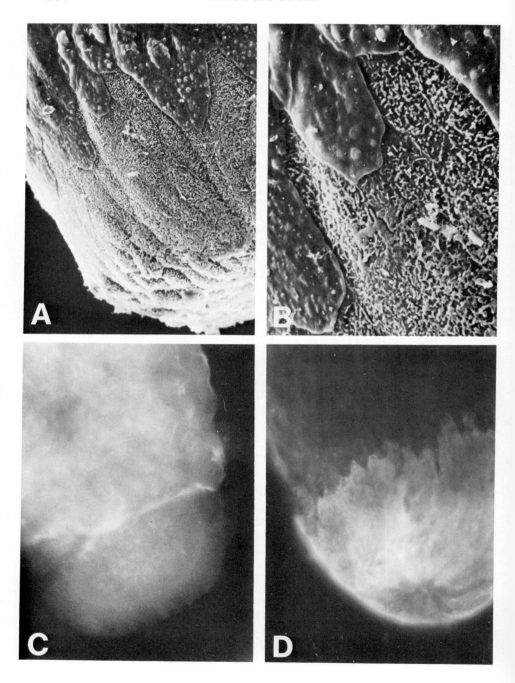

Of the epithelial cells displaced in an apical direction, some end up on the hypostome instead of on the tentacles. The morphological differences between these cells and those on the body column are not dramatic; the cells in the hypostome are smaller and have a smaller vacuole. One might not necessarily consider them different cell types. However, there are antigenic differences. One antibody, TS12, does not bind to ectodermal epithelial cells of the hypostome, in contrast to its aforementioned binding to epitheliomuscular cells. Another antibody, CPB, stains granules near the apical surface of the ectodermal epithelial cells of the hypostome and tentacles, but not of the body column (Fig. 5). The density of labeled granules is very high throughout the hypostome, but drops steeply across the ring of tentacles so that the epithelial cells below the ring at the top of the body column are essentially devoid of such granules.

Thus, every epitheliomuscular cell that leaves the body column by displacement into either the head or foot undergoes morphological and antigenic changes. That these changes occur in response to a change in position and is not just a maturing process that is complete as the epithelial cell leaves the body column is evident from the following consideration. This is the case since epitheliomuscular cells from any axial level of the body column can be rapidly converted into epithelial cells of either extremity. For example, bisection of an animal anywhere along the body column results in the regeneration of a head at the apical end of the lower half with the corresponding conversion of many epitheliomuscular cells into battery cells. Similarly, in a regenerating basal disk, epitheliomuscular cells form basal disk gland cells very rapidly.

The transition from a epitheliomuscular cell to either a battery cell or a basal disk gland cell appears to be a change in the differentiated state of the epithelial cell, or a transdifferentiation. However, for two reasons it could be argued that the transitions are not true switches in differentiated state.

One reason is based on the relationship among the cell types. This group of epithelial cells has many characteristics of the cells of a vertebrate epidermis. In that tissue a layer of stem cells continually

---

FIG. 3. The body column/basal disk boundary as visualized with scanning electron microscopy (A and B) and monoclonal antibodies (C and D). In (C) TS12 binds only to the epitheliomuscular cells, while in (D) DB17 binds only to the basal disk gland cells. Visualization of the antibodies is with indirect immunofluorescence on whole mounts of adult hydra. The basal disk is toward the bottom in all photos. (A) ×900; (B)×2600; (C and D) ×300.

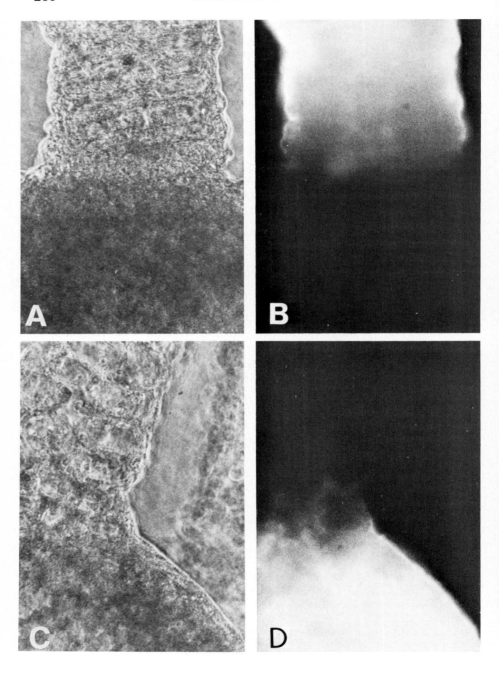

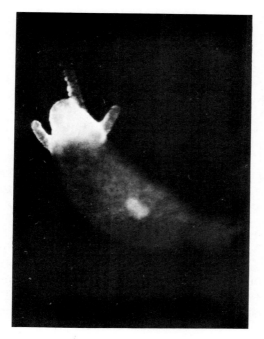

FIG. 5. A whole mount of a young hydra stained with the monoclonal antibody CP8. ×55.

gives rise to epidermal cells which no longer undergo mitosis, are terminally differentiated, and are short-lived. Analogously, the epitheliomuscular cells are constantly in the mitotic cycle and give rise to the short-lived, nondividing battery cells and basal disk gland cells. However, there are significant differences. The transition in the vertebrate epidermis involves the change from an undifferentiated cell, the stem cell, to a differentiated cell, the epidermal cell. Hence, this is simply the process of a cell traversing a differentiation pathway, and not a switch from one differentiated state to another. The situation is different in the ectoderm of hydra. The epitheliomuscular cell has stem-cell-like qualities in that it is always in the mitotic cycle and

FIG. 4. Visualization of the body column/tentacle boundary with two monoclonal antibodies. Phase (A) and fluorescent (B) photographs of the tentacle/body column boundary stained with TS19, which binds to epithelial cells only on the tentacles. Phase (C) and fluorescent (D) photographs of the same region stained with TS12, which binds only to cells on the body column. Visualization of the antibodies as in Fig. 6. ×600.

gives rise to the other cell types. But, the epitheliomuscular cell is not undifferentiated. It is fully differentiated with specific physiological functions in the animal. Hence, the transition is from one cell type with distinct functions, the epitheliomuscular cell, to a second cell type with other functions, the battery cell or basal disk gland cell. We consider this kind of change to be a transdifferentiation.

The second argument is that the changes are more quantitative than qualitative in nature. Epitheliomuscular cells do share some characteristics with battery cells and basal disk gland cells, such as epidermal and muscle cell functions. Further, one out of every five epithelial cells of the body column has a single nematocyte embedded within it (Bode and Flick, 1976). Embedded nematocytes are the distinguishing morphological feature of a battery cell. Similarly, epitheliomuscular cells have small numbers of granules that are characteristic of basal gland cells and secrete small amounts of material that contribute to an amorphous periderm surrounding the animal.

Nevertheless, there are sufficient differences to warrant calling the transitions a change in differentiated state. Two of the four types of nematocytes are necessary for the capture of prey (Ewer, 1947). Both are found in every battery cell, while only one type is found on the body column, and then only in 10% of the epitheliomuscular cells (Bode and Flick, 1976). Hence, the epithelial cells of the body column could not carry out the function of those on the tentacles. The basal disk gland cells have lost the large vacuole, and more prominently, they secrete mucus in quantities sufficiently large to allow the basal disk to serve as a holdfast. The amounts secreted by the epitheliomuscular cell are much too small to serve this purpose. In addition to the differences in function which are reflected in quantitative as well as qualitative differences in the structure of the three cell types, there are the aforementioned differences in the antigenicity of the cells.

In summary, it is reasonable to consider the transitions of the epitheliomuscular cells to be changes in its differentiated state. As any epitheliomuscular cell is capable of these switches at any time, the differentiated state of this cell type is, at most, metastable. Whether the differentiated state of the battery cells and basal disk gland cells is fixed, or also only metastable, is not known.

IV. The Nervous System

A. STRUCTURE AND DYNAMICS OF THE NERVE NET

A second example of change in state of differentiation of a cell is found in the nervous system. Since growth dynamics again play an important role, they will be considered in detail in order to put the differentiation behavior in perspective. As with other aspects of hydra, the structure of the nervous system is quite simple.

Ganglion cells are scattered throughout the animal (Hadzi, 1909; Semal-Van Gansen, 1952). The cell bodies lie close to the mesoglea (Hadzi, 1909; Lentz and Barrnett, 1965) with most located on the ectodermal side, and a smaller number on the endodermal side (Davis, 1972; Epp and Tardent, 1978). Neuronal processes extending from the cells along, or across, the mesoglea make contact with one another so that the entire population of nerve cells forms a loose net throughout the animal. A part of the net in the peduncle, stained with the monoclonal antibody, RC14, is shown in Fig. 6. The "mesh size" of the net is not uniform throughout the animal, as the density of cell bodies and processes is higher in the hypostome, tentacles, and basal disk compared to the body column. There is on average one nerve cell for every six epithelial cells in the body column, while the ratio is closer to 1:1 in the tentacles and basal disk and about 2:1 in the hypostome (Bode et al., 1973).

There are also sensory nerve cells whose cell bodies lie between the

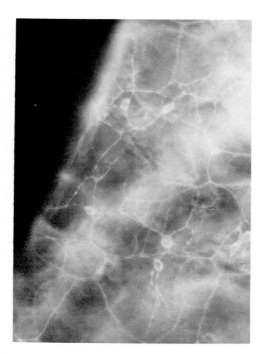

Fig. 6. Part of the nerve net of the peduncle, stained with the monoclonal antibody, RC14. ×400.

lateral surfaces of epithelial cells throughout the animal (Hadzi, 1909). Those in the ectoderm of the tentacles and hypostome, termed epidermal sensory cells (Westfall and Kinnamon, 1978), extend from the basal to the apical surface of the epithelium. In the body column the cell body does not reach the apical surface. Instead, a process extends from the apical end of the cell and protrudes toward, and sometimes into, the external environment (Lentz and Barrnett, 1965). Processes extending from the basal end of the sensory cells connect them into the net.

The nerve net is not a static structure. Since the epithelial layers within which it is embedded are constantly growing, the nerve net must grow in a corresponding fashion. This is supported by the fact that the nerve density is constant in this steady-state system (Bode *et al.*, 1973). The growth of the nerve cell population has also been directly demonstrated. Nerve cells are not capable of cell division, but arise by differentiation from interstitial cells (Campbell, 1967a; David and Gierer, 1974; Yaross and Bode, 1978). By continuously labeling hydra with [^{3}H]thymidine, three-quarters of the population of nerve cells in a body column becomes labeled within 2 weeks (David and Gierer, 1974). Thus, as the number of epithelial cells increases by cell division, new nerve cells are continuously inserted into the net.

Since the number of nerve cells per animal is a constant, neurons must be constantly lost from the animal to balance their continuous production. The growth dynamics of the epithelia provide a mechanism for such loss. As body column tissue is displaced up or down into an extremity, it is likely that the corresponding region of the nerve net is also displaced. As epithelial cells are sloughed at the tips of the extremities, or are displaced in sheets into a developing bud, the associated nerve cells are similarly lost from the body column. The implication is that each nerve cell, like the epithelial cells, is continuously changing its position on the body column by displacement. In one instance this has been directly demonstrated.

The epidermal sensory cells are restricted to the ectoderm of the hypostome and tentacles (Westfall and Kinnamon, 1978). They arise by differentiation from interstitial cells, since elimination of the interstitial cells by hydroxyurea resulted in the absence of any new sensory cells (Yaross *et al.*, 1985). The treatment left the existing nerve net and tissue dynamics unaffected. Thus, these animals presented us with an opportunity to examine the displacement behavior of a cohort of epidermal sensory cells and their associated battery cells. The cohort could be studied without the complications of the continual addition of new sensory cells. Since new sensory cells are continually added at the base of the tentacle, they would obscure the size and location of a particular cohort.

Epidermal sensory cells can be specifically labeled with the monoclonal antibody, JD1, and the entire pattern of stained cells visualized in a whole mount (Dunne et al., 1985). Stained sensory cells of the hypostome are shown in Fig. 7A, while the pattern of JD1-positive nerve cells on part of the tentacle is shown in Fig. 7B. Hence, the location of the cohort of sensory cells could be determined at any time after treatment, by staining animals with JD1 using indirect immunofluorescence. The displacement of the battery cells was followed by marking the ectoderm of the base of a tentacle with India ink (Campbell, 1973).

Shortly after treatment the patterns were the same as in untreated animals. The marked battery cells were at the base, and JD1-positive cells were found along the entire length of a tentacle. Five days later, the marked battery cells had been displaced from the base to a point halfway up the tentacle. Correspondingly, the outer half of the tentacle contained JD1-positive cells, but the inner half did not. This indicated that the cohort of sensory cells had been displaced along the tentacle toward the tip, and that many had been sloughed. By day 10, only a few JD1-positive cells remained, and these were located at the tip of the tentacle (Fig. 7C and D). The marked battery cells had been displaced equally far (Yaross et al., 1985). As the JD1-positive nerve cells were clearly shifted with the epithelial cells in this experiment, it is likely that the remainder of the net undergoes comparable movements.

B. NEURONAL PLASTICITY

Since individual neurons are continually changing their axial location, the question arises as to whether a given neuron maintains the same differentiated state throughout its existence. Until recently this has been difficult to approach since specific types of neurons were not well defined. Distinctions based on morphology or ultrastructure were not always clear-cut, leading to estimates ranging from 1 to 11 types of nerve cells (Davis et al., 1968; Westfall, 1973; Epp and Tardent, 1978). In addition, for the few that are well defined there was no convenient assay to probe the question.

Using antisera to a number of neuropeptides, Grimmelikhuijzen and colleagues provided a new approach (reviewd by Grimmelikhuijzen, 1984). They found that each antiserum bound a subset of the nerve cells in the net that was restricted to one or two specific regions of the animal. For example, the antiserum to cholecystokinin bound to

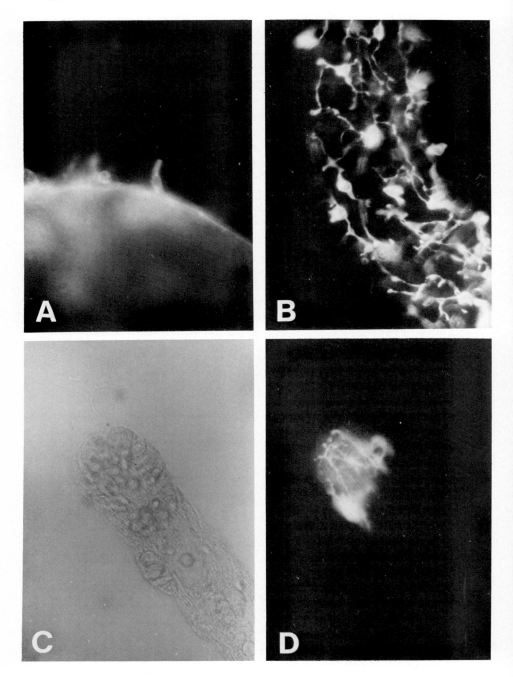

nerves in the hypostome (Grimmelikhuijzen *et al.*, 1980), while that against substance P bound to nerve cells in the tentacles and basal disk (Grimmelikhuijzen *et al.*, 1981). Five of the six subsets were restricted to the extremities or the lower peduncle. The last, a subset that bound an antiserum to oxytocin/vasopressin, was distributed throughout the body, but clearly did not bind all neurons (Grimmelikhuijzen, 1984). For the most part the subsets bound by any two of the antisera do not appear to be identical although some overlap may exist.

More recently, we have isolated a number of monoclonal antibodies that also bind to subsets of nerves. One subset, defined by JD1 as described above, is in the head and consists almost entirely of epidermal sensory cells (Dunne *et al.*, 1985). Several others, including RC14, bind some of the ganglion cells of the peduncle, tentacles, and hypostome. Although only preliminary analyses have been carried out, none of the subsets defined by the antisera to neuropeptides appear to be equivalent to those identified with the monoclonal antibodies.

With these kinds of reagents a clearer definition of types of neurons can be established, and the question of the stability of the differentiated state of a neuron can be addressed with greater precision. As an individual neuron is continuously changing its location, does it change the subset to which it belongs? For example, there are no neurons with FRMFamide-like immunoreactivity (FLI) in the body column, but there are large numbers of them in the hypostome, tentacles, and lower peduncle (Fig. 8A and C, and Grimmelikhuijzen *et al.*, 1982a). As the nerve net at the upper end of the body column is displaced into the head, FLI-positive neurons appear in the net. Is this due to FLI-negative neurons switching subsets and becoming part of the FLI-positive subset? If so, it would indicate that a neuron can shift from one type to another. There are, however two alternate explanations.

In one, the nerve cells of the body column could be considered as immature, or not having fully differentiated, and they would express the neuropeptide only upon maturation in the head. This is unlikely for three reasons. The ganglion and sensory cells of the nerve net in the body column have the morphology and ultrastructure of fully ma-

---

FIG. 7. Binding characteristics of the monoclonal antibody JD1. (A) Epidermal sensory cells of the hypostome. The cell bodies are between the epithelial cells of the ectoderm, which are not visible in the fluorescent photograph. (B) Binding pattern of sensory cells and their processes in a portion of a normal tentacle. Phase (C) and fluorescent (D) images of a tentacle tip in animals in which the nerve precursors had been removed 10 days earlier. Visualization of the antibodies was by indirect immunofluorescence on whole mounts. ×480.

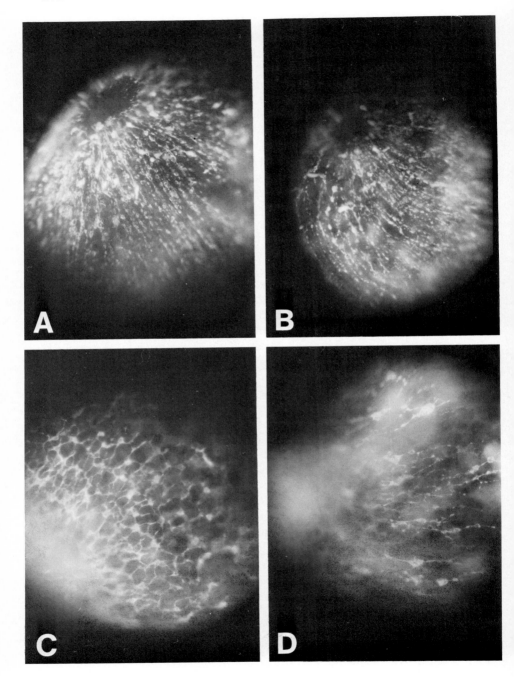

ture nerve cells (e.g., Davis *et al.*, 1968). Also, the net is functional in coordinating the movements of the body column as well as responding to external stimuli. Finally, one subset of body column neurons apparently contains oxytocin- or vasopressin-like immunoreactivity (Grimmelikhuijzen *et al.*, 1982b), indicating these cells are sufficiently mature to express a neuropeptide.

Another possibility is that the FLI-positive neurons arise directly by differentiation from interstitial cells in the head and peduncle. This possibility was tested in the following way. Exposure to hydroxyurea or nitrogen mustard removed all the interstitial cells and nerve differentiation intermediates from 60 to 90% of the treated animals (Koizumi *et al.*, 1985). The remainder contained only 1–10 interstitial cells each. As described above, the nerve net was unaffected by this treatment. The head or peduncle of such animals was removed, and the animals allowed to regenerate. Once regeneration was complete, the animals were stained as whole mounts with the antiserum to FMRFamide. More than 90% of the animals regenerating a peduncle showed the typical FLI-positive staining pattern (Fig. 8D), as did 80–90% of the regenerated heads (Fig. 8B; Koizumi *et al.*, 1985). Although >50% of the animals had no interstitial cells to differentiate into new nerves, FLI-positive cells still appeared. We conclude that the FLI-positive neurons arose by conversion from FLI-negative neurons. As animals totally devoid of interstitial cells can be maintained for periods of weeks or months and remain nerve free, the possibility that the FLI-positive cells arose by conversion from another cell type can be excluded (Marcum and Campbell, 1978). Whether the plasticity in expression of neuropeptides is a general phenomenon in hydra is under study.

It is plausible that the observed neuronal plasticity is a response due to the experimental intervention of decapitation and the subsequent regeneration. The following consideration and experiment suggest this is not so. If neurons were continuously converted from FLI-negative to FLI-positive cells in the intact animal, one would expect the regional staining pattern to remain unchanged in the absence of new nerve differentiation. If new differentiation were required to produce FLI-positive cells in the head, then with time after removal of the nerve precursors one would expect to see the changing staining

---

FIG. 8. Distribution of nerve cells exhibiting FMRFamide-like immunoreactivity in whole mounts. (A) Hypostome and (C) peduncle of normal hydra; (B) hypostome and (D) peduncle of regenerated hydra that were devoid of nerve precursors before amputation. ×200.

pattern described above for the JD1-positive sensory cells. The stained cells would decrease in number and be found progressively farther and farther from the base of the tentacle. The experiment using JD1 indicated that a cell traverses the length of the tentacle of a hydroxyurea- or nitrogen mustard-treated animal in about 8–10 days.

To determine the behavior of neurons with respect to FLI during tissue movements, animals devoid of nerve precursors were maintained and stained 13 or 20 days after treatment. The FLI-staining pattern was the same as in the normal animal, although the number of FLI-positive nerve cells per animal was somewhat fewer (Koizumi *et al.*, 1985). In this period of time the complement of nerve cells on the tentacle was replaced at least once, indicating that FLI-negative neurons of the body column are continuously converted into FLI-positive cells. Thus, the switch occurs in intact as well as regenerating animals.

Not all subsets of neurons in the head arise by conversion from subsets of the body column displaced in an apical direction. There are two arguments for thinking that a large number must arise by differentiation in the hypostome. One is simply arithmetic. The density of nerve cells, as measured in number of neurons per epithelial cell, is 6-fold higher in the head than in the body column (Bode *et al.*, 1973). The increase in density must be due to the production of new nerve cells as there is no reason to believe that a major loss of epithelial cells, a possible alternative explanation, occurs.

The other evidence for the direct differentiation of particular subsets of neurons concerns the subset defined by JD1. Removal of the nerve precursors results in the absence of new JD1-positive cells in newly formed tentacle tissue as described above. Further, if animals devoid of nerve precursors are decapitated and allowed to regenerate, they do not form JD1-positive cells (Yaross *et al.*, 1985). Hence, none of the body column JD1-negative cells were converted into JD1-positive cells when they became part of the head during tissue reorganization.

The evidence accumulated so far suggests that some neurons are capable of switching subsets, whereby a subset is defined by the binding of antiserum or monoclonal antibody. Whether this is a sufficient definition for different types of nerve cells is probably a matter of taste. Switching a neurotransmitter substance could be considered modulation within a cell type, or it might be considered a switch in differentiated phenotype, as the physiology of the cell has changed. In terms of the latter view, and assuming the neuropeptides act as neurotransmitters in hydra, there are neurons in hydra which change their differentiated state as they change their axial location. Because of the growth dynamics of the animal, such changes take place continuously throughout the animal's existence. This implies that for many neurons in hydra the differentiated state is only metastable, which stands in contrast to the conventional view that the differentiated state of a neuron is fixed in adult animals.

V. Conclusion

There are at least two populations of cells in hydra which always contain cells that are undergoing changes in their differentiated state: the epithelial cells of the ectoderm and some of the nerve cells. In fact, there may be a few more. For example, gland cells are distributed among the epithelial cells of the endoderm in the body column. They occur in large numbers at the apical end of the column, but are absent in the head (Bode et al., 1973). When endodermal tissue is displaced into the head, the gland cells must either be lost from the tissue or differentiate into another cell type. A reasonable possibility is that they are converted into mucous cells which exist in large numbers in the hypostome, but only in very small numbers in the body column (Bode et al., 1973; Wanek, personal communication).

The extent of the changes a cell undergoes when switching from one differentiated state to another varies in the systems studied. Some changes, as in Wolffian lens regeneration, are large and dramatic. A pigmented iris epithelial cell is converted into a cell type, the lens cell, that has a completely unrelated function and is of different embryological origin. Other transitions are more restricted in that cell types within a class are converted into one another. The conversion of one connective tissue cell type, the fibroblast of the dermis, into another, the chondroblast, during limb regeneration in newts is an example. The changes in differentiated state in hydra are analogous to the latter instance. The epitheliomuscular cell is converted into another type of epithelial cell, while the switches in the nerve cells are from one subset of neurons to another. More extreme switches involving a conversion from one class to another, for example, an epithelial cell type changing into a cell of the interstitial cell system, has not been observed (Sugiyama and Fujisawa, 1978; Marcum and Campbell, 1978).

That transitions from one differentiated state to another take place continuously in an adult animal is unusual. One reason this occurs in hydra and infrequently elsewhere is the constant change of location of every cell in a hydra. Similar changes occur in only a few tissues in vertebrates, such as in the intestinal villi of mammals. Cells are continuously displaced but these movements are not known to be accompanied by switches in cell type.

Another reason has to do with hydra being a primitive organism of simple construction. Although it has a small number of cell types, the animal must carry out many of the same functions that more complicated metazoans perform using many more cell types. For example, the epitheliomuscular cell serves as an epidermal cell, a muscle cell, and a stem cell for the other ectodermal epithelial cells in the tentacles and basal disk, and it conducts electrical signals. In higher organisms these several functions are usually each allotted to a different cell type.

Is the metastability of the differentiated state observed for the hydra cell types described here something peculiar to this primitive metazoan? It is plausible, but there is more reason to believe that there is a great deal of common ground concerning the stability of the differentiated state between hydra and vertebrates. In higher animals most cells do not normally alter their position within a tissue or organ. Hence, their differentiated state appears stable or fixed. Where metaplasia or transdifferentiation has been demonstrated, it always involves one of two conditions. In one, the position of some cells change within a tissue or organ. For example, during the regeneration of a newt limb, dermal cells that become part of the blastema are often shuffled to more interior locations and are converted into chondrocytes. In the second, removal of a tissue or organ can cause neighboring cells to alter their differentiated state to compensate for the missing structures. In *Podocoryne* when the striated muscle cells lose all their usual neighbors, many are converted into the missing cell types.

These are the same two conditions in which a switch in cell type occurs in hydra: change in position or a change in the cell composition of the immediate environment. Removal of the head is an example of the latter. The apical end of the remaining animal reorganizes into a head which is accompanied by the rapid conversion of epithelial cells and neurons typical of the body column into related cell types of the head. The only difference between hydra and most other animals is that in the latter a cell can only change its position by experimental intervention, whereas in hydra it also occurs normally.

As the conditions constraining the stability of the differentiated state are similar in hydra as elsewhere, the behavior of the ectodermal epithelial cells and neurons in hydra serves to demonstrate the normal variation in the stability of the differentiated state in organisms. In many cell types the differentiated state is clearly fixed and probably unalterable. An extreme example is a red blood cell that has lost its nucleus. However, the number of cell types in which the state of differentiation is only metastable may be higher than suspected. In fact the degree of finality of the differentiated state of a cell in an adult organism may vary from completely undifferentiated, as in a stem cell, to the terminally differentiated state. A condition of metastability may be sufficient for most cell types since they do not alter their location during the lifetime of the organism.

ACKNOWLEDGMENTS

We thank Patricia Bode, Lynne Littlefield, and Richard Campbell for their critical reading of the manuscript, and Margaret Chow for some of the photography. Some of the research described was supported by research grants from the National Institutes of Health (HD 08086 and HD 16440) to H. B., by fellowships (American Cancer Society: PF 21410 and NIH: GM 08513) to L.J., and by a research grant from the National Science Foundation (PCM-83-02581) to M.Y.

REFERENCES

Bode, H. R., and David, C. N. (1978). *Prog. Biophys. Mol. Biol.* **33**, 189–206.
Bode, H. R., and Flick, K. M. (1976). *J. Cell Sci.* **21**, 15–34.
Bode, H., Berking, S., David, C., Gierer, A., Schaller, H., and Trenkner, E. (1973). *Wilhelm Roux's Arch. Entwicklungsmech. Org.* **181**, 269–285.
Bosch, T., and David, C. N. (1984). *Dev. Biol.* **104**, 161–171.
Brien, P. (1961). *Bull. Biol. Fr. Belg.* **95**, 301–364.
Brien, P., and Reniers-Decoen, M. (1949). *Bull. Biol. Fr. Belg.* **83**, 293–386.
Burnett, A. L. (1961). *J. Exp. Zool.* **146**, 21–83.
Campbell, R. D. (1965). *Science* **148**, 1231–1232.
Campbell, R. D. (1967a). *Dev. Biol.* **15**, 487–502.
Campbell, R. D. (1967b). *J. Morphol.* **121**, 19–28.
Campbell, R. D. (1967c). *J. Exp. Zool.* **164**, 379–392.
Campbell, R. D. (1973). *J. Cell Sci.* **13**, 651–661.
Campbell, R. D., and Bode, H. R. (1984). *In* "Hydra: Research Methods" (H. M. Lenhoff, ed.), pp. 5–14. Plenum, New York.
Campbell, R. D., Josephson, R. K., Schwab, W. E., and Rushforth, N. B. (1976). *Nature (London)* **262**, 388–390.
David, C. N., and Campbell, R. D. (1972). *J. Cell Sci.* **11**, 557–568.
David, C. N., and Gierer, A. (1974). *J. Cell Sci.* **16**, 359–375.
Davis, L. E. (1972). *Z. Zellforsch.* **123**, 1–17.
Davis, L. E., Burnett, A. L., and Haynes, J. F. (1968). *J. Exp. Zool.* **167**, 295–332.
Dunis, D. A., and Namenwirth, M. (1977). *Dev. Biol.* **56**, 97–109.
Dunne, J. F., Javois, L. C., Huang, L. W., and Bode, H. R. (1985). *Dev. Biol.* **109**, 41–53.
Eguchi, G. (1964). *Embryologia (Nagoya)* **8**, 247–287.
Epp, L., and Tardent, P. (1978). *Wilhelm Roux's Arch.* **185**, 185–193.
Ewer, R. F. (1947). *Proc. Zool. Soc. London* **117**, 365–376.
Grimmelikhuijzen, C. J. P. (1984). *In* "Evolution and Tumour Pathology of the Neuroendocrine System" (S. Falkmer, R. Hakanson, and F. Sundler, eds.), pp. 39–58. Elsevier, Amsterdam.
Grimmelikhuijzen, C. J. P., Sundler, F., and Rehfeld, J. F. (1980). *Histochemistry* **69**, 61–68.
Grimmelikhuijzen, C. J. P., Baife, A., Emson, P. C., Powell, D., and Sundler, F. (1981). *Histochemistry* **71**, 325–333.

Grimmelikhuijzen, C. J. P., Dockray, G. J., and Schot, L. P. C. (1982a). *Histochemistry* **73**, 499–508.

Grimmelikhuijzen, C. J. P., Dierickx, K., and Boer, G. J. (1982b). *Neuroscience* **7**, 3191–3199.

Hadzi, J. (1909). *Arb. Zool. Inst. Univ. Wien* **17**, 225–268.

Koblick, D. C., and Yu-tu, L. (1965). *J. Exp. Zool.* **166**, 325–330.

Koizumi, O., and Bode, H. R. (1985). Submitted.

Lentz, T. L., and Barrnett, R. (1965). *Am. Zool.* **5**, 341–356.

Marcum, B. A., and Campbell, R. D. (1978). *J. Cell Sci.* **29**, 17–33.

Namenwirth, M. (1974). *Dev. Biol.* **41**, 42–56.

Otto, J. J., and Campbell, R. D. (1977). *J. Cell Sci.* **28**, 117–132.

Semal van Gansen, P. (1952). *Acad. R. Belg. Bull. Classe Sci.* **38**, 718–735.

Slautterback, D. B., and Fawcett, D. W. (1954). *J. Biophys. Biochem. Cytol.* **5**, 441–452.

Steward, F. C., Kent, A. E., and Mapes, M. O. (1966). *Curr. Top. Dev. Biol.* **1**, 113–154.

Sugiyama, T., and Fujisawa, T. (1978). *J. Cell Sci.* **29**, 35–52.

Takebe, I., Labib, G., and Melchers, G. (1971). *Naturwissenschaften* **58**, 318–320.

Tripp, K. (1928). *Z. Wiss. Zool.* **132**, 476–525.

Westfall, J. A. (1973). *J. Ultrastruct. Res.* **42**, 268–282.

Westfall, J. A., and Kinnamon, J. C. (1978). *J. Neurocytol.* **7**, 365–379.

Yamada, T. (1967). *Curr. Top. Dev. Biol.* **2**, 247–283.

Yaross, M. S., and Bode, H. R. (1978). *J. Cell Sci.* **34**, 1–26.

Yaross, M. S., Westerfield, J., Javois, L. C., and Bode, H. R. (1985). Submitted.

CHAPTER 19

NEMATOCYTE DIFFERENTIATION IN HYDRA

Toshitaka Fujisawa, Chiemi Nishimiya, and Tsutomu Sugiyama

DEPARTMENT OF DEVELOPMENTAL GENETICS
NATIONAL INSTITUTE OF GENETICS
MISHIMA, SHIZUOKA-KEN, JAPAN

I. Introduction

In a variety of developmental systems, the fate of cells in differentiation is specified by the position within the system. How cells recognize their relative position and respond with a specific behavior is one of the fundamental problems in developmental biology today.

The differentiation of four types of nematocytes in hydra, which arise from interstitial stem cells, offers an excellent model system to study this subject. These differentiation processes occur in a highly position-dependent manner along the body axis (Bode and Smith, 1977). For example, stenoteles, one of the nematocyte types, are produced predominantly in the proximal regions and in sharply decreasing amounts in more distal regions. In contrast, another type, desmonemes, are produced largely in the distal and middle regions and in gradually decreasing amounts in the more proximal regions.

In this chapter, we summarize the results of our recent studies which were carried out with the aim of understanding the underlying mechanisms responsible for position-dependent nematocyte differentiation in hydra. The first two studies address to the question of whether the position-dependent nematocyte differentiation is correlated in any way to the "developmental gradients." This was examined by comparing the nematocyte differentiation patterns in animals in

281

which the developmental gradients are altered by various means. The next part deals with a biochemical study on the isolation and characterization of a factor which specifically inhibits stenotele differentiation. This factor is present in a high concentration in the head region of hydra, and in gradually decreasing concentrations down the body column. Finally, we discuss the alteration of the normal differentiation pathway of nematocytes. Nematoblasts which normally differentiate into one type of nematocyte can be diverted into the other pathways by treating animals with the stenotele-specific factor just described, or by other means. This alteration of the differentiation pathway occurs near the S/G_2 boundary in the terminal cell cycle prior to the final nematocyte differentiation (Fujisawa and David, 1981, 1982; Fujisawa 1984). This suggests that until this critical period the fate of nematoblasts is undetermined, or alternatively the fate which is already determined can be changed under the influence of their environment (transdifferentiation).

II. Position-Dependent Nematocyte Differentiation

Nematocytes of hydra are stinging cells which are involved in the capture of prey, defense, and locomotion. Based on the distinct morphology of large complex organelles, nematocysts, four types of nematocytes have been characterized in hydra: stenoteles, desmonemes, holotrichous isorhiza, and atrichous isorhiza (Weil, 1934). Since nematocytes are terminally differentiated cells which have no mitotic activity and turn over rapidly, they are constantly supplied from interstitial stem cells by differentiation (Lehn, 1951; Slautterback and Fawcett, 1959). Stem cells entering the nematocyte differentiation pathway undergo two to five rounds of synchronous cell division (Lehn, 1951; Rich and Tardent, 1969). The daughter cells remain connected to each other by cytoplasmic bridges to form clusters or nests (Slautterback and Fawcett, 1959). Following the final cell division, all cells in a nest differentiate synchronously into the same type of nematocytes (Lehn, 1951; Rich and Tardent, 1969).

A striking feature of nematocyte differentiation is its position dependency along the body axis (Fig. 1; Bode and Smith, 1977). For example, desmonemes are produced largely in the distal and middle regions and in gradually decreasing amounts in the more proximal regions. In contrast, stenoteles are produced predominantly in the proximal regions, and in sharply decreasing amounts in the more distal regions of the body column. Head and foot tissues do not support nematocyte differentiation. This regional pattern of nematocyte differentiation may be the result of regional commitment of nematocyte

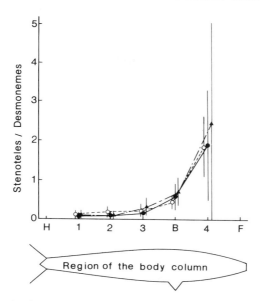

Fig. 1. Changes in the ratio of differentiating stenoteles to desmonemes along the body column. Differentiating nematocytes were stained with thiolacetic acid and lead nitrate (David and Challoner, 1974) and analyzed. The body column of an animal with its first bud protrusion was divided into seven regions from the head (H) to the foot (F), and the ratio of stenoteles to desmonemes in each region was measured. The values for the head and foot regions are not shown since the total numbers of nematocytes in these regions are too low to obtain meaningful values. B designates the budding region. (●) 105; (○) reg-16; (▲) mh-1. Vertical bars represent standard deviations.

precursors to a particular differentiation pathway or the regional migration of committed cells. So far, however, the selected migration of committed cells has not been known. Thus, we will confine our discussion to the question of what nematocyte precursors "measure" to recognize their relative position along the body column and differentiate accordingly.

In view of the parallelism between the gradients of morphogenetic factors and the gradients of nematocyte differentiation along the body column, Bode and David (1978) have suggested that morphogenetic factors are also involved in the regional control of nematocyte differentiation. Hydra tissue exhibits gradients of four types of morphogenetic potentials which are thought to play crucial roles in pattern formation (see Bode and Bode, 1984, for recent review). The head-activation and the head-inhibition potentials are involved in governing head formation, and their levels are both high near the head and decrease monotonically toward the foot, forming gradients down the body column. The foot-activation and the foot-inhibition potentials are involved in governing foot formation, and they exist in gradients in the foot-to-head direction. The levels of these potentials are determined presumably by the concentration of the specific morphogenetic factors present in hydra tissue (Schaller, 1973; Berking, 1977; Schaller *et al.*, 1979). To test the hypothesis of Bode and David (1978) we examined position-dependent nematocyte differentiation in animals in which the levels of morphogenetic potentials were artificially altered by various means. If morphogenetic potentials are indeed involved in the regional control of nematocyte differentiation, one may expect to see alterations in the regional pattern of nematocyte differentiation along the body axis in these animals.

A. Nematocyte Differentiation in Regenerating Animals

One way to artificially alter the levels of morphogenetic potentials is to remove a part of the body and allow the remaining body to regenerate. For example, after head removal the level of the head-inhibition potential falls drastically very soon, and the level of the head-activation potential rises gradually (Webster and Wolpert, 1966; Mac-Williams, 1983a,b). These events occur long before any morphological changes of regeneration process become visible at the tissue level in the regenerating animals.

Fujisawa and David (1981) excised a distal portion of the body column and allowed it to regenerate a whole hydra. They then examined whether the regional pattern of nematocyte differentiation was altered in any way in the regenerating animal. Normally most of nematoblasts present in the distal regions differentiate into desmonemes and rarely into stenoteles (Bode and Smith, 1977). In an isolated distal portion, however, a significant fraction of the nematoblasts was found to have shifted from the desmoneme to the stenotele differentiation pathway.

The kinetics of the appearance of these newly differentiated stenoteles indicated that this shift occurred soon after the isolation of a distal portion long before any morphological changes of regeneration process became visible (see below). Moreover, new stenotele differentiation occurred predominantly in the proximal regions of the regenerating pieces (Fujisawa and David, unpublished results). These results were consistent with the idea that the regional changes in nematocyte differentiation were induced by the changes in the levels of morphogenetic factors and support the hypothesis of Bode and David (1978). These results, however, did not reveal which of the morphogenetic factors were involved (if involved at all) in the regional control of nematocyte differentiation.

B. Nematocyte Differentiation in Mutants Which Show Altered Levels of Morphogenetic Potentials

In the experiments just described we altered the morphogenetic potential levels in normal animals by amputating them, and then examined the effects on nematocyte differentiation. In the next experiment we examined the nematocyte differentiation patterns in mutant strains in which the morphogenetic potential levels were altered by genetic defects.

Sugiyama and Fujisawa (1977a) isolated and characterized a number of mutants defective in developmental processes. Of these, two strains, reg-16 and mh-1, were used for the present studies (Fig. 2). Reg-16 regenerates a head very poorly after removal of the original head (Sugiyama and Fujisawa, 1977a). However, foot regeneration and budding occur normally in this strain. When compared with a wild-type standard strain (105), reg-16 has a significantly lower head-activation potential and a significantly higher head-inhibition potential (Sugiyama and Fujisawa, 1977b; Achermann and Sugiyama, 1984). In contrast, mh-1 forms extra heads outside of the normal budding zone, resulting in the multiheaded morphology (Sugiyama and Fujisawa,

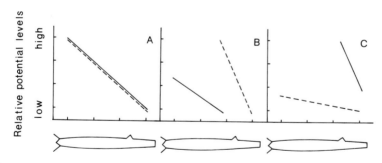

Fig. 2. Schematic representation of the gradients of the head-activation and head-inhibition potentials in 105 (A), reg-16 (B), and mh-1 (C). Strain 105 is used as the standard to express the gradients of the two potentials, employing two straight lines to represent the gradients of this strain. As compared to 105, reg-16 has a significantly lower activation and a significantly higher inhibition potential (Achermann and Sugiyama, 1984), whereas mh-1 has a significantly higher activation and a significantly lower inhibition potential (Sugiyama, 1982). The levels of the potentials in the three strains were compared by lateral grafting of tissue (Webster and Wolpert, 1966), using as hosts animals which had just produced their first bud protrusion after detachment from parents.

1977a). It has a significantly higher head-activation potential and a significantly lower head-inhibition potential than 105 (Sugiyama, 1982). If the morphogenetic factors governing head formation are responsible for position-dependent nematocyte differentiation, the regional pattern of nematocyte differentiation should be altered in some way in these mutants.

Nishimiya et al. (1984) carefully examined the ratios of stenoteles to desmonemes produced in the different regions along the body column in these mutants (Fig. 1). In 105, this ratio was 0.1 near the head and became gradually higher reaching a value of 2.0 in the proximal regions. The ratio is, therefore, a useful and sensitive indicator of position-dependent nematocyte differentiation along the body axis for these two nematocyte types. Comparison of these indicator values along the body axis obtained in the mutants with those obtained in 105 revealed no significant difference which could be attributed to the altered levels of morphogenetic factors. These results suggest that morphogenetic factors involved in the head formation are uninvolved in position-dependent nematocyte differentiation. Instead, some other parameter(s) of axial location not directly associated with those potentials is presumably responsible for the positional influence of interstitial cell differentiation into the nematocytes. One possibility is the morphogenetic factors involved in foot formation (MacWilliams et al., 1970; Cohen and MacWilliams, 1975; Schmidt and Schaller, 1976). At present, however, no mutant strains are available which have alterations in these potentials. This possibility, therefore, could not be examined by the same approach used by Nishimiya et al. (1984).

C. Biochemical Studies on the Factor Responsible for Position-Dependent Nematocyte Differentiation

An alternative and more direct approach to understanding the underlying mechanisms of the regional control of nematocyte differentiation is to look for and identify the factors which specifically influence position-dependent nematocyte differentiation. In view of the regional differences of nematocyte differentiation along the body column, one may expect to find a factor which suppresses stenotele differentiation and/or enhances desmoneme differentiation in distal tissue, or a factor which enhances stenotele differentiation or suppresses desmoneme differentiation in proximal tissue. Homogenates of hydra tissue were assayed for such factors (Fujisawa, 1984). A factor was found which specifically inhibits stenotele differentiation. This factor, termed the stenotele inhibitor, is probably a nonpeptide (insensitive to proteases) and has a molecular weight of about 500–1000 (determined by differential membrane filtration). When applied

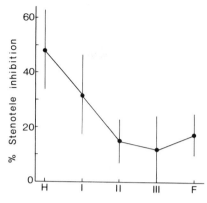

FIG. 3. Axial distribution of the stenotele inhibitor. The body columns of hydra which showed no bud protrusion were divided into five regions from head (H) to foot (F), and a homogenate was prepared from each region to determine the stenotele inhibitor level. (From Fujisawa, 1984.)

externally to intact hydra, it specifically inhibits stenotele differentiation but has no effect on proliferation or differentiation of other cell types. It is contained in a high level in the head region and in the gradually decreasing levels in the more proximal regions (Fig. 3). It appears to be different from any of the known morphogenetic factors involved in head or foot formation (Schaller, 1973; Schaller and Gierer, 1973; Berking, 1977; Schmidt and Schaller, 1976). Fractionating the macerated hydra cells (David, 1973) by the glycerol gradient centrifugation (Schaller and Gierer, 1973) and assaying each fraction for the stenotele inhibitor showed that it is located in epithelial cells but not in the nerve cells where the head activator and the head inhibitor are presumably stored under normal conditions (Schaller and Gierer, 1973; Berking, 1977; Schmidt and Schaller, 1980).

From these results a tentative conclusion has been drawn. Hydra has a gradient of the stenotele-specific inhibitor along the body column and this gradient appears to be responsible, at least in part, for the position-dependent stenotele differentiation along the body axis. How differentiation of other nematocyte types is regulated and whether any of the morphogenetic factors are involved in their regulation are not clear at present.

III. The Instability of Commitment to Nematocytes

As already described, the majority of nematoblasts present in the distal regions of the body column normally turn into desmonemes

(Bode and Smith, 1977). However, when the distal portion is isolated and allowed to regenerate, a significant fraction of these nematoblasts turn into stenoteles (Fujisawa and David, 1981).

Fujisawa and David (1981) carried out an experiment to determine at which step in the nematocyte differentiation pathway the decision for this alteration occurs. By comparing the time required for the appearance of the new stenoteles in the isolated distal pieces and the kinetics of the normal nematocyte differentiation process (David and Gierer, 1974; Fujisawa and Sugiyama, 1978), it was shown that the nematoblasts in the terminal cell cycle of the differentiation pathway responded to the environmental change and shifted their pathway from desmoneme to stenotele differentiation. Furthermore, the standard cell cycle analysis using [^{3}H]thymidine labeling and hydroxyurea treatment in the isolated distal pieces showed that this alteration of the pathway occurs near the S/G_2 boundary in the terminal cell cycle (Fujisawa and David, 1982).

It was also found that the alteration of the nematocyte differentiation pathway by means of treatment with the stenotele inhibitor occurs at the same step (Fujisawa, 1984). Intact hydra were treated with the stenotele inhibitor and the kinetics of the inhibition of stenotele differentiation was examined. Comparison of the inhibition kinetics with the kinetics of the normal stenotele differentiation process showed that stenotele differentiation was suppressed for only a 6-hour period near the S/G_2 boundary in the terminal cell cycle. When treated within this period, the nematoblasts, which otherwise differentiate into stenoteles, presumably differentiate into other types (Fig. 4). Outside of this period, the nematoblasts were insensitive to the inhibitor treatment.

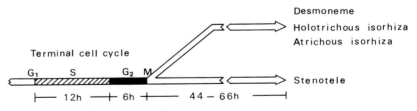

FIG. 4. Schematic representation of the nematocyte differentiation pathway altered by means of treatment with the stenotele inhibitor. Nematoblasts which normally differentiate into stenoteles can be altered to differentiate into the other types when animals are treated with the stenotele inhibitor. This alteration occurs during a 6-hour period near the S/G_2 boundary (filled bar) in the terminal cell cycle in the nematocyte differentiation pathway.

These results have suggested strongly that there is a distinct step in the nematocyte differentiation pathway at which alteration of the normal pathway occurs. Two possibilities can be considered. First, commitment to differentiate a specific nematocyte type occurs at this step. Second, commitment occurs at an earlier step but it can be changed by the environmental influence (transdifferentiation) at this step.

IV. Summary

Nematocyte differentiation from interstitial stem cells in hydra occurs in a highly position-dependent manner along the body axis. The results of the studies summarized here have shown that the morphogenetic factors involved in head formation are probably not responsible for this. Whether the morphogenetic factors involved in foot formation are responsible has not been determined.

A new factor, presumably unrelated to any of the known morphogens, has been identified which specifically inhibits the developing nematoblasts to differentiate into stenoteles. This factor is present in a gradient along the body column, and appears to be responsible, at least in part, for producing position-dependent nematocyte differentiation.

Nematoblasts which normally differentiate into one nematocyte type can be altered to differentiate into another by means of regeneration or treatment with stenotele inhibitor. This alteration occurs near the S/G_2 boundary in the terminal cell cycle in the nematocyte differentiation pathway. It appears that either the nematoblasts are not committed to any specific nematocyte pathway until this critical time, or the nematoblasts committed to differentiate into a specific type can transdifferentiate into another type at this step.

REFERENCES

Achermann, J., and Sugiyama, T. (1985). *Dev. Biol.* **107**, 13–27.
Berking, S. (1977). *Wilhelm Roux's Arch.* **181**, 215–225.
Bode, H. R., and David, C. N. (1978). *Prog. Biophys. Mol. Biol.* **33**, 189–206.
Bode, H. R., and Smith, G. S. (1977). *Wilhelm Roux's Arch.* **181**, 203–213.
Bode, P. M., and Bode, H. R. (1984). *In* "Pattern Formation, a Primer in Developmental Biology" (G. M. Malacinski and S. V. Bryant, eds), pp. 213–241. Macmillan, New York.
Cohen, J. E., and MacWilliams, H. K. (1975). *J. Theor. Biol.* **50**, 87–105.
David, C. N. (1973). *Wilhelm Roux's Arch.* **171**, 259–268.
David, C. N., and Challoner, D. (1974). *Am. Zool.* **14**, 537–542.
David, C. N., and Gierer, A. (1974). *J. Cell Sci.* **16**, 359–375.
Fujisawa, T. (1985). In preparation.
Fujisawa, T., and David, C. N. (1981). *J. Cell Sci.* **48**, 207–222.
Fujisawa, T., and David, C. N. (1982). *Dev. Biol.* **93**, 226–230.

Fujisawa, T., and Sugiyama, (1978). *J. Cell Sci.* **30,** 175–185.

Lehn, H. (1951). *Z. Naturforsch.* **6b,** 388–391.

MacWilliams, H. K. (1983a). *Dev. Biol.* **96,** 217–238.

MacWilliams, H. K. (1983b). *Dev. Biol.* **96,** 239–257.

MacWilliams, H. K., Kafatos, F. C., and Bossert, W. H. (1970). *Dev. Biol.* **23,** 380–398.

Nishimiya, C., Fujisawa, T., and Sugiyama, T. (1985). In preparation.

Rich, F., and Tardent, P. (1969). *Rev. Suisse Zool.* **76,** 779–789.

Schaller, H. C. (1973). *J. Embryol. Exp. Morphol.* **29,** 27–38.

Schaller, H. C., and Gierer, A. (1973). *J. Embryol. Exp. Morphol.* **29,** 39–52.

Schaller, H. C., Schmidt, T., and Grimmelikhuijzen, C. J. P. (1979). *Wilhelm Roux's Arch.* **186,** 139–149.

Schmidt, T., and Schaller, H. C. (1976). *Cell Differ.* **5,** 151–159.

Schmidt, T., and Schaller, H. C. (1980). *Wilhelm Roux's Arch.* **188,** 133–139.

Slautterback, D. B., and Fawcett, D. W. (1959). *J. Biophys. Biochem. Cytol.* **5,** 441–452.

Sugiyama, T. (1982). *Am. Zool.* **22,** 27–34.

Sugiyama, T., and Fujisawa, T. (1977a). *Dev. Growth Differ.* **19,** 187–200.

Sugiyama, T., and Fujisawa, T. (1977b). *J. Embryol. Exp. Morphol.* **42,** 65–77.

Webster, G., and Wolpert, L. (1966). *J. Embryol. Exp. Morphol.* **16,** 91–104.

Weill, R. (1934). *Trav. Sta. Zool. Wimereux* **10/11,** 1–701.

CHAPTER 20

THE MICROENVIRONMENT OF T AND B LYMPHOCYTE DIFFERENTIATION IN AVIAN EMBRYOS

N. Le Douarin

INSTITUT D'EMBRYOLOGIE
ANNEXE DU COLLÈGE DE FRANCE
CENTRE NATIONAL DE LA RECHERCHE SCIENTIFIQUE
NOGENT-SUR-MARNE, FRANCE

I. Introduction

How the microenvironment of the primary lymphoid organs influences the differentiation of lymphocytes, the effector cells of the immune function, is an important problem in developmental immunology. In all vertebrates, the site where cells responsible for cell-mediated immunity are produced is the thymus, hence their designation as T lymphocytes (for thymus-dependent lymphocytes). Identification of the organ in which the B lymphocytes, which ensure humoral immunity and antibody production, differentiate has long been controversial in all vertebrates except birds. It is now generally accepted that, in mammals at least, B lymphocytes develop in the bone marrow. Birds in this respect are unique; they possess a lymphoepithelial organ appended to the cloaca, the function of which long remained an enigma. This organ is the bursa of Fabricius, which in principle is the source of

CURRENT TOPICS IN
DEVELOPMENTAL BIOLOGY, VOL. 20

all the B lymphocytes of the body. This is why the avian embryo is a particularly appropriate model for studying the differentiation of lymphocytes. During recent years, the origin and identification of the various cell types in the thymus and the bursa of Fabricius of the avian embryo have been studied with the quail–chick marker system, in combination with various monoclonal and polyclonal antibodies that identify cell-specific-surface antigens and exhibit, at least for some of them, a specificity restricted to the quail or the chick species. This chapter will review some of the data gathered so far on the ontogeny of the two primary lymphoid organs of the avian embryo as well as the embryonic origin of their different cellular components, which are currently being identified through their surface antigens.

II. Cellular Composition of the Early Thymic and Bursal Primordia

The primary thymic rudiment is formed in the avian embryo by endodermal and mesenchymal cells of the third and fourth pharyngeal pouches. The thymic mesenchyme, like most mesenchymal component cells of vertebrate branchial arches, is derived from the ectodermal germ layer via the neural crest. Rhombencephalic crest cells (also called mesectodermal cells) invade the ventral area of the pharynx during the second and third days of incubation in both chick and quail embryos. This was demonstrated by constructing chimeric thymuses by means of heterospecific grafting of a quail rhombencephalon into a chick embryo, prior to the onset of neural crest cell migration. As a result, the branchial arch mesenchyme (except for a central core of mesodermal cells that yield striated muscle) of the host was of the quail type, whereas the endothelial cells of the blood vessels were mesodermal in origin and therefore of the chick host type (Le Douarin and Jotereau, 1975).

The fate of the mesectodermal cells that take part in thymus histogenesis could be followed at later stages, due to the stability of the quail nuclear marker. They formed only a very small contingent of the thymic cortical and medullary tissue, the pericytes located along the blood vessels.

When the thymic endoderm detaches from the pharynx, it becomes a cord of epithelial cells located along the jugular vein. The differentiated thymic epithelial cells form a network in which the lymphocytes develop. In spite of the fact that they lose their epithelial arrangement and acquire long processes during histogenesis, the thymic reticuloepithelial cells remain closely joined by desmosomes and exhibit a basal lamina at the interface with the surrounding connective tissue.

The bursa of Fabricius develops as a dorsal appendage of the cloaca; it arises at 5 days as an endodermal bud surrounded by a mesenchymal component of mesodermal origin.

One of the particularities of the development of the bursa of Fabricius is that it associates, for a certain period during embryonic life, two hemopoietic organs. One, myeloid in nature, develops in the mesenchyme and is mainly the site of granulopoiesis. The other consists in the epithelial follicles in which the B lymphocytes differentiate.

III. The Origin of Lymphocytes in Thymus and Bursa of Fabricius

A. RAPID HISTORICAL OVERVIEW

The histogenetic processes that give rise to the lymphoid differentiation of these primary lymphoid organs have been the subject of controversial interpretations for many years, although it has long been recognized that the early precursor of lymphocytes is a large cell with a strongly basophilic cytoplasm that appears more or less simultaneously in the epithelial and mesenchymal parts of the organ. For some researchers, the lymphocyte precursors were endodermal in origin, whereas others considered them to arise from the mesenchyme and to migrate secondarily into the endoderm, which was supposed to trigger their multiplication and differentiation into lymphocytes. It was in 1965 (see Metcalf and Moore, 1971; Le Douarin et al., 1984a,b, for reviews) that, for the first time, the problem of the ontogeny of the thymus and the bursa of Fabricius was approached through an appropriate experimental technique, i.e., the use of a cell marker.

Moore and Owen (1965, 1967a,b), using sex chromosomes as cell markers, demonstrated that an immigration of cells into the thymus and the bursa of Fabricius via the circulation takes place during development. They concluded from this observation that the lymphocytes do not originate in the lymphoid organs themselves but from blood-borne stem cells. This and other experiments suggested that in fact all the hemopoietic cells, including those that develop in the mammalian fetal liver, the bone marrow, and the spleen, arise from precursors that originate in embryonic areas, which are different from the primordia of the blood-forming organs. Thus was formulated the hematogenous theory (Metcalf and Moore, 1971) of hemopoietic organ development.

Although very attractive, Moore and Owen's views were not readily accepted, since a limitation of the labeling technique relying on mitotic chromosomes as markers is that it identifies only cells dividing at the time of observation. Therefore, although an influx of cells into the thymus was demonstrated beyond doubt by the experiments mentioned above, it could not be stated with certainty that all the lymphocytes that developed in the thymus and the bursa of Fabricius were of extrinsic origin. It was quite conceivable at that time that certain subpopulations of lymphocytes could be derived from one or the other of the original components of the rudiments.

Thus it seemed worthwhile to reinvestigate this question by means of the quail–chick marker system, which provides information on the entire cell population of the organ at any time.

B. Application of the Quail–Chick Marker System to Tracing the Early Colonization of the Thymus by Hemopoietic Cells

Our first attempt was to see whether the endoderm and the mesenchyme of the early thymus and bursa rudiments were totally devoid of lymphopoietic capacities in the early embryo.

The presumptive territories of the thymus and bursa were taken from quail or chick embryos several days before the appearance of the basophilic cells, i.e., at 3 days of incubation (E3) for the thymus and E5 for the bursa. They were grafted in the somatopleure of a 3-day host of the other species (Fig. 1). An extra thymus or bursa developed in this site, and chimerism analysis could be performed after Feulgen staining or at the electron microscopic level. If either the endoderm or the mesenchyme of the rudiment had the capacity to differentiate into lymphocytes, lymphocytes with the nuclear marker of the graft would have been found in the organs. This was not the case. For example, in a quail thymic rudiment grafted at 3 days into a chick host, the lymphoid population was entirely of the chick host type, while the reticuloepithelial cells and the connective tissues were derived from the grafted quail thymus. Similar results were found for the bursa.

These results therefore confirmed Moore and Owen's findings and showed that lymphoid differentiation of thymus and bursa was totally dependent on a seeding by stem cells of extrinsic origin. By grafting thymic and bursic explants at increasingly late stages of development, it was possible to find out exactly when they are colonized by hemopoietic cells in both chick and quail species.

In all these experiments, the thymuses were observed when they had reached 14 days of age and the bursas 19 days. The times of colonization of chick and quail thymuses and bursas are indicated in Fig. 2. Several periods in the development of the prelymphoid organs, in terms of their relationships with the hemopoietic cells, could then be

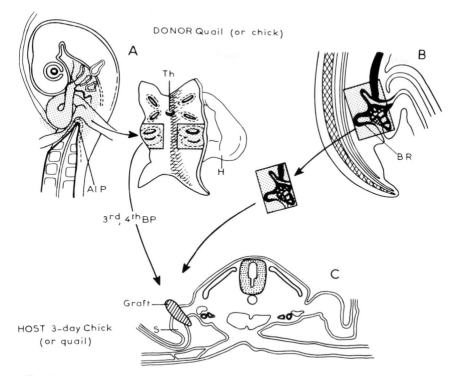

FIG. 1. Interspecific graft of the thymic and bursal rudiments into the somatopleure of a 3-day-old host. (A) Removal of the thymic rudiment from a 30-somite embryo from which the ventral side of the pharynx, comprising the branchial arches and the cardiac primordium, has been removed. Bilateral areas corresponding to the third and fourth pharyngeal pouches (endoderm + mesenchyme) have been selectively taken and grafted into the somatopleure of a 3-day-old host. (B) Removal of the bursal rudiment from a 5-day-old embryo. (C) Graft of the thymic or bursal rudiments into the somatopleure of the host embryo at 3 days of incubation. The donor and the graft were always of different species, i.e., quail or chick. AIP, Anterior intestinal portal; BP, branchial pouches; BR, bursal rudiment; H, heart; S, somatopleure; Th, thyroid rudiment.

identified. For the thymus, a *prereceptive* period during which hemopoietic cells do not enter the rudiment precedes a fast influx of cells that takes place during a *receptive period,* itself followed by a *postreceptive,* or *refractory,* period characterized by a shutoff of hemopoietic cell seeding (Fig. 3).

The influx of hemopoietic cells in the bursa was found to have a longer duration than that for the thymus. It occurs between E7 and E11 for the quail and from E8 to E14 for the chick embryos.

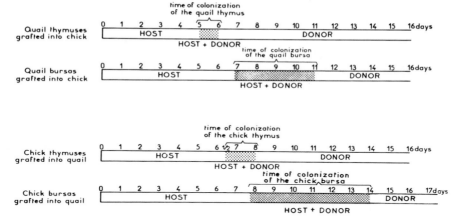

Fig. 2. Diagram showing the timing of the early colonizations of bursal and thymic rudiments by hemocytoblasts in quail and chick embryos.

C. APPLICATION OF THE MONOCLONAL ANTIBODY α-MB1 TO THE STUDY OF THYMUS AND BURSA HISTOGENESIS

A monoclonal antibody (α-MB1) was prepared in our laboratory by Peault *et al.* (1983) by immunizing a mouse with the quail immunoglobulin μ chain purified from the plasma. Besides several monoclonal antibodies actually directed against the quail μ chain, the fusion yielded a particular clone that exhibited a broad immunoreactivity toward hemopoietic and endothelial cells of the embryonic and adult quail. The surface marker, referred to as MB1, is expressed early

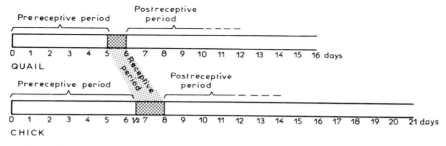

Fig. 3. Four periods can be identified in the early phases of thymus development according to its receptivity to lymphocyte precursor cells (LPC): a prereceptive period during which the thymus primordium is unable to receive LPC; a receptive period during which the seeding of the organ takes place; a postreceptive period of 4–5 days; and eventual LPC reentry to the thymus.

on both intra- and extraembryonic mesodermal cell precursors from which vascular endothelial and blood cells develop, designated as *hemangioblasts*. This marker is transmitted to the whole progeny of these hemangioblastic precursors with the exception of the erythrocytes. Moreover, it is a constant feature of the endothelial cell surface throughout ontogenesis and adult life, irrespective of their localization and specialization.

The epitope recognized by α-MB1 is shared by several molecular entities, among which is the immunoglobulin μ chain both on B lymphocyte surfaces and in its soluble plasmatic form. In addition, MB1 is present in another quail plasma component of MW 160K (Fig. 4).

Staining of serial sections of the thymic and bursal rudiments by α-MB1 at increasing stages of development in the quail embryo confirms

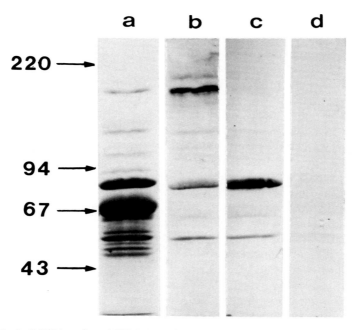

FIG. 4. Anti-MB1 and anti-MD2 (specific anti-μ monoclonal antibody) binding to quail plasma components. Three-day quail plasma was submitted to SDS–electrophoresis in a 6.5% acrylamide gel, in reducing conditions, and the material was blotted on nitrocellulose. Lane a: Ponceau S staining of transferred proteins. After destaining, individual lanes were treated with anti-MB1 antibody (lane b) or anti-MD2 (lane c), and reactions were visualized by the peroxidase anti-peroxidase/diaminobenzidine (PAP–DAB) technique. On lane d, as a control, the first antibody was omitted. Arrows indicate standard MW markers ($\times$ 10^3) run in the same gel. (From Peault *et al.*, 1983.)

the hemopoietic origin of the lymphocytes. The blood vessels surround the thymic rudiment, which at the time of its first colonization is still avascular (Fig. 5). The bursal mesenchyme, in contrast, is already irrigated at 7 days before the hemopoietic cell seeding has started (Fig. 6). In both organs, immigration of hemopoietic cells in the epithelium component is easily followed with α-MB1 immunostaining.

D. Periodicity of Hemopoietic Cell Seeding of the Thymus

The next problem was whether the influx of cells that reaches the thymus and the bursa at the early stages of their development is the only one occurring during embryonic life or whether it is followed by other inflows of cells later in development. In this respect, the thymus and bursa behave differently.

For the thymus, our investigations (so far devoted especially to the quail species) revealed that several waves of hemopoietic cell influx reach the thymus during embryonic and early postnatal life. The principle of the experiments was to submit the thymic rudiment, taken from the quail after completion of the first colonization, to two successive grafts, the first host being a 3-day chick embryo and the second a 3-day quail embryo.

By varying the duration of the first and second grafts as explained in detail in Jotereau and Le Douarin (1982) and in Le Douarin et al. (1984a,b), we could establish that colonization of the quail thymus by lymphocyte precursors during embryonic and postnatal life appears as a cyclic process with a sequence of receptive periods lasting about 24 hours, separated by refractory phases of about 5 days, as represented in Fig. 7. As a corollary, the proliferative capabilities of the hemopoietic cells that colonize the young thymus appear to be limited to a definite and probably relatively small number of cell cycles.

The cyclic renewal pattern of the quail embryo thymic lymphocytes was further confirmed by experiments involving homospecific cellular combinations based on sex chromosome markers (Le Douarin, Coltey, and Guillemot, unpublished; see Le Douarin et al., 1984a,b).

Fig. 5. (a) Transverse section in the thymus (Th) of a quail embryo during the sixth day of incubation. Immunoreactivity with MB1 shows the blood vessels surrounding the thymus rudiment. The endothelial cells (En) of these blood vessels are MB1-positive. A few hemopoietic cells (HC) MB1-positive have already homed to the thymus epithelium. ×300. (b) At day 7 the thymic rudiment is completely colonized by immunoreactive HC. ×300. JV, Jugular vein. (c) At 9 days of incubation, thymus histogenesis is in progress. ×170.

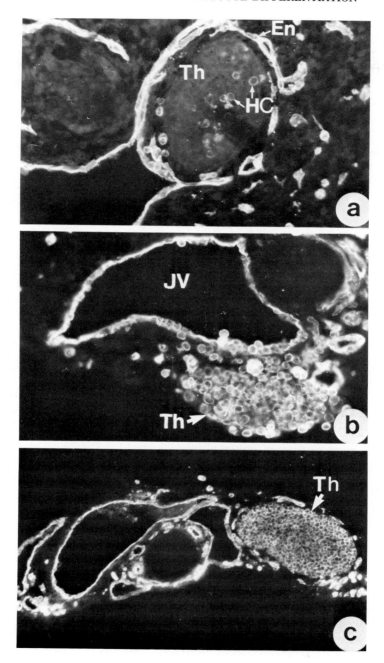

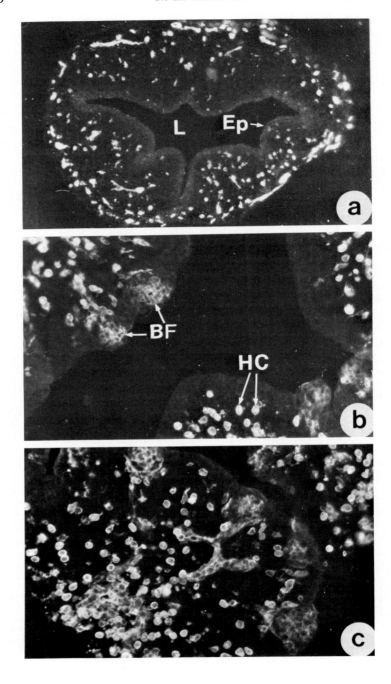

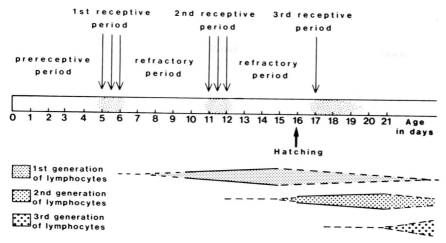

FIG. 7. Schematic representation of the three successive waves of HC entering the thymus in the embryo and in the early postnatal quail. The entry of HC is restricted to short periods separated by several days (five in our experimental conditions) during which no significant seeding takes place. The precise timing of the third wave of LPC into the thymus could not be determined precisely in our experimental conditions.

E. SEEDING OF THE BURSA AS A SINGLE DEVELOPMENTAL EVENT

In both quail and chick embryos, seeding of the bursa starts after the completion of thymus colonization and lasts several days: from E7 to E11 in the quail and from E8 to E14 in the chick.

Whether the stem cells that colonize the epithelium and those that remain in the mesenchyme belong to the same homogeneous initial population of hemopoietic cells is a crucial, still unanswered question. Their respective fate is, in any case, totally different in these two instances.

In all these experiments, the grafted bursas were analyzed for chimerism when they had reached the total age of 19–20 days. The grafting time was then prolonged in order to see whether an eventual renewal of the lymphocyte precursors took place in the bursa as it does for the thymus. Colonized quail bursas (taken at 11 days) were grafted

FIG. 6. (a) Bursa of Fabricius of a 9-day quail embryo. Numerous MB1-positive blood vessels are seen in the mesoderm. ×300. L, Lumen of the bursa; Ep, bursal epithelium. (b) At E11 numerous hemopoietic cells (HC) are present in the mesenchyme of the bursa. Some have penetrated the epithelium, where bursal follicles (BF) are in formation. ×300. (c) At E12 follicle formation has progressed. ×300.

into 3-day chick embryos either for 13 days in a single host or for 20 days in two successive hosts. No renewal was ever observed in any of the explants that still contained quail lymphocytes, even at 31 days of total age (i.e., 7 days after hatching) (Le Douarin et al., 1984a).

It seems, therefore, that embryonic seeding of the bursa is unique, meaning that in this respect the bursa behaves differently from the thymus. This is in agreement with the previous studies by Weber and co-workers (Weber and Mausner, 1977; Weber and Foglia, 1980) and Toivanen et al. (1972), who showed that cells able to seed the bursa of Fabricius were no longer present in the peripheral blood or in bone marrow of chickens around hatching time.

Along this line, Houssaint et al. (1983) demonstrated that an empty embryonic bursal rudiment implanted into an histocompatible hatched chicken never became colonized, whereas the thymic rudiment, in a similar situation, was seeded and developed normally.

IV. The Role of the Bursa of Fabricius in B Lymphocyte Production

The principal observation that indicated the role of the bursa of Fabricius in B lymphocyte production was that of Bruce Glick, who showed that removal of the bursa resulted in impaired antibody formation against bacterial antigens (Glick et al., 1956). Various bursectomy methods have since been devised that cause more or less complete B cell defects and agammaglobulinemia (see Le Douarin et al., 1984a, for a review). In fact, it seems that it is only when several methods are combined that lifelong agammaglobulinemia is achieved; such is the case with anti-μ chain treatment followed by surgical bursectomy at hatching time (Kincade et al., 1970; Leslie and Martin, 1973) or hormonal bursectomy with testosterone plus cyclophosphamide in early life (Weidanz et al., 1971).

Some investigators have proposed that, although the bursa is certainly the central organ for B lymphocyte development, other sites exist in the embryo and early posthatched chicken that are also able to fulfill this function (see Le Douarin et al., 1984b, for review).

Recently, surgical bursectomy performed at 60 hours of incubation in chickens was followed by a thorough analysis of immune function in these birds (Jalkanen et al., 1983a). The operated chickens nevertheless produced immunoglobulins of all three classes: IgM, IgG, and IgA. The IgG level was decreased, but the levels of IgM and IgA were normal. However, and interestingly, no specific antibodies were obtained even after heavy immunizations with a variety of antigens. Moreover, the bursectomized chickens lacked natural antibodies against phosphorylcholine, fecal bacteria, rabbit red blood cells and autoantibodies to the liver, kidney, and thyroid (Jalkanen *et al.*, 1983b). Another abnormality of these birds was that immunoglobulin-bearing lymphocytes (s-Ig$^+$) of the IgG$^+$ type were markedly lower in number than in controls in the spleen and peripheral blood, s-IgM$^+$ cells were decreased in the peripheral blood, and s-IgA$^+$ cells were decreased in the spleen. In contrast, s-Ig$^+$ cells of the three classes were present in normal frequency in the bone marrow. Moreover, no effect on the frequency of major histocompatibility complex (MHC) class II antigen-bearing cells in different organs was observed, nor was any effect observed on the total lymphocyte or white cell counts in the peripheral blood. The incapacity of the bursectomized chickens to produce specific antibodies against definite antigens shows that the bursa of Fabricius is the site where the antibody repertoire is generated. This was further confirmed by the demonstration of a defect in generation of light-chain diversity in bursectomized chickens (Jalkanen *et al.*, 1984).

V. Mechanisms of the Seeding of the Primary Lymphoid Organs by Hemopoietic Cells

The fact that hemopoietic cells able to seed the thymus and the bursa of Fabricius are present in the bloodstream well before these organs are actually colonized was in my view suggestive of an intrinsically regulated mechanism for their receptivity. I put forward the hypothesis that, at least for the seedings that occur in the embryo, the hemopoietic cells are attracted to the thymic and bursal anlagen through the production by these organs of factors triggering (1) their adhesion to the endothelial wall of the blood vessels located in the vicinity of the thymic and bursal epithelium and (2) their migration toward an increasing gradient of chemoattractant secretion due to the developmentally programmed activity of the thymic and bursal epithelia (Le Douarin, 1978).

Since the attractivity of the thymic epithelium is maintained at least several days if its seeding by hemopoietic cells is prevented, I proposed that the arrest of thymic and bursal attractivity was due to a negative retrocontrol exerted by these cells on chemoattractant production. Receptivity was resumed when the hemopoietic cells had, after proliferating and differentiating into lymphocytes, changed their cell-surface properties and lost their capacity to inhibit the thymic epithelial cells to produce the chemoattractant. The latter, reappearing at E11 in the quail, is responsible for the second influx of hemopoietic cells into the organ (Fig. 8).

This hypothesis has been substantiated by a series of experiments carried out either *in vivo* or *in vitro* in my laboratory and others (Le

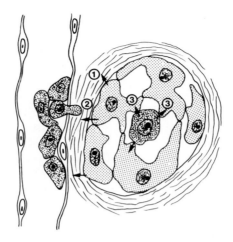

① Production of a diffusible "attractive" substance

② Invasion of the thymic rudiment by hemocytoblasts

③ Feedback inhibition of the production of the

diffusible "attractive" substance by the hemocytoblasts

Fig. 8. Hypothesis for hemopoietic cell seeding in the thymic rudiment. The endodermal epithelial cells are supposed to produce a chemoattractant that leads the blood-borne hemopoietic cells (HC) to agglomerate in the vessels located close to the thymus anlage and thereafter to migrate and home to the thymus. A negative feedback induced by the HC stops the production of the chemoattractant, resulting in the refractory periods that separate the first from the second HC influx to the thymus.

Douarin and Jotereau, 1980; Jotereau *et al.*, 1980; Pyke and Bach, 1979). Ben Slimane *et al.* (1983) set up an experimental system derived from the chemotactic chamber described by Zigmond (1977) and could demonstrate that the embryonic chick thymus produces a low-molecular-weight factor (not retained by a 12-kDa-cutoff dialysis membrane) that has a strong chemoattractive action on bone marrow cells. (On the other hand, these have been shown capable of homing to a thymus and of differentiating into T lymphocytes.) Another factor can be evidenced in the culture medium of thymuses. This factor has a molecular weight higher than 50K and a chemokinetic effect on the same cells. Interestingly, an iterative production of the chemoattractant could be demonstrated and was shown to coincide with the phases of thymic receptivity for hemopoietic cells, whereas the larger chemokinetic molecule appeared permanently produced. The latter substance considerably enhances hemopoietic cell adhesion to glass *in vitro*. It is proposed that it triggers the arrest of circulating hemopoietic cells by making them adhere to the blood vessel endothelia in the vicinity of the attractive thymus and bursa. The purification of these two substances has not yet been achieved.

Although these experiments have been performed on the thymus, evidence has accumulated suggesting that similar mechanisms are probably operative for the bursa as well. This evidence relies on the fact that hemopoietic cells that have homed to a bursa can be displaced to an attractive thymus if the two organs are placed side by side *in vitro* or *in vivo* (see Le Douarin *et al.*, 1982, 1984a,b). Thus the cells that have been able to seed the bursa can also respond to the attractive signals emanating from the thymus.

VI. Expression of Class II Antigens of the Major Histocompatibility Complex in the Thymus as Studied by the Quail–Chick System

It is generally accepted that the thymus plays a central role not only in the differentiation of the thymocytes but also in the "education" of T cells and the restriction of T cell specificities.

During their interaction with the thymic stroma, thymocytes become restricted for immunological responses involving cytotoxicity against modified self-determinants, delayed-type hypersensitivity, and interaction with Ia-positive accessory cells. In addition, the thymus appears to be a site for tolerance induction (see Le Douarin *et al.*, 1984b; Guillemot *et al.*, 1984, for references).

The cellular diversity of the thymic stroma and the expression of the antigens encoded by the major histocompatibility gene complex by thymic cells thus become important to consider. It is already established that cells from the three initial embryonic germ layers contribute to thymus ontogeny: the *endoderm* provides the reticuloepithelial cells; the *ectoderm* yields the pericytes, the connective cells, and the myoid elements via the neural crest, as shown by Le Lièvre and Le Douarin (1975) and H. Nakamura (for the myoid elements; personal communication); and the *mesoderm* gives rise to the lymphocyte precursors and to the endothelium of all the thymic blood vessels (Le Douarin and Jotereau, 1975; Peault *et al.*, 1983).

Evidence has progressively emerged that nonlymphoid cells of extrinsic origin also take part in thymus histogenesis and might, with the other cell types mentioned above, play a role in thymocyte processing (Barclay and Mayerhofer, 1981; Kyewski *et al.*, 1982). These thymic "accessory cells" include macrophages and interdigitating reticular cells, originally described in the human by Kaiserling *et al.* (1974) and corresponding to the dendritic cells isolated from rodent thymuses (Beller and Unanue, 1980; Wong *et al.*, 1982). Dendritic cells (DCs), first described in splenic cell cultures by Steinman and Cohn (1973), have since been shown to share certain characteristics with macrophages; for example, they have a low buoyant density, are glass adherent, and, like macrophages, express class II (Ia) and class I antigens encoded by the MHC. However, they differ from macrophages in that they lack Fc receptors and phagocytic properties (Steinman *et al.*, 1980).

The avian MHC, or *B* locus (essentially studied in the chicken), is thought to be analogous to the mammalian MHC (Briles *et al.*, 1982). Within the *B* locus, the *B–L* subregion is the counterpart of murine *H-2I* and human *HLA-D* (Pink *et al.*, 1977; Ewert and Cooper, 1982). B–L antigens are known to be distributed on B cells and also on monocytes, macrophages, and DC, including those found in the thymus (Ewert and Cooper, 1978; Peck *et al.*, 1982; Oliver and Le Douarin, 1984).

We have recently isolated macrophages and DCs from quail and chick thymuses and used those cells as immunogens to prepare monoclonal antibodies directed against the Ia-like antigens of the avian MHC (Oliver and Le Douarin, 1984; Le Douarin *et al.*, 1983; Guillemot *et al.*, 1984). The macrophages and DCs were isolated by means of their property of adhering to glass.

Three interesting reagents recognizing Ia-like determinants of chicken and quail have been obtained. All three stain specifically both DCs and macrophages of thymus, bursa, and spleen as well as B lymphocytes. They do not stain thymocytes or postthymic lymphocytes. One is restricted to quail cells and does not react with chick cells. Another has a strict chick species specificity, whereas the third one reacts equally well with chick and quail cell-surface determinants. The three reagents have the ability to immunoprecipitate cell-surface antigens of approximately 30 kDa in reducing conditions (Le Douarin *et al.*, 1983; Guillemot *et al.*, 1984). These reagents have been applied on sections of chick and quail thymuses taken around or after hatching. The anti-quail Ia antibodies stain small cells with intricated processes in the quail thymic cortex and, in the medulla, large areas of cells that appear brightly fluorescent (Fig. 9a). No immunoreactivity was found in the chick with the anti-quail Ia antibody. In contrast, the anti-chick Ia reagent stains chicken thymuses according to a similar pattern (Fig. 9b).

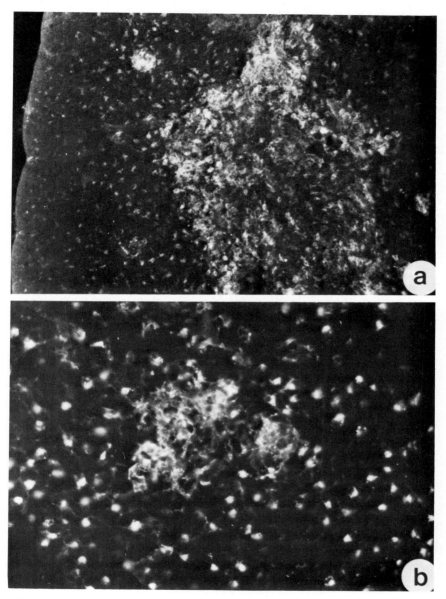

FIG. 9. Distribution of Ia-positive cells in the avian thymus. (a) Normal thymus section from a 10-day quail labeled with anti-quail–chick Ia. In the cortex, scattered cells with long processes are labeled, whereas in the medulla the antibody stains very brightly compact cell aggregates. ×155. (b) A higher magnification of anti-quail–chick Ia monoclonal antibody staining on the normal late embryonic quail thymus reveals the network of processes joining the Ia-positive cortical cells and shows the densely packed cells of the medulla. ×480. (From Guillemot *et al.*, 1984.)

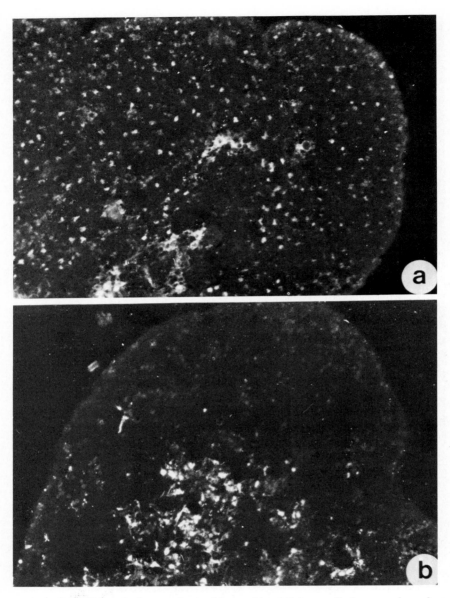

Fig. 10. (a) Chimeric thymus obtained by grafting 3.5-day quail pharyngeal pouches into a chick embryo for 17 days. The monoclonal antibody, specific for quail Ia antigen, reacts only with the reticuloepithelial cells of the cortex and the corticomedullary margin. Note that the center of the medulla is practically devoid of immunoreaction. ×150. (b) The same chimeric thymus as in (a) on which the anti-chick Ia monoclonal antibody has been applied. Now, the medullary Ia-positive cells are brightly stained, whereas the epithelial cells of quail endodermal origin are negative. ×155. (From Guillemot *et al.*, 1984.)

Distinction between intrinsic and extrinsic Ia-positive cells in the avian thymus was achieved in a series of experiments in which quail thymic primordia (i.e., the third and fourth pharyngeal pouches removed from a 3-day quail embryo) were grafted into the somatopleure of a 3.5-day chick for 17 days (total age of the thymus at the time of observation, 20 days). As demonstrated before, all the hemopoietic cells developing in these thymuses were of chick host origin (Le Douarin and Jotereau, 1973, 1975; Le Douarin et al., 1984a,b). All the Ia-specific reagents reacted with certain cell components of these chimeric thymuses. The anti-quail Ia antibody stained the small cells with processes in the cortex, but not the compactly arranged cells of the medulla (Fig. 10a). Immunoreactivity of those elements was observed with the anti-chick Ia antibody, which in contrast did not stain the cortical epithelial cells. Due to its double chick and quail specificity, the third reagent exhibited a positive immune reaction with both cortical and medullary Ia-bearing cells of these chimeric thymuses (Fig. 10b).

This demonstrated that while the Ia-positive cells of the cortex are derived from the thymic primordium itself, those of the medulla are of extrinsic hemopoietic origin.

If the quail thymuses were removed from the donor embryo after the first colonization by hemopoietic cells (e.g., at 7 or 8 days of incubation) and grafted into a 3.5-day chick host, it was found that, though the first generation of lymphocytes belonged to the quail, it was progressively replaced by lymphocytes originating from the second stem cell influx according to the pattern indicated in Fig. 7. It could also be shown that the precursors for DCs and macrophages reached the thymus along with the lymphocyte precursors during the first wave of stem cell entry and that they were subsequently replaced according to a timing comparable to that of lymphocytes (Guillemot et al., 1984).

However, we have not yet performed a detailed study of their turnover in the thymus, and the data presently available do not exclude the possibility that the precursors of thymic accessory cells penetrate the thymus in a continuous rather than in a cyclic way. More work is required to clarify this problem.

VII. Concluding Remarks

One of the issues raised by these studies is the embryonic origin of the hemopoietic cells that colonize the primary lymphoid and the other hemopoietic organs of the body.

In a series of studies based on the use of the quail–chick marker system, Dieterlen-Lièvre and co-workers have significantly en-

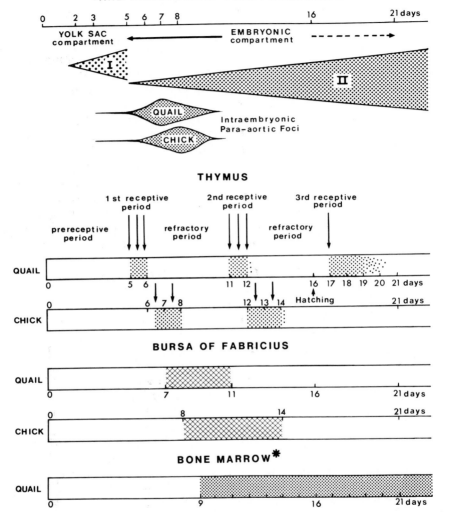

FIG. 11. Timetable of the major events in the ontogeny of the hemopoietic system in the quail and chick embryo. (I) Primitive erythrocytes (hemoglobins P and E); (II) secondary erythrocytes (hemoglobins A, D, and H) (studied in the chick; see Bruns and Ingram, 1973, for a review). The yolk sac is the major erythropoietic site during most of incubation. During phase I, it functions with its own stem cells; during phase II, it receives stem cells from the embryo. (From Le Douarin *et al.*, 1984a.)

lightened us on this question (Dieterlen-Lièvre, 1975, 1986; Dieterlen-Lièvre *et al.*, 1976; Dieterlen-Lièvre and Martin, 1981; Beaupain *et al.*, 1979; Martin *et al.*, 1978, 1979, 1980). They have found that in the avian embryo at least two generations of stem cells endowed with different fates are produced at different times and in different locations, first in the yolk sac and then the embryonic mesoderm. The particular stem cells responsible for the first seeding of the thymus and the colonization of the bursa arise from the intraembryonic mesoderm and not from the yolk sac. Another important notion to come out of the studies carried out on bird hemopoiesis is that the development of the hemopoietic system appears as a sequence of precisely timed recurring events involving the inflow of successive erythroid lineages, the functional succession of hemopoietic organs, and the periodic colonization of the thymus. Figure 11 is a summary of the data accumulated so far on the principal landmarks of hemopoietic system ontogeny of the bird embryo.

REFERENCES

Barclay, A. N., and Mayerhofer, G. (1981). *J. Exp. Med.* **153**, 1666–1671.
Beaupain, D., Martin, C., and Dieterlen-Lièvre, F. (1979). *Blood* **53**, 212–225.
Beller, D. I., and Unanue, E. R. (1978). *J. Immunol.* **121**, 1861–1864.
Beller, D. I., and Unanue, E. R. (1980). *J. Immunol.* **124**, 1433–1440.
Ben Slimane, S., Houllier, F., Tucker, G., and Thiery, J. P. (1983). *Cell Differ.* **13**, 1–24.
Briles, W. E., Bumstead, N., Ewert, D. L., Gilmour, D. G., Gogusev, J., Hala, K., Koch, C., Longenecker, B. M., Nordskog, A. W., Pink, J. R. L., Schierman, L. W., Simonsen, M., Toivanen, A., Toivanen, P., Vaino, O., and Wick, G. (1982). *Immunogenetics* **15**, 441–448.
Bruns, G. A. P., and Ingram, V. M. (1973). *Philos. Trans. R. Soc. London Ser. B* **226**, 225–305.
Dieterlen-Lièvre, F. (1975). *J. Embryol. Exp. Morphol.* **33**, 607–619.
Dieterlen-Lièvre, F. (1984). *Dev. Comp. Immunol.* **8** (Suppl. 3), 75–80.
Dieterlen-Lièvre, F., and Martin, C. (1981). *Dev. Biol.* **88**, 180–191.
Dieterlen-Lièvre, F., Beaupain, D., and Martin, C. (1976). *Ann. Immunol. (Paris)* **127c**, 857–863.
Ewert, D. L., and Cooper, M. D. (1978). *Immunogenetics* **7**, 521–535.
Ewert, D. L., and Cooper, M. D. (1982). *In* "Ia Antigens and their Analogs in Man and Other Animals" (S. David and S. Ferrone, eds.), p. 1. CRC Press, Boca Raton, Florida.
Glick, B., Chang, T. S., and Jaap, R. G. (1956). *Poultry Sci.* **35**, 224–245.
Guillemot, F. P., Oliver, P. D., Péault, B. M., and Le Douarin, N. M. (1984). *J. Exp. Med.* **160**, 1803–1819.
Houssaint, E., Torano, A., and Ivanyi, J. (1983). *Eur. J. Immunol.* **13**, 590–595.
Jalkanen, S., Granfors, K., Jalkanen, M., and Toivanen, P. (1983a). *J. Immunol.* **130**, 2038–2041.
Jalkanen, S., Granfors, K., Jalkanen, M., and Toivanen, P. (1983b). *Cell. Immunol.* **80**, 363–373.

Jalkanen, S., Jalkanen, M., Granfors, K., and Toivanen, P. (1984). *Nature (London)* **311,** 69–71.

Jotereau, F. V., and Le Douarin, N. M. (1982). *J. Immunol.* **129,** 1869–1877.

Jotereau, F. V., Houssaint, E., and Le Douarin, N. M. (1980). *Eur. J. Immunol.* **10,** 620–627.

Kaiserling, E., Stein, H., and Müller-Hermelink, H. K. (1974). *Cell Tissue Res.* **155,** 47.

Kincade, P. W., Lawton, A. R., Bockman, D. E., and Cooper, M. D. (1970). *Proc. Natl. Acad. Sci. U.S.A.* **67,** 1918–1925.

Kyewski, B. A., Rouse, R. V., and Kaplan, H. S. (1982). *Proc. Natl. Acad. Sci. U.S.A.* **79,** 5646–5650.

Le Douarin, N. M. (1978). "Differentiation of Normal and Neoplastic Hematopoietic Cells," pp. 5–31. Cold Spring Harbor Laboratory, Cold Spring Harbor, New York.

Le Douarin, N. M., and Jotereau, F. V. (1973). *Nature (London) New Biol.* **246,** 25–27.

Le Douarin, N. M., and Jotereau, F. V. (1975). *J. Exp. Med.* **142,** 17–40.

Le Douarin, N. M., and Jotereau, F. V. (1980). *In* "Immunology 1980" (M. Fougereau and J. Dausset, eds.), pp. 285–302. Academic Press, London.

Le Douarin, N. M., Jotereau, F. V., Houssaint, E., Martin, C., and Dieterlen-Lièvre, F. (1982). *In* "The Reticuloendothelial System" (N. Cohen and M. Siegel, eds.), Vol. 3, pp. 589–616. Plenum, New York.

Le Douarin, N. M., Guillemot, F. P., Oliver, P., and Peault, B. (1983). *In* "Progress in Immunology V" (Y. Yamamura and T. Tada, eds.), pp. 613–631. Academic Press, New York.

Le Douarin, N. M., Dieterlen-Lièvre, F., and Oliver, P. D. (1984a). *Am. J. Anat.* **170,** 261–299.

Le Douarin, N. M., Jotereau, F. V., Houssaint, E., and Thiery, J. P. (1984b). *In* "Chimeras in Developmental Biology" (N. M. Le Douarin and A. Mc Laren, eds.), pp. 179–215. Academic Press, London.

Le Lièvre, C. S., and Le Douarin, N. M. (1975). *J. Embryol. Exp. Morphol.* **34,** 125–154.

Leslie, G. A., and Martin, L. N. (1973). *J. Immunol.* **110,** 959–967.

Martin, C., Beaupain, D., and Dieterlen-Lièvre, F. (1978). *Cell Differ.* **7,** 115–130.

Martin, C., Lassila, O., Nurmi, T., Eskola, J., Dieterlen-Lièvre, F., and Toivanen, P. (1979). *Scand. J. Immunol.* **10,** 333–338.

Martin, C., Beaupain, D., and Dieterlen-Lièvre, F. (1980). *Dev. Biol.* **75,** 303–314.

Metcalf, D., and Moore, M. A. S. (1971). "Haemopoietic Cells." North Holland Publ., Amsterdam.

Moore, M. A. S., and Owen, J. J. T. (1965). *Nature (London)* **208,** 956, 989–990.

Moore, M. A. S., and Owen, J. J. T. (1967a). *J. Exp. Med.* **126,** 715–723.

Moore, M. A. S., and Owen, J. J. T. (1967b). *Lancet* **2,** 658–659.

Oliver, P. D., and Le Douarin, N. M. (1984). *J. Immunol.* **132,** 1748–1755.

Peault, B. M., Thiery, J. P., and Le Douarin, N. M. (1983). *Proc. Natl. Acad. Sci. U.S.A.* **80,** 2976–2980.

Peck, R., Murthy, K. K., and Vaino, O. (1982). *J. Immunol.* **129,** 4–5.

Pink, J. R. L., Droege, W., Hala, K., Miggiano, V. C., and Ziegler, A. (1977). *Immunogenetics* **5,** 203–216.

Pyke, K. W., and Bach, J. F. (1979). *Eur. J. Immunol.* **9,** 317–323.

Steinman, R. M., and Cohn, Z. A. (1973). *J. Exp. Med.* **137,** 1142–1162.

Steinman, R. M., Chen, L. L., Witmer, M. D., Kaplan, G., Nussenzweig, M. C., Adams, J. C., and Cohn, Z. A. (1980). *In* "Mononuclear Phagocytes Functional Aspects, Part II" (R. Van Furth, ed.), pp. 1781–1807. Martinus-Nijhoff, The Hague.

Toivanen, P., Toivanen, A., and Good, R. A. (1972). *J. Immunol.* **109,** 1058–1070.

Weber, W. T., and Foglia, L. M. (1980). *Cell. Immunol.* **52,** 84–94.

Weber, W. T., and Mausner, R. (1977). *In* "Avian Immunology" (A. A. Benedict, ed.), pp. 47–59. Plenum, New York.

Weidanz, W. P., Konietzko, D., and Lerman, S. P. (1971). *J. Reticuloendothel. Soc.* **9,** 635.

Wong, T. W., Klinkert, W. E. F., and Bowers, W. E. (1982). *Immunobiology* **160,** 413–423.

Zigmond, S. H. (1977). *J. Cell Biol.* **75,** 606–616.

CHAPTER 21

DIFFERENTIAL COMMITMENT OF HEMOPOIETIC STEM CELLS LOCALIZED IN DISTINCT COMPARTMENTS OF EARLY *XENOPUS* EMBRYOS

Chiaki Katagiri, Mitsugu Maéno, and Shin Tochinai

ZOOLOGICAL INSTITUTE
FACULTY OF SCIENCE
HOKKAIDO UNIVERSITY
SAPPORO, JAPAN

I. Introduction

The generally accepted concept regarding the differentiation of hemopoietic cells in vertebrates is that pluripotent stem cells that are initially differentiated in the yolk sac or its phylogenetic equivalent migrate via the circulation to colonize receptive organs, where they undergo subsequent differentiation into discrete lineages under the influence of specific microenvironments (Metcalf and Moore, 1971). Although this concept of hematogenous metastasis may be correct in general terms, the exact site of the origin of the stem cells as well as their state of commitment before and during migration in early embryos are still a matter of conjecture. In fact, more recent studies on the avian model (e.g., Le Douarin *et al.*, 1982; chapter by Le Douarin, this volume) have emphasized that definitive hemopoiesis is dependent not on the yolk sac blood islands, as had previously been believed, but on the intraembryonic stem cells localized in the mesenchyme associated with the dorsal aorta. A number of experiments made by Turpen and co-workers (Turpen *et al.*, 1982) on anuran amphibians, particularly *Rana pipiens,* seem to argue against the concept of the common hemopoietic stem cells that originate from the ventral blood islands of

315

early embryos, the amphibian counterpart of yolk sac blood islands in mammals.

In this brief review, we will summarize knowledge of the early differentiation of lymphocytes mainly derived from our recent studies of the African clawed frog *Xenopus laevis,* with particular attention to the site of the origin of thymocytes in early embryos and its relevance to the differentiation capacity as stem cells.

II. Interstitial Immigration of Lymphoid Stem Cells in Early Thymus Rudiments

As in other anuran amphibians, the thymus rudiments in *X. laevis* emerge as a pair of buddings from the pharyngeal epithelium (second visceral pouch; Manning and Horton, 1982). These epithelial buds first become obvious 2 days after fertilization (st. 40; Nieuwkoop and Faber, 1956). By st. 45 (4 days old) the rudiments have detached themselves from the pharyngeal epithelium and contain two clearly discernible types of cells: epithelial and lymphoid precursor cells (Fig. 1). By st. 47 (6 days old) the latter type of cells have increased greatly in number to give rise to large lymphocytes.

The evidence that the lymphocytes in the amphibian thymus are exclusively of extrinsic origin came from experiments grafting the gill region that contains the thymus primordia between diploid and triploid or tetraploid tailbud embryos. All experiments that have been made with the anurans *R. pipiens* and *X. laevis* (Volpe *et al.,* 1979) and an urodele, *Pleurodeles waltlii* (Deparis and Jaylet, 1976), have agreed that the thymocytes are of the ploidy of host tissues. To determine at what stage of its histogenesis the thymus rudiment is colonized by lymphoid stem cells, the diploid thymus rudiments from 3- to 11-day-old larvae were grafted under the tail dermis of histocompatible, triploid tadpoles at the 40-day-old stage (Tochinai, 1978). Cytophotometric determinations of the lymphocyte ploidy when the thymus was 18 days chronologically revealed that the host-derived lymphocytes occur in the grafted thymuses taken from 3- to 4-day-old (sts. 42–45) larvae, and the grafted thymuses taken from 5-day-old (st. 46) and later larvae contained mostly donor-derived lymphocytes. Colonization by thymocyte precursor cells evidently occurs, therefore, during the limited period of 3–4 days of age (sts. 42–45). This cell immigration will certainly occur via the mesenchyme, since the vascularization in the thymus occurs in the 12-day-old (st. 49) larvae (Manning and Horton, 1982).

Stage-by-stage electron microscopic (TEM) observations of the early thymus histogenesis by Nagata (1977) have revealed that, in confir-

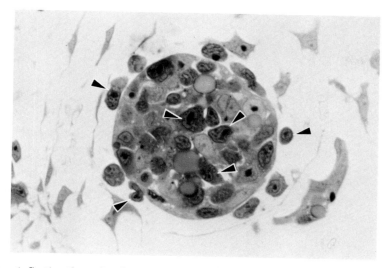

FIG. 1. Section through 4-day-old thymic rudiments of *Xenopus laevis* (st. 44), showing lymphoid stem cells (arrowheads) in and outside the rudiment. ×750.

mation of the classical notion advanced by Sterba (1950) with regard to the basophilic "amöboide Wanderzellen," putative lymphoid precursor cells possessing a high content of free ribosomes frequently attach to and pass through the basal lamina immediately surrounding the thymic rudiments to become engulfed in the rudiments. Strong confirmation of this finding was obtained by an observation in which 4-day-old thymus rudiments with their surrounding mesenchyme were cultured *in vitro* (Tochinai, 1980). In the mesenchyme of larvae at this particular stage, the cells possessing conspicuous blebs or lobopodia were frequently observed to move toward the rudiments (Fig. 2). These cells attach themselves to the surface of the organ rudiments in a cone-shaped form, finally entering the rudiments. Analyses of cell movements on several occasions indicated that the pathway of their migration is by no means random, always moving toward and never away from the rudiments. In agreement with TEM studies as well as experimental results with the grafted organ rudiments just described, this cell invasion into the rudiments was no longer observed in the preparation taken from 5-day-old larvae (st. 46).

Following this initial colonization by thymocyte stem cells in 3- to 4-day-old larvae (sts. 42–45), there is no further opportunity for cell invasion in the thymus until the commencement of vascularization in 12-day-old (st. 49) larvae. Therefore, the burst of lymphoid cell pro-

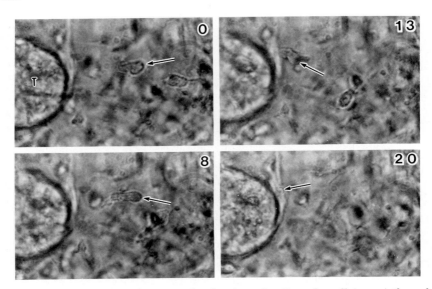

Fig. 2. Time-lapse photomicrographs showing migration of a cell (arrow) through mesenchyme toward thymus rudiment (T) in a 4-day-old larva (st. 44). Live specimen. Numerals in each photograph indicate time in minutes when photographed. ×450.

liferation in this organ during the prevascularization period is apparently dependent on the interstitially immigrated stem cells. It should be mentioned that these initially immigrated stem cells are sufficient by themselves to support the differentiation of a wide variety of T lymphocyte clones, since thymectomizing larvae after the commencement of thymic vascularization have been shown to be capable of eliminating the least number of the committed lymphocytes functioning in cellular immune response in adult frogs (Horton and Manning, 1974; Tochinai and Katagiri, 1975).

III. Localization of Hemopoietic Stem Cells in Embryos

In their experiments designed to investigate the exact source of thymocyte stem cells in amphibian embryos, Volpe *et al.* (1979) produced chimeras by joining the anterior and posterior halves of diploid and triploid tailbud embryos. Determination of the ploidy of cells in such postmetamorphic chimeras in both *R. pipiens* and *X. laevis* revealed that the thymocytes are derived exclusively from the posterior half of the embryos. More recently, extensive studies are being carried out on *R. pipiens* to localize in embryos the source of differentiating hemocytes. The major findings by Turpen *et al.* (1979, 1981) are that the dorsolateral plate (DLP) mesoderm in the prospective mesonephric region gives rise to all classes of hemopoietic cells. In addition, it is emerging that, in contradiction to the previous belief (Finnegan, 1953), the ventral blood island (VBI) mesoderm is solely responsible for the transient population of embryonic erythrocytes that declines during larval development. The most recent report by Flajnik *et al.* (1984) on *X. laevis* seems to substantiate this view by showing that the adult lymphocytes are exclusively derived from the DLP mesoderm of tailbud embryos. A series of experiments on *X. laevis* carried out in our laboratory (Maéno *et al.*, 1985a) indicates, however, that the above straightforward conclusion may be only partially correct for the cell population that contributes to the definitive hemopoiesis. Our results as summarized below demonstrate that the DLP and VBI mesoderms of tailbud embryos constitute two different compartments, which supply hemopoietic cells functioning at different developmental stages.

The experimental design and techniques we employed were essentially the same as those used by previous workers (Volpe *et al.*, 1979; Flajnik *et al.*, 1984). Thus, in agreement with previous reports, the precursor cells of all the hemopoietic cells were localized at the posterior one-half to three-quarters level of the embryos, and the DLP mesoderm (indicated as D in Fig. 3a) contributed to the major but not to the whole hemopoietic population when the chimeric animals were sacrificed at the late larval (>30 days old; st. 54) through postmetamorphic stages (60–100 days old). An intriguing finding was that none of the hemopoietic cells in early larvae (<30 days old) was derived from the same DLP mesoderm grafts. In contrast, when the cytogenetically labeled VBI mesoderm (indicated as V in Fig. 3a) was grafted and its contribution to hemopoietic cells was evaluated at various developmental stages, the graft-derived cells comprised the major population

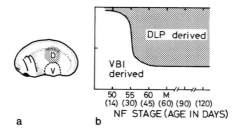

a   b

FIG. 3. (a) The dorsolateral plate (D) and ventral blood island (V) regions of st. 22 *Xenopus* embryo. (b) The rate of their contribution in supplying hemopoietic cell populations at various developmental stages. (Data based on Maéno *et al.*, 1985a.)

of hemopoietic cells in the early larvae and the minor population of such cells in advanced larvae and adults (Fig. 3b). In all the chimeric individuals examined, there was no significant difference in the ratio of distribution of donor and host cells between thymocytes, splenocytes, and erythrocytes, suggesting a lack of preferential localization of hemopoietic stem cell lines in the mesoderm of st. 22 embryos, or their pluripotency as hemopoietic stem cells.

Our results thus indicate the existence of distinct compartments (the DLP and VBI) in the tailbud embryos for the hemopoietic precursor cells that commence to function at different developmental stages. In their recent similar grafting experiments, Kau and Turpen (1983) demonstrated that in embryos at the neural fold stage of X. laevis (sts. 14–19) the thymocyte precursor cells for the 12-day-old larvae (st. 49) are localized in both the VBI and DLP mesoderms, whereas the erythrocyte precursors are localized solely in the VBI region. It is plausible that the location of lymphoid precursor cells becomes increasingly restricted to the ventral region during the 8 hours between the neural fold and tailbud stages. On the other hand, the preferential localization of erythrocyte precursors in the VBI of such early embryos is of particular interest in that it suggests a possible relevance to the occurrence of transient, embryo-specific hemoglobin expression that is soon replaced by the tadpole-specific one (Kobel and Wolff, 1983).

IV. Differential Commitment of VBI and DLP Mesoderms as Hemopoietic Stem Cells

The finding of the distinct compartments described above raises questions as to whether there will also be a difference in the state of commitment between the DLP and VBI mesoderms with regard to their capacity to differentiate into hemopoietic cells. To try to answer this question, diploid VBI tissue (region V in Fig. 3a of an embryo at st. 22) was grafted heterotopically onto the triploid embryos at the DLP region (region D in Fig. 3a) and vice versa (Maéno et al., 1985a). The results clearly showed that the VBI mesoderm contributes largely to the hemopoietic cells in *both* early and late larvae, but that none of the cells grafted from the DLP to the VBI region differentiated into hemopoietic cells at any developmental stage of the hosts examined. Control heterotopic grafting comprising the more anteriorly located, prospective pronephric region grafted onto either the DLP or VBI region did not give rise to the graft-derived hemopoietic cells at all. We suggest that the precursor cells in the VBI region that differentiate into a hemopoietic lineage have already been determined by st. 22, thus giving rise to a transient hemopoietic population functioning in the early larval stages. Assuming that the precursors in the DLP mesoderm are not yet determined in this sense, they will require additional inductive influences from surrounding tissues in order to differentiate into the hemopoietic lineage. We therefore argue that the two compartments demonstrated here signify not only the topography of constituent cells but also the differential states of commitment as hemopoietic precursor cells.

The next question of interest concerns how the lymphoid precursor cells in the VBI region migrate toward the primary lymphoid organs in early embryos. Previous study (Mangia *et al.*, 1970) has shown that the primitive blood cells first appear at st. 31 throughout the VBI region, accompanied by the establishment of blood circulation around sts. 33–39 (2 days old). Consequently, the thymocyte precursor cells may find a way of vascular migration as soon as the vasculature is established in the embryo. This has indeed been substantiated by our own recent observations (Maéno *et al.*, 1985b) employing the unique stainability of *Xenopus borealis* nuclei as a marker (cf. Thiébaud, 1983). The VBI or DLP tissue from *X. borealis* tailbud embryos was grafted orthotopically to stage-matched *X. laevis* embryos, followed by histological observations on the quinacrine-stained sections of host larvae at various developmental stages. The results showed that the VBI-derived cells occur in all blood capillaries throughout the body as soon as the circulatory system is established. Remarkable in sts. 43–45 larvae was that besides blood vessels, the thymus rudiments and their adjacent mesenchyme as well as the initial primordia of mesonephros at the DLP region were the only sites where the VBI-derived cells were localized (Fig. 4). The DLP-derived cells, however, did not show any migratory pattern during these early larval stages, although they were actively engaged in hemopoiesis in later larval stages.

Our observations have thus established that the lymphoid precursor cells emerging in the VBI first enter the circulation through the vitelline veins in association with differentiating erythrocytes. In contrast with the erythrocytes, these precursor cells leave the blood vessels around the jugular vein and/or internal carotid artery, and move interstitially toward the thymic rudiments in order to colonize them. Similarly, other populations of VBI-derived cells evidently leave the circulation around the mesonephric primordia to become involved in hemopoiesis in conjunction with the DLP-residing stem cells that differentiate into definitive hemocytes. The possible relevance of the late-differentiating, definitive population of DLP-derived hemopoietic cells is the observation (Tochinai, unpublished) that there is a large scale of renewal in thymus lymphocyte populations during the late larval

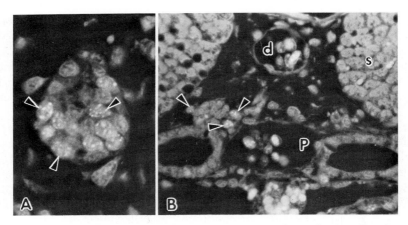

Fɪɢ. 4. Sections through right thymus rudiment (A) and initial primordia of mesonephros (B) of st. 44 *X. laevis,* which received ventral blood island (VBI) graft (cf. Fig. 3)
from *X. borealis* at st. 22, showing the brightly stained graft-derived cells (arrowheads)
in the thymus rudiment (A) and mesonephros (B). Graft-derived cells are found also in
dorsal aorta (d) and posterior vena cava (P); s, somite. (A) ×550; (B) ×200.

stages. We emphasize, in partial agreement with a previous notion by
other workers (Turpen *et al.,* 1982), that the larval mesonephros is the
site of active hemopoiesis involving the stem cells derived both from
the VBI and DLP regions of tailbud embryos.

V. Concluding Remarks

The VBI-derived lymphoid precursor cells that colonize the early
thymic rudiments provide an intriguing model through which the
mechanism of interaction between particular types of cells and the
organ primordia at particular stages may be analyzed. In some preliminary experiments by one of us (Tochinai, unpublished observation), 3-
to 5-day-old (sts. 42–45) thymic primordia were implanted into advanced larvae at the area close to the well-differentiated host thymuses, which had been broken *in situ* to allow a release of the constituent thymocytes. Examination of the organ rudiments 24 hours after
the implantation revealed an extremely large number of thymocytes
that had colonized or gathered around the rudiments of 3- to 4-day-old
(sts. 42–45) larvae, but not those of 5-day-old larvae (st. 46). Evidently
the VBI-derived precursor cells colonizing the organ rudiments in normal thymus development do share with the lymphocytes in the well-
differentiated thymus the behavioral property of reacting against thymic rudiments. Besides these properties, however, nothing has currently been shown about the state of commitment of these stem cells in

molecular terms. Employing this experimental model, further efforts are warranted to try to detect the expression of cell-surface properties that are likely to be associated with commitment to a certain repertoire of hemopoietic cell lineages.

ACKNOWLEDGMENT

This study was supported by Grants-in-Aid from the Ministry of Education, Science and Culture (Project Nos. 58480412 and 59480023, Special Project Research, Multicellular Organization).

REFERENCES

Deparis, P., and Jaylet, A. (1976). *Ann. Immunol.* **123**, 827–831.
Finnegan, M. F. (1953). *J. Exp. Zool.* **123**, 371–395.
Flajnik, M. F., Horan, P. K., and Cohen, N. (1984). *Dev. Biol.* **104**, 247–254.
Horton, J. D., and Manning, M. J. (1974). *Immunology* **26**, 797–807.
Kau, C. L., and Turpen, J. B. (1983). *J. Immunol.* **131**, 2262–2266.
Kobel, H. R., and Wolff, J. (1983). *Differentiation* **24**, 24–26.
Le Douarin, N. M., Jotereau, F. V., Houssaint, E., and Dieterlen-Lièvre, F. (1982). *In* "The Reticuloendothelial System" (N. Cohen and M. M. Sigel, eds.), pp. 589–616. Plenum, New York.
Maéno, M., Todate, A., and Katagiri, Ch. (1985a). *Dev. Growth Differ.* **27**, 137–148.
Maéno, M., Tochinai, S., and Katagiri, Ch. (1985b). *Dev. Biol.* **110**, 503–508.
Mangia, F., Procicchiani, G., and Manelli, H. (1970). *Acta Embryol. Exp.* **1970**, 163–184.
Manning, M. J., and Horton, J. D. (1982). *In* "The Reticuloendothelial System" (N. Cohen and M. M. Sigel, eds.), pp. 423–495. Plenum, New York.
Metcalf, D., and Moore, M. A. S. (1971). "Haemopoietic Cells." North-Holland Publ., Amsterdam.
Nagata, S. (1977). *Cell Tissue Res.* **179**, 87–96.
Nieuwkoop, P. D., and Faber, J. (1956). "Normal Table of *Xenopus laevis* (Daudin)." North-Holland Publ., Amsterdam.
Sterba, G. (1950). *Abh. Sachs. Akad. Wiss.* **44**, 1–54.
Thiébaud, C. H. (1983). *Dev. Biol.* **98**, 245–249.
Tochinai, S. (1978). *Dev. Comp. Immunol.* **2**, 627–635.
Tochinai, S. (1980). *Dev. Comp. Immunol.* **4**, 273–282.
Tochinai, S., and Katagiri, Ch. (1975). *Dev. Growth Differ.* **17**, 383–394.
Turpen, J. B., Turpen, C. J., and Flajnik, M. (1979). *Dev. Biol.* **69**, 466–479.
Turpen, J. B., Knudson, C. M., and Hoefen, P. S. (1981). *Dev. Biol.* **85**, 99–112.
Turpen, J. B., Cohen, N., Deparis, P., Jaylet, A., Tompkins, R., and Volpe, E. P. (1982). *In* "The Reticuloendothelial System" (N. Cohen and M. M. Sigel, eds.), pp. 569–588. Plenum, New York.
Volpe, E. P., Tompkins, R., and Reinschmidt, D. (1979). *J. Exp. Zool.* **208**, 57–66.

CHAPTER 22

PROBABLE DEDIFFERENTIATION OF MAST CELLS IN MOUSE CONNECTIVE TISSUES

Yukihiko Kitamura, Takashi Sonoda, Toru Nakano, and Yoshio Kanayama

INSTITUTE FOR CANCER RESEARCH AND DEPARTMENT OF MEDICINE
OSAKA UNIVERSITY MEDICAL SCHOOL
KITA-KU, OSAKA, JAPAN

I. Introduction

Mast cells are a ubiquitous constituent of connective tissues in most mammalian species. Their morphological identification is easy because of the prominent basophilic granules in their cytoplasm. These basophilic granules contain many pharmacologically active substances such as histamine and heparin. Immunological functions of mast cells are accomplished through the immunoglobulin E (IgE) receptors on the surface (reviewed by Ishizaka and Ishizaka, 1984). The stimulated mast cells release histamine and also synthesize prostaglandins or leukotrienes (reviewed by Schwartz and Austen, 1984). Although mast cells were considered to be proper connective tissue cells, recent *in vivo* and *in vitro* studies have clearly shown that mast cells are derived from hematopoietic tissues (reviewed by Kitamura *et al.*, 1983; Galli *et al.*, 1984). However, the differentiation process of mast cells is different from other blood cells such as erythrocytes and neutrophils. In this chapter, recent studies on mast cell differentiation will be reviewed. It is expected that they will provide a potential system for analyzing the mechanism of cell commitment and the instability in phenotypic expression of differentiated cells.

325

II. Useful Mutant Mice

A. THE bg^J/bg^J MOUSE

The bg^J/bg^J mouse (bg locus; chromosome 10) has a disorder similar to the human Chediak–Higashi syndrome, which is characterized by marked enlargement of lysosomes and specific granules of various types of cells. For example, the C57BL/6–bg^J/bg^J mouse is easily distinguished from its normal litter mates ($bg^J/+$ or $+/+$) by its diluted coat color, which is attributed to giant melanosomes in hairs (Silvers, 1979). By using giant basophilic granules of bg^J/bg^J mice as a marker, Kitamura et al. (1977) demonstrated the bone marrow origin of mast cells.

B. THE W/W^v MOUSE

A double gene dose of mutant alleles at the W locus (chromosome 5) has been known to produce the pleiotropic effects of macrocytic anemia, sterility, and lack of hair pigmentation (reviewed by Russell, 1979). In addition to the above-mentioned abnormalities, Kitamura et al. (1978) found that W mutant alleles affected the production of mast cells. The number of mast cells in the skin of adult W/W^v mice is less than 1% observed in the congenic $+/+$ mice. No mast cells appeared in other tissues of W/W^v mice. The histamine concentration in the skin of W/W^v mice is about 2% of the value detected in the $+/+$ mice (Yamatodani et al., 1982). Although the content of glycosaminoglycans other than heparin (hyaluronic acid, chondroitin sulfate, dermatan sulfate, and heparan sulfate) in the skin of W/W^v mice is comparable to that of the $+/+$ mice, heparin was not detectable in the skin of W/W^v mice (Nakamura et al., 1981).

The depletion of erythrocytes and melanocytes has been demonstrated to be due to a defect in their precursor cells (reviewed by Russell, 1979). The depletion of mast cells is also attributable to a defect in precursor cells. In fact, the intravenous injection of bone marrow cells from the congenic $+/+$ mice normalized the number of mast cells in the W/W^v mice (Kitamura et al., 1978).

C. THE Sl/Sl^d MOUSE

A double gene dose of mutant alleles at the Sl locus (chromosome 10) produces pleiotropic effects that cannot be distinguished from those produced by a double gene dose of W mutant alleles (reviewed by Russell, 1979). Mice of Sl/Sl^d genotype show macrocytic anemia, sterility, lack of hair pigmentation, and depletion of mast cells (Kitamura and Go, 1979). In spite of the similarity of phenotypes, the underlying

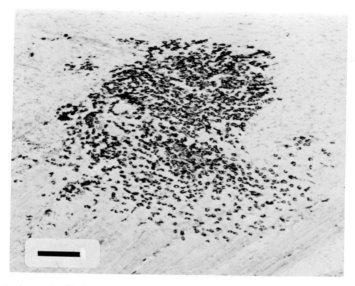

FIG. 1. A mast cell cluster which developed in the skin of a genetically mast cell-deficient W/W^v mouse after the injection of a cell suspension containing mast cell precursor(s). Stained with toluidine blue. Bar, 100 μm.

mechanisms are quite different. The defects in the W/W^v mice are attributed to precursor cells, although precursors of erythrocytes, melanocytes, and mast cells of the Sl/Sl^d mice are apparently normal. In fact, the intravenous injection of bone marrow cells from Sl/Sl^d mice cured the anemia and mast cell depletion of W/W^v mice (Kitamura and Go, 1979). In contrast, the injection of bone marrow cells from $+/+$ to Sl/Sl^d mice did not increase the number of erythrocytes and mast cells. The direct injection of a cell suspension containing $+/+$ mast cell precursors into the skin of W/W^v mice resulted in the development of a mast cell cluster at the injection site (Fig. 1). In contrast, no mast cells were observed at the injection sites in the skin of Sl/Sl^d mice (Sonoda et al., 1982).

III. In Vitro Cultures

Trials to develop in vitro systems for studying the differentiation of mast cells are not new. Ginsburg and Sachs (1963) cultured thymus cells of mice on a feeder layer composed of mouse skin fibroblasts and then observed the development of mast cells. Ishizaka et al. (1976, 1977) obtained a similar result by using the rat thymus, and they demonstrated the content of histamine and the presence of IgE recep-

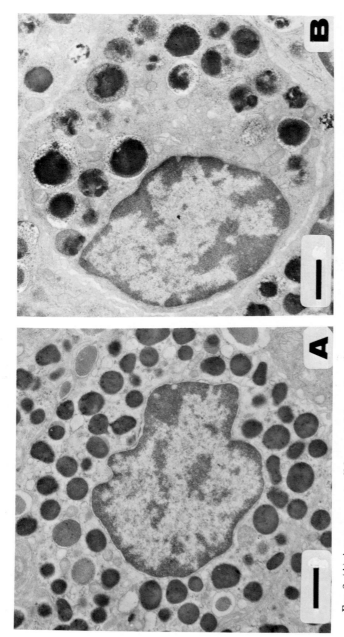

FIG. 2. (A) A mouse mast cell harvested from the peritoneal cavity. (B) A mouse mast cell which developed in a suspension culture of bone marrow cells. Granules are fewer in the cell developing *in vitro* than in the cell harvested from the peritoneal cavity. Mitosis of *in vitro* developing mast cells is frequent but that of peritoneal mast cells is rare. Bars, 1 μm.

tors in mast cells differentiated in such a system. These results were used by Burnet (1977) as evidence supporting his hypothesis that some or all mast cells might be derived from T cells. However, recent detailed studies have shown that T cells themselves are not precursors of mast cells (reviewed by Galli et al., 1984). On the other hand, T cells produce a factor that stimulates in vitro growth and differentiation of mast cells. The factor has been purified and designated as interleukin 3 (Ihle et al., 1983). The cDNA sequence coding interleukin 3 has been cloned and characterized (Fung et al., 1984; Yokota et al., 1984). Methods of making mast cell colonies in semisolid culture media have also been developed (Zucker-Franklin et al., 1981; Nakahata et al., 1982). As reviewed by Galli et al. (1984), mast cells produced in vitro are less differentiated than typical peritoneal mast cells (Fig. 2).

IV. Differentiation Process

The relation of mast cells to the multipotential hematopoietic stem cell was investigated by Kitamura et al. (1981), Shrader et al. (1981), and Nakahata et al. (1982). All groups obtained the same conclusion that mast cells are a progeny of the hematopoietic stem cell. Most progenies of the stem cell, such as erythrocytes, neutrophils, and platelets, leave hematopoietic tissues after maturation. In contrast, mast cells do not complete differentiation in hematopoietic tissues, although they are progeny of the stem cell as well. No mast cells are detectable in the blood. However, when blood mononuclear cells of the +/+ mice were directly injected into the skin of the congenic W/W^v mice, mast cell clusters appeared at the injection sites (Sonoda et al., 1982). This indicates that mast cell precursors migrate in the bloodstream.

When mast cell precursors arrive in connective tissues, they proliferate and then differentiate into mast cells (Hatanaka et al., 1979; Kitamura et al., 1979). When the mixture of bone marrow cells from bg^J/bg^J mice and those from +/+ mice was injected intravenously into W/W^v mice, clusters of mast cells appeared in the skin and mesentery of the W/W^v recipients. Since each of the clusters consisted of either bg^J/bg^J-type mast cells alone or +/+ mast cells alone, the mast cell cluster seemed to originate from a single precursor cell. In other words, a single precursor proliferates and forms a mast cell colony after its arrival in connective tissues.

A portion of the descendants of blood-migrating precursors retain the ability to proliferate and differentiate into mast cells after localization in the skin (Matsuda et al., 1981). Bone marrow cells from the +/+ mice were injected intravenously into the congenic W/W^v

mice. On various days after the bone marrow transplantation, pieces of the skin were removed from the W/W^v recipients and grafted onto the back of the $+/+$ mice that had been irradiated and injected with bone marrow cells from the bg^J/bg^J mice. Proliferation of $+/+$-type mast cell precursors, which had been derived from the injected bone marrow cells, was demonstrated by radioautography. Since preparation of single-cell suspensions from the skin of adult mice is difficult, further characterization of localized mast cell precursors was done using peritoneal cells as the starting material. The concentration of mast cell precursors in the peritoneal cavity was about five times as great as the concentration in the bone marrow. Although peritoneal mast cell precursors were also shown to originate from the bone marrow, physical characterization revealed that the peritoneal precursors differed from the marrow precursors. The peritoneal precursors were less susceptible to irradiation than the marrow precursors; the former were heavier than the latter. When a 95% mast cell suspension was prepared from the peritoneal cells by the removal of phagocytes and density gradient centrifugation, 1 out of 16 cells had the potentiality to make a mast cell cluster, as shown in Fig. 1. Moreover, when a single mast cell was identified under the phase-contrast microscope and picked up with the micromanipulator, 1 out of 17 mast cells made a cluster (Sonoda et al., 1984). This indicates that some peritoneal mast cells keep extensive proliferative potentiality even after morphological differentiation. In other words, some peritoneal mast cells themselves may function as the localized precursors.

V. Probable Dedifferentiation

Although the division of mast cells grown in vitro is always observed, the division of mast cells is rarely detectable in either the skin (Matsuda et al., 1981) or the peritoneal cavity (Padawer, 1974). Even when production of mast cells is stimulated in the skin, mitotic figures of granular mast cells are rarely observed, although a considerable proportion of mast cells was labeled with [³H]thymidine (Matsuda et al., 1981). This result suggests that in vivo production of mast cells results from proliferation and differentiation of nongranular cells rather than from mitosis of granular mast cells. In contrast, a portion of morphologically identifiable mast cells may function as localized mast cell precursors (Sonoda et al., 1984). In order to resolve this problem, we recently carried out an experiment. One hundred morphologically identifiable peritoneal mast cells of the $+/+$ mice were collected and directly injected into the skin of the congenic W/W^v mice; the recipient W/W^v mice were killed at various days after the injection. No

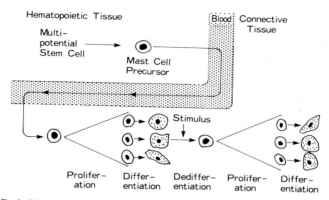

Fig. 3. Probable dedifferentiation of mast cells in the connective tissue. When local production of mast cells is demanded, mast cells with proliferative potentiality may dedifferentiate, proliferate, and differentiate again into mast cells.

mast cells were observed at the injection sites on the tenth day after the injection (unpublished data). Moreover, division of granular mast cells was rarely observed even after the development of mast cell clusters. This result suggests that mast cells may lose their granules before proliferation. Probably, such dedifferentiated mast cells divide and differentiate again into mast cells (Fig. 3). We speculate that this process may be similar to the blastogenesis of small lymphocytes.

ACKNOWLEDGMENTS

This work was supported by grants for Special Project Research and Cancer Research from the Ministry of Education, Science and Culture, a grant for Research of Intractable Diseases from the Ministry of Health and Welfare, and a grant from the Princess Takamatsu Cancer Research Fund.

REFERENCES

Burnet, F. M. (1977). *Cell. Immunol.* **30,** 358–360.
Fung, M. C., Hapel, A. J., Ymer, S., Cohen, D. R., Johnson, R. M., Campbell, H. D., and Young, I. G. (1984). *Nature (London)* **307,** 233–237.
Galli, S. J., Dvorak, A. M., and Dvorak, H. F. (1984). *Prog. Allergy* **34,** 1–141.
Ginsburg, H., and Sachs, L. (1963). *J. Natl. Cancer Inst.* **31,** 1–40.
Hatanaka, K., Kitamura, Y., and Nishimune, Y. (1979). *Blood* **53,** 142–147.
Ihle, J. N., Keller, J., Oroszlan, S., Henderson, L. E., Copeland, T. D., Fitch, F., Prystowsky, M. D., Goldwasser, E., Schrader, J. W., Palaszynski, E., Dy, M., and Lebel, B. (1983). *J. Immunol.* **131,** 282–287.
Ishizaka, T., and Ishizaka, K. (1984). *Prog. Allergy* **34,** 188–235.
Ishizaka, T., Okudaira, H., Mauser, L. E., and Ishizaka, K. (1976). *J. Immunol.* **116,** 747–754.

Ishizaka, T., Adachi, T., Chang, T. H., and Ishizaka, K. (1977). *J. Immunol.* **118,** 211–217.

Kitamura, Y., and Go, S. (1979). *Blood* **53,** 492–497.

Kitamura, Y., Shimada, M., Hatanaka, K., and Miyanao, Y. (1977). *Nature (London)* **268,** 442–443.

Kitamura, Y., Go, S., and Hatanaka, K. (1978). *Blood* **52,** 447–452.

Kitamura, Y., Matsuda, H., and Hatanaka, K. (1979). *Nature (London)* **281,** 154–155.

Kitamura, Y., Yokoyama, M., Matsuda, H., Ohno, T., and Mori, K. J. (1981). *Nature (London)* **291,** 159–160.

Kitamura, Y., Sonoda, T., and Yokoyama, M. (1983). *In* "Hemopoietic Stem Cells" (S. A. Killmann, E. P. Cronkite, and C. N. Muller-Berat, eds.), pp. 350–361. Munksgaard, Copenhagen.

Matsuda, H., Kitamura, Y., Sonoda, T., and Imori, T. (1981). *J. Cell. Physiol.* **108,** 409–415.

Nakahata, T., Spicer, S. S., Cantey, J. R., and Ogawa, M. (1982). *Blood* **60,** 352–361.

Nakamura, N., Kojima, J., Okamoto, S., and Kitamura, Y. (1981). *Biochem. Int.* **5,** 449–456.

Padawer, J. (1974). *Exp. Mol. Pathol.* **20,** 269–280.

Russell, E. S. (1979). *Adv. Genet.* **20,** 357–459.

Schrader, J. W., Lewis, S. J., Clark-Lewis, I., and Culvenor, J. G. (1981). *Proc. Natl. Acad. Sci. U.S.A.* **78,** 323–327.

Schwartz, L. B., and Austen, K. F. (1984). *Prog. Allergy* **34,** 271–321.

Silvers, W. K. (1979). "The Coat Color of Mice." Springer-Verlag, Berlin and New York.

Sonoda, T., Ohno, T., and Kitamura, Y. (1982). *J. Cell. Physiol.* **112,** 136–140.

Sonoda, T., Kanayama, Y., Hara, H., Hayashi, C., Tadokoro, M., Yonezawa, T., and Kitamura, Y. (1984). *J. Exp. Med.* **160,** 138–151.

Yamatodani, A., Maeyama, K., Watanabe, T., Wada, H., and Kitamura, Y. (1982). *Biochem. Pharmacol.* **31,** 305–309.

Yokota, T., Lee, F., Rennick, D., Hall, C., Arai, N., Mosmann, T., Nabel, G., Cantor, H., and Arai, K. (1984). *Proc. Natl. Acad. Sci. U.S.A.* **81,** 1070–1074.

Zucker-Franklin, D., Grusky, G., Hirayama, N., and Schnipper, E. (1981). *Blood* **58,** 544–551.

CHAPTER 23

INSTABILITY AND STABILIZATION IN MELANOMA CELL DIFFERENTIATION

Dorothy C. Bennett

DEPARTMENT OF ANATOMY
ST. GEORGE'S HOSPITAL MEDICAL SCHOOL
LONDON, ENGLAND

I. Introduction: Cell Types

Can cell differentiation be unstable or reversible? The answer depends partly on how one defines cell differentiation. If one chooses a standard definition, "the production of different cell types in development" (Balinsky, 1981), then the answer depends on how one defines cell types. Now the idea of different cell types is not as simple as it may appear. Experimental embryology and cell biology have led to several different, and even conflicting, concepts or definitions of "cell types," and different uses of the same term can cause misunderstandings. Some of these definitions will therefore be considered briefly. Afterward evidence will be reviewed from mouse melanoma (pigment tumor) cells and other systems, leading to the idea that a change in cell type (differentiation) may be unstable before becoming stable. The emphasis will be on mammalian cells.

The disparate definitions of "cell type" apply less to mature cells (like adult liver or nerve cells) than to immature cell types (like ectoderm) and to cell *lineages,* i.e., the branching successions of immature cell types through which mature types develop. A *lineage step* is defined here as a change in cell type. The problem is to detect

333

such a step in practice, when one cell type changes into the next. First immature cell types are sometimes classified by their physical (immediately detectable) properties, like size, shape, or the capacity to bind specific antibodies. However in developing embryos these properties can change continuously rather than stepwise. It is then easier to classify cells by either their developmental *fate* or their *potency* (which may correlate to physical properties). Potency will be defined below; *fate* is defined as the range of mature cell (or tissue) types found among the progeny of an embryonic cell (or tissue) type after undisturbed development (Balinsky, 1981; Slack, 1983). For example, the fate of the neural plate of a vertebrate embryo is to form the nervous system. The neural plate first develops into specific parts which have narrower fates. In this way one can map a fate lineage in which the successive cell types are defined by their different fates (Fig. 1a).

Differentiation is sometimes altered in cells transplanted or explanted to a new environment. A cell is said to be *committed* to a lineage step, *with respect to a given environment,* if this step is not prevented or reversed by a change in the specified environment. A test of commitment is *potency,* defined as all the cell types a given cell can engender if exposed to different environments (in practice, those environments tested so far) (Slack, 1983). Classifying cells by potency gives potency lineages, as usually used in hematology (Till and McCulloch, 1980). Here a lineage step is the same as commitment; this is not necessarily the case with lineages other than potency lineages, as will be discussed.

Potency may be defined to include only mature cell types or also immature types—the convention adopted in this chapter. For example, explanted human myoblasts (muscle precursor cells) proliferate to give more myoblasts in one culture medium, but fuse and differentiate into striated muscle in another (Blau and Webster, 1981). Their differentiation is not reversed by restoring the first medium. If potency is defined to include immature cell types, then the potency to produce myoblasts had been lost here, and commitment had occurred: this illustrates an unbranching lineage step (Fig. 1b). Otherwise, if potency is taken to include only mature cell types, then the change from myoblast to muscle does not in itself constitute commitment. The latter view seems unduly narrow, although it is possible that loss of potency to remain immature differs at a molecular level from loss of potency to produce a mature cell type. This may soon be clarified by the research in progress on commitment in both branching and unbranching lineage steps. The step now to be discussed is unbranching in the differentiation of cultured mouse melanoma cells. This step involves pigment synthesis and other features visible in single living cells, which

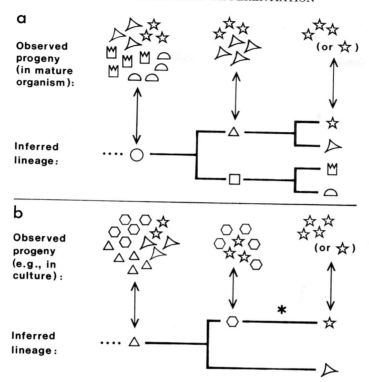

FIG. 1. Two common methods of mapping cell lineage which give different outcomes. Each symbol indicates a cell type. Branches show restrictions of lineage and are not necessarily connected with cell divisions. (a) Lineage based on fate. Here an embryonic cell or tissue is vitally marked and the organism is left to develop. At the end of development, the marked mature cells reveal the fate of the precursor cells (see also text). (b) Lineage based on potency. Here cells are cloned in culture or marked and transplanted to show potency (see also text). Even if only mature cells are included in the potency the results can differ from a fate lineage. For example, 16-cell mouse embryos contain two layers of cells which differ in their fates, but have the same potency (Ziomek et al., 1982; Section III,B). The case shown differs doubly from (a) because immature as well as mature cells are included in the potency. The asterisk (∗) represents an unbranching lineage step.

prove useful for comparing the timing of expression (physical changes) with that of commitment.

II. Commitment in Mouse Melanoma and Other Pigment Cells

A. DIFFERENTIATION OF MAMMALIAN PIGMENT CELLS

A brief background on the biology and development of melanocytes (pigment cells) is appropriate here (reviewed in detail by Fitzpatrick et *al.,* 1979 and Le Douarin, 1982). Mammalian epidermal melanocytes are stellate or bipolar cells which normally lie beneath the basal epithelial cells of the skin and within the hair follicles and produce melanosomes. These are membranous organelles containing melanin (pigment) and proteins including tyrosinase, a specific enzyme of melanin biosynthesis.

The embryonic lineage of mammalian melanocytes, like most lineages in vertebrate embryos, has been mapped in terms of fate, but not of potency. Melanocytes develop from melanoblasts, cells which migrate from the neural crest to the embryonic skin and other organs (Rawles, 1947; Le Douarin, 1982). Melanoblasts are usually defined as lacking tyrosinase, but they may have premelanosomes (melanosome precursors without melanin) (Fitzpatrick *et al.,* 1979; Hirobe, 1982). Much is known about the potency of neural crest (Le Douarin, 1982) but little about that of melanoblasts. There is weak evidence that abnormal melanoblasts or melanocytes can form neuron-like cells: the common skin mole, a benign proliferation of melanocytes, often contains an unpigmented portion described as neurofibroma (Willis, 1967). Alterations to neural or glial histology have been seen in malignant melanoma (reviewed by Di Maio *et al.,* 1982).

B. Evidence on a Commitment Step in This Lineage

The strongest evidence on pigment cell commitment comes from cell lines derived from melanomas. It has long been argued that melanomas contain incompletely differentiated pigment cells (Gordon, 1959). Proliferating, clonal mouse melanoma lines consisting only of unpigmented or only of pigmented cells have been reported (Silagi, 1969; Pawelek, 1979). Note that "unpigmented" here means at the light microscope level. In some cases the absence of pigment is due to a metabolic defect (Pawelek, 1979) or to culture conditions (Oikawa *et al.,* 1972); in other cases both pigmented and unpigmented cells are found under the same conditions, suggesting two cell types. Kreider and Schmoyer (1975) showed that the B16C3 melanoma line had cells of both types, and that pigmented cells could arise spontaneously from unpigmented cells. The change appeared irreversible under their conditions. The pigmented cells were also larger than the unpigmented, often dendritic, and less proliferative on average. Bennett (1983) confirmed these findings for well-pigmented B16C3 cells. Hence the change from an unpigmented to a pigmented cell has the properties of a change in cell type with commitment (see also Section II,C). It may not be equivalent to the differentiation from melanoblast to melanocyte, as some unpigmented cells in B16C3 (and related lines) contain tyrosinase unlike melanoblasts (White *et al.,* 1983 and author's unpublished data).

A closer normal counterpart to unpigmented B16C3 cells is found in cultures from neonatal human skin. Human melanocyte strains can reproducibly be isolated from normal skin (Eisinger and Marko, 1982; Halaban and Alfano, 1984). Melanocytes freshly cultured from adult skin were predominantly larger than those from neonates, more dendritic, more pigmented, and expressed numerous different antigens (Houghton *et al.,* 1982). Cloning studies showed that proliferating, pigmented cells with the "adult" morphology could develop spontaneously from unpigmented neonatal cells, and both types could be cloned, indicating commitment (Bennett *et al.,* 1985). The unpigmented neonatal cells contained tyrosinase, like unpigmented B16C3 cells.

C. "Switching On" and Stochastic Control

It has repeatedly been suggested that the initiation of a lineage step, commitment to that step, or both, may be a stochastic (probabilistic) event (Levenson and Housman, 1981). The kinetics of commitment of hemopoietic stem cells in mice appeared to fit those of a single stochastic event (Till *et al.,* 1964). A similar idea was proposed for the commitment of cultured mouse erythroleukemic cells (a tumor of the red blood cell lineage) (Gusella *et al.,* 1976). These cells will proliferate indefinitely as cells resembling early erythroid stem cells, but can be induced to mature to nonproliferative normoblasts (Friend *et al.,* 1971) or reticulocytes (Volloch and Housman, 1982) by dimethyl sulfoxide (DMSO) among other agents. The matured cells contain hemoglobin (Hb). Gusella *et al.* (1976) cultured erythroleukemic cells with DMSO for varying times, removed it before any increase in Hb was observable, cultured the resulting cells singly, and then examined the size and Hb content of the colony formed by each cell in 5 days. Virtually all colonies were either large and without Hb or small (up to 16 cells) and contained Hb. The proportion of small colonies increased with time in DMSO with approximately first-order kinetics. The conclusion was that the small colonies arose from cells committed to differentiate and that commitment resembled a stochastic event. This event was postulated also to initiate expression of Hb and other erythroid cell products and to be a controlling step, its probability being sensitive to extracellular signals. Theories on stochastic commitment in other lineages are summarized by Levenson and Housman (1981).

These ideas of stochastic initiation and commitment were reexamined using B16C3 melanoma cells. As will be shown, the results

broadly supported the concept, but led to a different view of the relation between commitment and the initiation of expression (Section
II,D).

The apparent commitment, when B16C3 cells become pigmented,
was discussed above (Section II,B). Pigmentation per se can be induced
in this and related cell lines by a number of stimuli, such as cyclic
AMP (Kreider *et al.,* 1973), melanocyte-stimulating hormone (Laskin
et al., 1982), and high extracellular or intracellular pH (Saeki and
Oikawa, 1978, 1983; Laskin *et al.,* 1980). For further analysis of this
change (Bennett, 1983), B16C3 cells were grown sparsely at an extracellular pH of 6.9; under these conditions only 0–5% of the cells were
detectably pigmented. Cultures were then transferred to a higher pH,
such as 7.7, at which usually all cells became pigmented. Time-lapse
films showed that the onset of pigmentation in a given cell was associated with a change in shape, size, and movement, consistent with the
morphological changes described by Kreider and Schmoyer (1975).

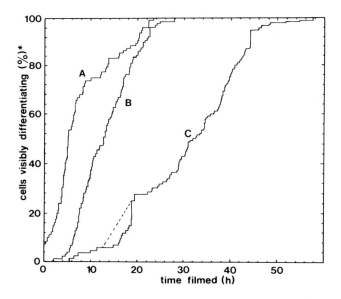

FIG. 2. Asynchrony of induced differentiation in B16C3 melanoma cells. Time-lapse
films were made of cultures after a change from standard growth medium (pH 6.9) to
media with higher pH of (A) approximately 8, (B) 7.9, and (C) 7.7. Using an analytical
projector the onset of visible differentiation was scored for each filmed cell and appears
as a vertical step in the graph. The broken line represents when the film was out of focus.
Data were corrected by a computer for cell deaths and divisions. Visible differentiation is
highly asynchronous within each culture, and the degree of asynchrony (or apparent
rate of recruitment) can change with conditions. From Bennett (1983) (copyright,
M.I.T.), where full methods are given.

However, this linked set of changes occurred at widely different times in different cells of the same culture (Fig. 2).

This variability was consistent with the idea of a stochastic initiation event. Alternatively such variability could arise from a genetic heterogeneity, which is common in mouse cell lines. However the asynchrony persisted in freshly cloned populations and the possibility of a major genetic component was ruled out by selection experiments (Bennett, 1983), based on the design of Gusella *et al.* (1976). The timing of pigmentation was thus not genetically predetermined, but indeterminate (stochastic) in the broadest sense.

D. "SWITCHING OFF": EARLY INSTABILITY AND GRADUAL COMMITMENT

Commitment in B16C3 cells was examined (Bennett, 1983) in cloning studies similar to those of Gusella *et al.* (1976). Melanoma cells were exposed to alkaline medium until some of them were pigmented, then cloned at a pH (6.9) which does not induce pigmentation. The size and pigmentation of individual clones were recorded. The results resembled those obtained with erythroleukemic cells in that nearly all the colonies formed in 14 days could be classified as large and unpigmented or small (1–2 cells) and pigmented (Table I, first half). However a different interpretation emerged when the colony characteristics on day 14 were also compared with the pigmentation of each progenitor cell on day 1 (Table I, second half). Unexpectedly a proportion of cells which had begun to form pigment on day 1 ceased to do so, and proliferated to give large, unpigmented colonies by day 14.

This loss of pigmentation was insufficient to prove that melanin synthesis had stopped. Whittaker (1963) observed that pigment epithelium cells of the retina lost pigmentation during rapid proliferation, but regained it as growth became inhibited in dense cultures, suggesting dilution by other cell products at high rates of cell growth. However some B16C3 cells lost pigment without proliferating (probably by secretion) (Laskin *et al.,* 1982). Moreover if the proliferating cells remained committed to pigment synthesis, they should have become repigmented rapidly and synchronously at high pH (which inhibits proliferation). This was tested when pigmented cells that had become depigmented after several days of growth at low pH were reinduced with alkaline medium. However these cells became pigmented just as slowly and asynchronously as cells in never-induced cultures (Bennett, 1983). The cell size and shape also reverted to that of uninduced cells in clones which became depigmented. Dilution of pigment thus does not seem a sufficient explanation in this case. Cases of similar reversion have now been observed in cloning studies with diploid human melanocytes (Bennett *et al.,* 1985).

TABLE I

Melanoma Clone Size Related to
Final and Initial Pigmentation[a]

Day	Pigmentation class[b]	Clone size on day 14 (number of cells)	
		1–2	Over 2
14[c]	U	8[a]	75
	VL	5	4
	L–W	27	1
1[d]	U	9	42
	VL	2	26
	L–W	29	12

[a] B16C3 melanoma cells were induced for a limited period and then cloned (see also text). Shown are the numbers of living clones in each size and pigmentation category.

[b] The five arbitrary pigmentation classes (criteria in Bennett, 1983) are unpigmented (U), very lightly (VL), lightly (L), moderately (M), and well pigmented (W).

[c] Pigment on day 14 related to size on day 14. Nearly all large colonies (three or more cells) were unpigmented.

[d] Pigment on day 1 related to size on day 14. Many cells pigmented on day 1 formed large colonies (mostly unpigmented) by day 14. Data are simplified considerably from Figs. 4 and 5 of Bennett (1983), where experimental details are given.

The simplest interpretation was that differentiating B16C3 cells had a finite chance of reverting to the pigment non-synthesizing cell type under the cloning conditions used. The proportion of reverting cells fell with increasing pigmentation (Table I), and no cells in the highest pigmentation class reverted (Bennett, 1983). These well-pigmented cells thus appeared to be committed (to pigment synthesis). However this commitment evidently did not precede pigment synthesis; instead it appeared as the end result of a gradual stabilization.

III. Perspectives

A. Summary and Rationale: Melanoma Cells

Cultured B16C3 mouse melanoma cells can be induced to change from a precursor cell type to a more mature type, as defined by a coordinated set of visible changes including pigment synthesis. The initiation of this change has properties suggesting a stochastic control. The change can be initiated without commitment; instead there seems to be a finite probability of *reversion* (to the precursor type), which depends on the extracellular and intracellular milieux. The probability of reversion falls as the new cell state is expressed (*stabilization*) and can reach zero, constituting *commitment,* under the experimental conditions examined.

B. OTHER MAMMALIAN CELL LINEAGES

The present volume contains evidence for initiations of changes in cell type (however defined) without or before commitment in a variety of nonmammalian organisms. There is also evidence for this from mammalian lineages other than pigment cells.

Sixteen-cell mouse embryos contain large, polar, microvillous "outer" cells and small, apolar "inner" cells with different developmental fates. However, if such embryos are disaggregated, then reaggregates of either 16 inner or 16 outer cells can generate a complete embryo (Ziomek *et al.*, 1982). In comparison, the cells of 8-cell embryos seem to be of only one type. This clearly illustrates a divergence of cell type, as defined by both fate and expression of physical characters at the 16-cell stage, which precedes commitment.

Erythroleukemic cells lend support to the idea of commitment before expression (Gusella *et al.*, 1976), because after a limited period of induction nearly all cloned cells form either small, differentiated or large, undifferentiated colonies (Section II,C). This shows early commitment or rapid stabilization under these conditions. Commitment must occur before detectable Hb synthesis and generally within the same cell cycle as the initiation of Hb synthesis, otherwise colonies with mixed sectors of Hb-positive and Hb-negative cells would have been common in such experiments. However it is now known that other inducers (or DMSO applied to other cell lines) can elicit increases in the earliest erythroid markers, like globin mRNA, without detectable commitment (Gusella *et al.*, 1980; Marks *et al.*, 1983; review by Nover, 1982, page 213).

These and further examples of unstable differentiation in mammalian lineages are summarized in Table II. A number of these examples come from tumor cells. This could be because instability is most readily observed in cultured lines and most mammalian cell lines come from tumors. However metaplasia (changed differentiation in an intact tissue) is relatively common in tumors. Metaplasia may be due to proliferation of a minor cell type, but an interesting alternative is that differentiation may be particularly unstable in tumor cells (Uriel, 1976).

TABLE II

INSTABILITY AND STABILIZATION IN OTHER MAMMALIAN LINEAGES[a]

Cell type	Evidence for instability[b]	Evidence for stabilization[c]
Inner, outer cells of early mouse embryo (see text)	Totipotency of both at 16-cell stage (1)	Yes (2)
Human pigmented retina	Transdifferentiation to lens in culture (3)	Inferred (3)
Rat myoblasts and muscle fibroblasts	Transdifferentiation to cartilage, given bone matrix	Possibly not (4)
Rat skeletal muscle	Muscle products reversibly expressed by mutant myoblast lines (so commitment not required for expression) (5)	Commitment is normal (5)[d]
Mouse fat cells, 3T3-T cell line	Reversion of differentiating fat cells to proadipocytes by retinoic acid (6)	Yes (6)
Human myeloid leukemic cells	Transdifferentiation of line with Philadelphia chromosome (marker of myeloid leukemia) to erythroid cells (7 and 8)	[d]
Mouse and human erythroid leukemic cells	Reversible induction of globin mRNA and other red cell products under some conditions (9–12 and see text)	Commitment is normal (12)
Rat neuroblastoma, cloned line	Partial differentiation to muscle (13)	[d]
Rat glial tumor line	Transdifferentiation to muscle (14)	[d]

[a] (1) Ziomek et al. (1982); (2) Slack (1983), review; (3) Yasuda et al. (1978); (4) Nathanson et al. (1978) and see Nathanson (present volume); (5) Nguyen et al. (1983); (6) Scott et al. (1982); (7) Lozzio and Lozzio (1975); (8) Andersson et al. (1979); (9) Gusella et al. (1980); (10) Marks et al. (1983); (11) Dean et al. (1981); (12) Nover (1982); (13) Brandt et al. (1976); (14) Lennon and Peterson (1979).

[b] Altered differentiation of a tissue (metaplasia) may be due to selection of preexisting cells, so examples have been chosen where the starting population is cloned or well characterized.

[c] Evidence that the instability is lost with maturation.

[d] This is not applicable to isolated instances or to mutants.

IV. Concluding Remarks

Evidence has been reviewed for the proposition that a change in cell type can be first initiated and later stabilized, and that this stabilization in melanoma cells is a gradual process rather than an abrupt event. Here the change in cell type would first be detected as changes in specific gene expression. However some investigators would assay (and define) a change in cell type as a change in potency (Section I; Till and McCulloch, 1980). Logically a cell cannot lose the potency to form some cell type before it is committed to lose that potency, so here the change in cell type is detected at the time of commitment. Consequently, if changes in expression can be initiated before commitment, investigators who use different criteria may disagree as to when the change in cell type occurs.

Perhaps the solution is to accept that the concept of "cell type," like that of "gene," has split into several different useful concepts. A change in cell type, differentiation, can be seen as comprising several component changes, including changed expression and commitment. There is, then, no contradiction in the possibility that these component changes may occur at different times. The study of this possibility may further our understanding of the mechanisms of cell differentiation.

ACKNOWLEDGMENTS

The author is grateful for many stimulating talks and conversations at the meeting which gave rise to this volume. The author's research is supported by the Cancer Research Campaign.

REFERENCES

Andersson, L. C., Jokinen, M., and Gahmberg, C. G. (1979). *Nature (London)* **278,** 364–365.

Balinsky, B. I. (1981). "An Introduction to Embryology," 5th Ed. Saunders, Philadelphia, Pennsylvania.

Bennett, D. C. (1983). *Cell* **34,** 448–453.

Bennett, D. C., Bridges, K., and McKay, I. A. (1985). *J. Cell. Sci.,* in press.

Blau, H. M., and Webster, C. (1981). *Proc. Natl. Acad. Sci. U.S.A.* **78,** 5623–5627.

Brandt, B. L., Kimes, B. W., and Klier, F. G. (1976). *J. Cell. Physiol.* **88,** 255–275.

Dean, A., Erard, F., Schneider, A. B., and Schechter, A. N. (1981). *Science* **212,** 459–461.

Di Maio, S. M., Mackay, B., Smith, J. L., and Dickersin, G. R. (1982). *Cancer* **50,** 2345–2354.

Eisinger, M., and Marko, O. (1982). *Proc. Natl. Acad. Sci. U.S.A.* **79,** 2018–2022.

Fitzpatrick, T. B., Szabó, G., Seiji, M., and Quevedo, W. C. (1979). *In* "Dermatology in General Medicine" (T. B. Fitzpatrick, A. Z. Eisen, K. Wolff, I. M. Freedberg, and K. F. Austen, eds.), 2nd Ed., pp. 131–163. McGraw-Hill, New York.

Friend, C., Scher, W., Holland, J. G., and Sato, T. (1971). *Proc. Natl. Acad. Sci. U.S.A.* **68,** 378–382.

Gordon, M. (1959). *In* "Pigment Cell Biology" (M. Gordon, ed.), pp. 215–239. Academic Press, New York.

Gusella, J., Geller, R., Clarke, B., Weeks, V., and Housman, D. (1976). *Cell* **9,** 221–229.

Gusella, J. F., Weil, S. C., Tsiftsoglou, A. S., Volloch, V., Neumann, J. R., Keys, C., and Housman, D. E. (1980). *Blood* **56,** 481–487.

Halaban, R., and Alfano, F. D. (1984). *In Vitro* **5,** 447–450.

Hirobe, T. (1982). *J. Exp. Zool.* **223,** 257–264.

Houghton, A. N., Eisinger, M., Albino, A. P., Cairncross, J. G., and Old, L. J. (1982). *J. Exp. Med.* **156,** 1755–1766.

Kreider, J. W., and Schmoyer, M. E. (1975). *J. Natl. Cancer Inst.* **55,** 641–647.

Kreider, J. W., Rosenthal, M., and Lengle, N. (1973). *J. Natl. Cancer Inst.* **50,** 555–558.

Laskin, J. D., Mufson, R. A., Weinstein, I. B., and Engelhardt, D. L. (1980). *J. Cell. Physiol.* **103,** 467–474.

Laskin, J. D., Piccinini, L., Engelhardt, D. L., and Weinstein, I. B. (1982). *J. Cell. Physiol.* **113,** 481–486.

Le Douarin, N. (1982). "The Neural Crest." Cambridge Univ. Press, London and New York.

Lennon, V. A., and Peterson, S. (1979). *Nature (London)* **281,** 586–588.

Levenson, R., and Housman, D. (1981). *Cell* **25,** 5–6.

Lozzio, C. B., and Lozzio, B. B. (1975). *Blood* **45,** 321–334.

Marks, P. A., Chen, Z.-X., Banks, J., and Rifkind, R. A. (1983). *Proc. Natl. Acad. Sci. U.S.A.* **80,** 2281–2284.

Nathanson, M. A., Hilfer, S. R., and Searls, R. L. (1978). *Dev. Biol.* **64,** 99–117.

Nguyen, H. T., Medford, R. M., and Nadal-Ginard, B. (1983). *Cell* **34,** 281–293.

Nover, L. (1982). *In* "Cell Differentiation. Molecular Basis and Problems" (L. Nover, M. Luckner, and B. Parthier, eds.), pp. 99–254. Springer-Verlag, Berlin and New York.

Oikawa, A., Nakayasu, M., Claunch, C., and Tchen, T. T. (1972). *Cell Differ.* **1,** 149–155.

Pawelek, J. (1979). *In* "Methods in Enzymology" (S. N. Timasheff and C. H. W. Hirst, eds.), Vol. 48, pp. 564–570. Academic Press, New York.

Rawles, M. E. (1947). *Physiol. Zool.* **20,** 248–266.

Saeki, H., and Oikawa, A. (1978). *J. Cell. Physiol.* **94,** 139–146.

Saeki, H., and Oikawa, A. (1983). *J. Cell. Physiol.* **116,** 93–97.

Scott, R. E., Hoerl, B. J., Wille, J. J., Florine, D. L., Krawisz, B. R., and Yun, K. (1982). *J. Cell Biol.* **94,** 400–405.

Silagi, S. (1969). *J. Cell Biol.* **43,** 263–274.

Slack, J. M. W. (1983). "From Egg to Embryo." Cambridge Univ. Press, London and New York.

Till, J. E., and McCulloch, E. A. (1980). *Biochim. Biophys. Acta* **605,** 431–459.

Till, J. E., McCulloch, E. A., and Siminovitch, L. (1964). *Proc. Natl. Acad. Sci. U.S.A.* **51,** 29–36.

Uriel, J. (1976). *Cancer Res.* **36,** 4269–4275.

Volloch, V., and Housman, D. (1982). *J. Cell Biol.* **93,** 390–394.

White, R., Hu, F., and Roman, N. A. (1983). *Stain Technol.* **58,** 13–19.

Whittaker, J. R. (1963). *Dev. Biol.* **8,** 99–127.

Willis, R. A. (1967). "Pathology of Tumours." 4th Ed. Butterworth, London.

Yasuda, K., Okada, T. S., Eguchi, G., and Hayashi, M. (1978). *Exp. Eye Res.* **26,** 591–595.

Ziomek, C. A., Johnson, M. H., and Handyside, A. H. (1982). *J. Exp. Zool.* **221,** 345–355.

CHAPTER 24

DIFFERENTIATION OF EMBRYONAL CARCINOMA CELLS: COMMITMENT, REVERSIBILITY, AND REFRACTORINESS

Michael I. Sherman

ROCHE INSTITUTE OF MOLECULAR BIOLOGY
ROCHE RESEARCH CENTER
NUTLEY, NEW JERSEY

I. Introduction

By the mid-blastocyst stage (approximately 64 cells) in the mouse embryo, the cells appear to have been segregated into two lineages: the outer trophectoderm layer will constitute the trophoblast and some other cells of the placenta, whereas the enclosed inner cell mass retains the potential to give rise to all of the embryo proper, as well as most of the extraembryonic membranes (for reviews, see Rossant and Papaioannou, 1977; Johnson *et al.,* 1977; Johnson, 1979; Sherman, 1981). Much attention has been given to the question of how and when cells of the early embryo become irreversibly committed to these lineages. Many experiments indicate that position is important in the establishment of the lineages, but the nature of the positional signal(s) is obscure.

It has been proposed that determinants for the trophectoderm lineage are present in the mouse embryo from very early stages, perhaps at fertilization (Sherman, 1981). If this is the case, then these determinants do not become adequately influential to fix cell fate for several cell generations. It seems clear that by the morula or even early blastocyst stage, most progeny of cells on the outside of the embryo are destined to participate in the trophectoderm lineage, yet the disposition of the majority of the outside cells is still adequately flexible at

345

those stages so that they can contribute to the inner cell-mass lineage if their position within the embryo is altered (reviewed by Sherman, 1981). It is important to note that even though cell fate is not fixed at the morula stage, inside and outside cells can be distinguished at the biochemical level, notably by protein synthetic profiles (e.g., Johnson *et al.*, 1977). This has led to the view that cells of the early mouse embryo begin to diverge biochemically ("differentiate") at very early stages, but that commitment only occurs when the cascade of biochemical changes reaches such a level of complexity that it cannot be reversed (Johnson *et al.*, 1977).

My objective in this chapter is to consider the issue of commitment of embryonal carcinoma (EC) cells to differentiate and, in keeping with the theme of this volume, to assess the question of whether differentiation of embryonal carcinoma cells can be reversed. I have begun with a discussion on commitment and differentiation in early mouse embryos because of the relationships between EC cells and cells of the early embryo and because of the possibility that factors controlling differentiation during embryogenesis will be operative in EC cells.

EC cells are stem cells of teratocarcinomas. Murine EC cells bear striking similarities to cells of early mouse embryos, and experiments illustrating this have been reviewed in detail elsewhere (e.g., Illmensee and Stevens, 1979; Martin, 1980; Jetten, 1986). In brief, both cell types share rare surface antigens and lack other antigens associated with most mature cells: both have rapid cell cycles with little or no G_1 phase; they are refractory to productive infection with several viruses; like early embryonic cells, EC cells have the potential to give rise to several different cell types; and, most compellingly, in some cases EC cells can participate in normal development, and even colonize the germ line, when injected into, or aggregated with, early embryos.

Many different EC cell lines have been established from teratocarcinomas. These EC lines retain tumorigenicity after culture for long periods; however, if the cells are induced to differentiate *in vitro*, tumorigenicity is reduced or lost when the resulting cells are injected into suitable hosts (Sherman *et al.*, 1981; M. Sherman, unpublished observations). Cells can be induced to differentiate in response to various stimuli, both physical and chemical. However, a survey of several "wild-type" and mutagenized EC lines reveals that cells from very few lines respond to all of these inducers (Sherman *et al.*, 1981; Jetten, 1985; Sherman, 1985). Furthermore, when they are induced to differentiate, cells from some EC lines give rise to many phenotypes (e.g., Evans and Martin, 1975), whereas cells from other lines appear to be substantially more limited in their repertoire (e.g., Sherman and Miller, 1978); such qualitative differences are also observed when tumors from different EC lines are analyzed (Sherman *et al.*, 1981; McCue *et al.*, 1983).

Recently, EC-like lines have been generated directly from embryo cultures (Evans and Kaufman, 1981; Martin, 1981). These lines might be even more similar to early embryonic cells than conventionally derived EC lines in that they show less tumorigenic potential when they are combined with embryos to generate chimeric offspring (see Bradley *et al.*, this volume).

It is very likely that principles of commitment and differentiation are at least qualitatively conserved among mouse embryos of different strains and even among different mammalian species. Because of disparities in potentiality and in responsiveness of EC lines to various inducers of differentiation, it would be incautious for one to generalize from restricted studies with selected inducers and a limited number of EC lines about such issues as patterns of commitment, mechanisms regulating differentiation, and the possibility of reversibility of differentiation.

II. Criteria for Assessing Differentiation of EC Cells

The phenotypes resulting from the induction of differentiation of EC cells can sometimes be identified as, e.g., parietal or visceral endoderm (Hogan *et al.*, 1981), neural (McBurney *et al.*, 1982; Liesi *et al.*, 1983), or muscle (McBurney *et al.*, 1982). In other instances, the phenotypes cannot be definitively identified, although they are clearly distinguishable from EC cells by morphology, surface or cytoskeletal antigenic properties, enzyme profiles, or the presence of mRNAs coding for these proteins. Jetten (1986) has recently elaborated upon these markers. As mentioned above, differentiated derivatives, unlike EC cells, are generally nontumorigenic. Similarly, when cultured in semisolid medium or when seeded at clonal density in liquid culture, EC cells survive and proliferate at much higher frequencies than their differentiated derivatives (Sherman, 1975; Wang and Gudas, 1984; Sherman *et al.*, 1985). In studies on commitment to, and irreversibility of, differentiation, it is not necessary to establish the identity of the differentiated phenotype, but it is obviously important to ascertain without doubt whether the cells are or are not differentiated at a given time.

III. Commitment of EC Cells

There are only limited published studies dealing with the question of EC cell commitment, and most involve the effects of retinoic acid (RA), a potent inducer of EC cell differentiation (see Jetten, 1986), upon proliferation. Rosenstraus *et al.* (1982) have reported that after 3 days of treatment of F9 EC cells with RA, the average cell-cycle time increased from 12 to 17 hours, the difference being due to the elongation or establishment of the G_1 phase and the prolongation of the S phase. Qualitatively similar observations were made when cells from another EC line, PSA-1, were induced to differentiate by aggregation (Rosenstraus *et al.*, 1982). Several cell cycles occurred within the 3-day period during which the cultures contained RA, and other investigators (Schindler *et al.*, 1981; Linder *et al.*, 1981; Ogiso *et al.*, 1982; Rayner and Graham, 1982; Mummery *et al.*, 1984) have provided evidence with various EC lines that RA does not alter the initial cell cycle times. In detailed analyses with synchronized clonal cultures, Rayner and Graham (1982) and Mummery *et al.* (1984) reported that RA appeared to slow growth of PC13 EC cells only beyond the second or third cell division, respectively.

If altered proliferation rate reflects differentiation of EC cells and if cell commitment precedes the appearance of the differentiated phenotype in the above experiments, then the latter must be elicited prior to the point at which cell cycle times for treated and untreated cells diverge. Rayner and Graham (1982) have provided evidence that this is so: in their studies, treatment of cloned PC13 EC cells with RA only interfered with proliferation after 48 hours. If cells were treated with RA for 24 hours and then exposed to RA-free medium, the clones showed the same growth patterns that were observed with continuous exposure to inducer, i.e., they grew essentially unimpeded for a further day, after which there was no increase in cell number per clone. Even a 6-hour exposure to RA led to an eventual diminution in clone size which was evidenced on or after 4 days of culture (Rayner and Graham, 1982). We have observed similar behavior of Nulli-SCC1 EC cells tested in a somewhat different way. Nonsynchronous cultures were exposed to RA for varying periods of time, cells were plated without RA at clonal densities in liquid culture or at higher densities in semisolid medium (survival and proliferation are much reduced under the latter conditions), and the number of clones or aggregates, respectively, was determined several days later. Some effects of RA on cloning efficiency were detectable following a 6-hour pulse, and suppression of cloning was close to maximal (85%) after a 48-hour exposure (Sherman *et al.*, 1985).

Although these results suggest that commitment precedes differentiation of EC cells, it must be noted that modulation of proliferation is only one of several indicators of differentiation and that RA can interfere with growth of many cell types without obviously influencing their state of differentiation (Lotan, 1980).

IV. Is Differentiation of EC Cells Reversible?

Sherman *et al.* (1981) have demonstrated that lengthy treatments of PCC4.aza1R cells with RA dramatically reduced their tumorigenicity, and consistent results have been obtained with PC13 cells by Rayner and Graham (1982). In the latter study, none of 17 clones treated with RA for 4 days and containing from 10 to 40 cells formed tumors, whereas 13 of 15 control clones (containing 30–48 cells) were tumorigenic (Rayner and Graham, 1982). In the former study by Sherman *et al.* (1981), the number of cells in the inoculum was many logs greater (10^6 cells) and the time of exposure to RA was 14–28 days; under these conditions the incidence of tumorigenicity was reduced from 100 to 25%. When Sherman *et al.* (1981) inspected cultures of PCC4.aza1R cells after exposure to RA, they found that within 2 weeks virtually all cells possessed the morphology of differentiated cells, but with further culture, foci of EC-like clones appeared. They proposed that cells had been sequestered within the cultures which initially neither proliferated nor differentiated in response to RA and that these cells eventually overcame the block to growth (Sherman *et al.*, 1981). However, an alternative interpretation of these results would be that a rare differentiated cell could revert to an undifferentiated EC cell.

Reversibility of EC cell differentiation was assessed more directly by Ogiso *et al.* (1982). These investigators studied the effects of RA on EC line 311. They utilized as indicators of differentiation the secretion of increased amounts of plasminogen activator, reactivity with an anti-F9 EC cell serum, and the loss of the ability of the cells to bind peanut agglutinin (PNA). Ogiso *et al.* (1982) observed clear alterations in these parameters 4 days after exposure to 10^{-7} M RA and demonstrated that most EC cells that had lost PNA receptors retained the PNA$^-$ phenotype 20 hours following removal of RA from the medium. When 311 cells were treated for only 2 days with RA, there was little if any increase in plasminogen activator secretion, but an obvious reduction was observed in the sensitivity of the cells to anti-F9 serum in a cytotoxicity assay. Most notable were effects on PNA binding: about half of the cells were unstained following challenge with fluorescein isothiocyanate-labeled PNA (compared with about 70–80% unstained cells in cultures treated with RA for 4 days and almost no PNA$^-$ cells in untreated cultures). Unlike cells treated for 4 days with RA, most 2-day-treated cells subsequently reacquired PNA receptors within 5 hours in RA-free medium. Ogiso *et al.* (1982) concluded from these results that differentiation of 311 EC cells begins prior to the second day of treatment with RA, progresses in a stepwise fashion (with e.g., PNA receptors being lost as a step preceding the increase in plasminogen activator secretion), and only becomes irreversible beyond the second day. In effect, within the context of the discussion in the preceding sections, one could interpret the results of Ogiso *et al.* as indicative of a rapid initiation of the differentiation cascade of 311 EC cells in response to RA with a protracted precommitment phase of at least 2 days. The proposed duration of the precommitment period is at variance with the suggestion from studies by Rayner and Graham (1982) that exposure to RA for 24 hours is adequate to irreversibly alter growth properties of most PC13 EC cells. However, it must be considered that the two groups were using different EC lines and different criteria to evaluate differentiation.

Our laboratory has been studying the reversibility of phenotypic changes in EC cells induced by retinol. Retinol is a less potent and less broadly effective inducer of differentiation than RA (Eglitis and Sherman, 1983; Jetten and DeLuca, 1983). We have studied in detail the response to retinol of a cell line named NR1 (Sherman *et al.*, 1986). This cell line, derived from EC line Nulli-SCC1, undergoes an alteration in morphology following treatment with retinol: the cells become flatter, more adherent to the substratum, and less interactive with each other. After treatment with retinol ($8.7 \times 10^{-7} M$) for 4 days, the cells acquire a morphological appearance intermediate between untreated cells and those exposed to RA (Fig. 1). Within 2 days of removal of the inducers, retinol-treated cells reacquire a typical EC-like appearance, whereas most RA-treated cells retain their altered morphology. NR1 cells secrete elevated levels of plasminogen activator following treatment for 4 days with RA, but not with retinol. Similarly, cloning efficiencies of these cells in semisolid medium are dramatically reduced with RA, but not with retinol. On the other hand, NR1 cells treated with either RA or retinol for 5 days have markedly elevated levels of surface-associated fibronectin, another commonly used indicator of differentiation (Sherman *et al.*, 1986).

Our results could be subject to the same kind of interpretation used by Ogiso *et al.* (1982), namely, that RA, as a potent inducer of NR1 cells, carries them through to irreversible stages of differentiation, whereas the less active retinol is only capable of eliciting some initial

---

F<small>IG</small>. 1. Effects of retinoids on morphology of cells from a clonal EC line, NR1. NR1 cells were derived from the EC cell line Nulli-SCC1 by M. Eglitis (Sherman *et al.*, 1986). (a) Untreated cells. (b) Cells treated for 4 days with $8.7 \times 10^{-7} M$ (0.25 μg/ml) all *trans*-retinol. (c) Cells treated for 4 days with $10^{-6} M$ all *trans*-retinoic acid. Bar (in c), 50 μm.

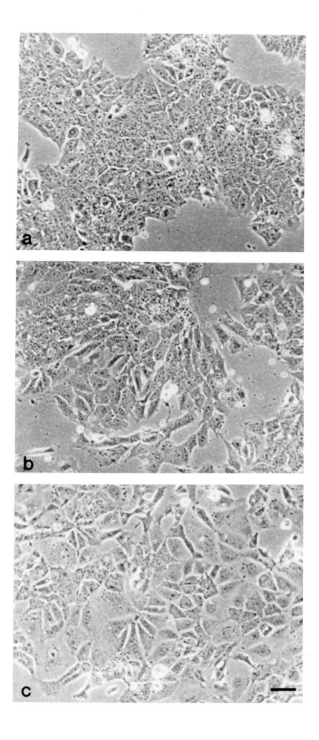

reversible changes, i.e., elevation of levels of surface-associated fibronectin and, perhaps resulting from this, alteration in morphology and adhesiveness.

One marked difference between our studies and the experiments reported by Ogiso *et al.* is that in the latter case the difference between reversibility and irreversibility of the response was time dependent, whereas in the former instance the difference was due to choice of the inducer. From the experiments described in the previous section, transient exposure to RA is adequate to commit some EC cells to a differentiated phenotype. Despite exposure of NR1 cells to retinol for as long as 2 weeks, the changes in phenotype are reversed within 2 days of removal of the retinol (Sherman *et al.*, 1986). If a pulse treatment with RA is adequate to trip off a cascade of events which lead eventually to an irreversibly differentiated phenotype, why are the changes induced in NR1 cells by retinol restricted in scope and reversible? Differential accessibility to the cells or stability of the two retinoids cannot readily explain the differences in their effectiveness; on the contrary, retinol appears to become associated with EC cells more efficiently than RA (Sherman *et al.*, 1983), and it is metabolically much more inert (Gubler and Sherman, 1985).

Unlike NR1 cells, cells from other EC lines (e.g., F9 and OC15) appear to be irreversibly altered by retinol (Eglitis and Sherman, 1983; Sherman *et al.*, 1986). It is conceivable, therefore, that there is a genetic block in the retinol-provoked differentiation pathway of NR1 cells. If this is so, it would be reasonable to conclude that retinol and RA induce differentiation of EC cells by different mechanisms or pathways (a possibility considered by Sherman, 1986), since NR1 cells respond normally to RA. Yet the retinol-induced changes in adhesiveness and fibronectin patterns are also observed when cells are treated with RA (e.g., McCue *et al.*, 1983). The RA-induced pathway to differentiation of EC cells thereby appears at least to include the changes elicited by retinol.

The studies described in this and the previous section can be taken to support the view that, as proposed for early embryonic cells (Johnson *et al.*, 1977), differentiation of EC cells proceeds as a cascade of initially reversible events which at some point leads to an irreversible step, after which the cells are considered to be committed. However, I believe that at least one other interpretation of the data is equally plausible, namely, that the appearance of some phenotypic changes resulting from retinoid treatment of EC cells, which we consider to be indicative of differentiation (Section II), does not necessarily mean that differentiation has occurred, or even begun. For example, the indicators of EC cell differentiation discussed in this review include suppression of pro-liferation, altered morphology, increased secretion of plasminogen ac-tivator, and increased deposition of cell-surface-associated fibronectin. Yet it has been documented that retinoids can elicit any or all of these effects on some cells without obviously affecting their state of differ-entiation (growth suppression: Lotan, 1980; Schindler et al., 1981; al-tered morphology: Jetten et al., 1979; increased secretion of plas-minogen activator: Schroder et al., 1980; McCue et al., 1983; increased surface deposition of fibronectin: Bolmer and Wolf, 1982). It must be stressed that such properties are clearly characteristic of many or most differentiated progeny of EC cells; however, it appears that, because of their pleiotropic nature, retinoids can alter phenotypic properties of cells differentially and in different ways (see Sherman, 1986). Indeed, such changes do not necessarily reflect alterations at the level of ge-nome: Bolmer and Wolf (1982) have reported that retinoid-induced increases in fibronectin deposition on the surface of fibroblasts can be observed with enucleated cells. The rapid reappearance (maximal with-in 5 hours) of surface PNA receptors in 311 EC cells following removal of RA after a 2-day treatment (Ogiso et al., 1982) raises the possibility that this effect might also be posttranscriptional.

The above argument has the unfortunate consequence of raising the question of how to define "differentiation" of EC cells. If the cells retain a non-EC phenotype after the inducer of differentiation has been withdrawn, then there is little confusion; however, if changes are transient, we cannot presently be certain that the observed alterations really had anything to do with differentiation of the cells.

V. Refractoriness to Induction of Differentiation

As mentioned in the previous section, there appears to be a small population of cells in PCC4.aza1R cultures which retain their EC-like phenotype (judged by their ability to produce teratocarcinomas) after several weeks of exposure to high concentrations of RA (Sherman et al., 1981). The refractory cells did not appear to be genetically altered since, after isolation and expansion in the absence of RA, many of the progeny cells underwent differentiation when reexposed to RA (un-published observations). In fact, we have only been able to isolate reproducibly differentiation-defective cells by mutagen treatment (Schindler et al., 1981; McCue et al., 1983), and even some of the mu-tants are leaky (McCue et al., 1983).

We have recently observed another example of "spontaneous" re-fractoriness of EC cells to differentiation by retinoids (R. Thomas and M. Sherman, unpublished observations). In our studies to assess the

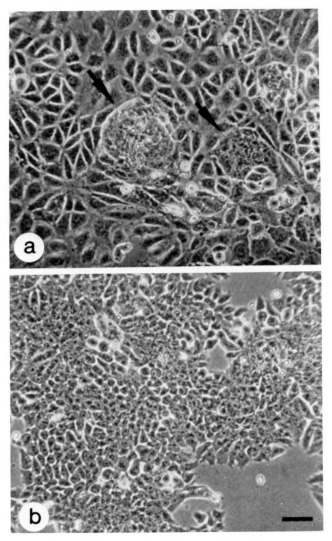

FIG. 2. Effects of long-term exposure of F9 EC cells to retinol. (a) Cells were treated continuously with $8.7 \times 10^{-7}\,M$ all *trans*-retinol for 51 days. Nests of persistent EC-like cells (arrows) are surrounded by epithelioid cell monolayers. (b) Control culture of F9 cells. Bar (in b), 50 μm.

extent to which retinol could induce differentiation of F9 EC cells, we maintained the cells for long periods of time in the presence of $8.7 \times 10^{-7} M$ retinol. We found that within several days, a condition was reached in which about half the cells possessed an altered morphology, whereas the remainder retained a typical EC phenotype. This situation persisted throughout many cell passages and for many weeks (e.g., Fig. 2). We have investigated by clonal analysis the possibility that we generated mixed cultures of genetically-altered, differentiation-defective EC cells and rapidly proliferating-differentiated cells. After lengthy exposure to retinol, cells seeded at clonal densities in the absence of retinol gave rise to clones that were EC-like, to pure clones of differentiated cells (which had a limited proliferative capacity), and to a large proportion of clones containing a mixture of EC and differentiated cells. All pure EC clones so far expanded and reexposed to retinol produced differentiated progeny in a proportion similar to that observed in the original mass culture. In other words, the ability of some F9 cells to resist the differentiation-inducing effects of retinol in these experiments appears to be due to epigenetic rather then genetic factors. An implication of our experiments is that the unresponsive state is retained more easily by cells in the presence of retinol than in its absence, and we are testing this directly. The nature of the epigenetic block is unclear. Since EC cells are derived from germ cells and early embryonic cells, it is interesting to speculate that this epigenetic refractoriness to differentiation might be a normal characteristic of, and a necessary feature to assure the integrity of, cells in the germ line and the early embryo.

ACKNOWLEDGMENTS

Some of the unpublished work described here was done in collaboration with Martin Eglitis and Richard Thomas. I wish to thank Alice O'Connor for secretarial assistance.

REFERENCES

Bolmer, S. D., and Wolf, G. (1982). *Proc. Natl. Acad. U.S.A.* **79**, 6541–6545.
Eglitis, M. A., and Sherman, M. I. (1983). *Exp. Cell Res.* **146**, 289–296.
Evans, M. J., and Kaufman, M. H. (1981). *Nature (London)* **292**, 154–156.
Evans, M. J., and Martin, G. R. (1975). *In* "Teratomas and Differentiation" (M. I. Sherman and D. Solter, eds.), pp. 237–250. Academic Press, New York.
Gubler, M. L., and Sherman, M. I. (1985). *J. Biol. Chem.* **260**, 9552–9558.
Hogan, B., Taylor, A., and Adamson, E. (1981). *Nature (London)* **291**, 235–237.
Illmensee, K., and Stevens, L. C. (1979). *Sci. Am.* **240**, 120–132.
Jetten, A. M. (1986). *In* "Retinoids and Cell Differentiation" (M. I. Sherman, ed.). CRC Press, Boca Raton, Florida, in press.
Jetten, A. M., and De Luca, L. M. (1983). *Biochem. Biophys. Res. Commun.* **114**, 593–599.

Jetten, A. M., Jetten, M. E. R., and Sherman, M. I. (1979). *Exp. Cell Res.* **124**, 381–391.

Johnson, M. H. (1979). *J. Reprod. Fertil.* **55**, 255–265.

Johnson, M. H., Handyside, A. H., and Braude, P. R. (1977). *In* "Development in Mammals" (M. H. Johnson, ed.), Vol. 2, pp. 67–97. North-Holland Publ., Amsterdam.

Liesi, P., Rechardt, L., and Wartiorvaara, J. (1983). *Nature (London)* **306**, 265–267.

Linder, S., Krondahl, U., Sennerstam, R., and Ringertz, N. R. (1981). *Exp. Cell Res.* **132**, 453–460.

Lotan, R. (1980). *Biochim. Biophys. Acta* **605**, 33–91.

McBurney, M. W., Jones-Villeneuve, E. M. V., Edwards, M. K. S., and Anderson, P. J. (1982). *Nature (London)* **299**, 165–167.

McCue, P. A., Matthaei, K. I., Taketo, M., and Sherman, M. I. (1983). *Dev. Biol.* **96**, 416–426.

Martin, G. R. (1980). *Science* **209**, 768–776.

Martin, G. R. (1981). *Proc. Natl. Acad. Sci. U.S.A.* **78**, 7634–7638.

Mummery, C. L., van den Brink, C. E., van der Saag, P. T., and de Laat, S. W. (1984). *Dev. Biol.* **104**, 297–307.

Ogiso, Y., Kume, A., Nishimune, Y., and Matsushiro, A. (1982). *Exp. Cell Res.* **137**, 365–372.

Rayner, M. J., and Graham, C. F. (1982). *J. Cell Sci.* **58**, 331–344.

Rosenstraus, M. J., Sundell, C. L., and Liskay, R. M. (1982). *Dev. Biol.* **89**, 516–520.

Rossant, J., and Papaioannou, V. E. (1977). *In* "Concepts in Mammalian Embryogenesis" (M. Sherman, ed.), pp. 1–36. MIT Press, Cambridge, Massachusetts.

Schindler, J., Matthaei, K. I., and Sherman, M. I. (1981). *Proc. Natl. Acad. Sci. U.S.A.* **78**, 1077–1080.

Schroder, E. W., Chou, I.-N., and Black, P. H. (1980). *Cancer Res.* **40**, 3089–3904.

Sherman, M. I. (1975). *In* "Teratomas and Differentiation" (M. I. Sherman and D. Solter, eds.), pp. 189–205. Academic Press, New York.

Sherman, M. I. (1981). *In* "Bioregulators of Reproduction" (G. Jagiello and H. J. Vogel, eds.), pp. 559–576. Academic Press, New York.

Sherman, M. I., ed. (1986). *In* "Retinoids and Cell Differentiation." CRC Press, Boca Raton, Florida, in press.

Sherman, M. I., and Miller, R. A. (1978). *Dev. Biol.* **63**, 27–34.

Sherman, M. I., Matthaei, K. I., and Schindler, J. (1981). *Ann. N.Y. Acad. Sci.* **359**, 192–199.

Sherman, M. I., Paternoster, M. L., Eglitis, M. A., and McCue, P. A. (1983). *Cold Spring Harbor Conf. Cell Prolif.* **10**, 83–95.

Sherman, M. I., Gubler, M. L., Barkai, U., Harper, M. L., Coppola, G., and Yuan, J. (1985). *Ciba Found. Symp.* **113**, 42–60.

Sherman, M. I., Eglitis, M. A., and Thomas, R. (1986). Submitted.

Wang, S.-Y., and Gudas, L. J. (1984). *J. Biol. Chem.* **259**, 5899–5906.

CHAPTER 25

EMBRYO-DERIVED STEM CELLS: A TOOL FOR ELUCIDATING THE DEVELOPMENTAL GENETICS OF THE MOUSE

Allan Bradley and Elizabeth Robertson

DEPARTMENT OF GENETICS
UNIVERSITY OF CAMBRIDGE
CAMBRIDGE, ENGLAND

I. Introduction

Fertilization or parthenogenetic activation of the mouse egg initiates a series of cell divisions which in the case of the fertilized egg ultimately results in the development of an adult individual. As the cell divisions proceed and cell numbers increase, differential gene expression is responsible for the phenotypic divergence of cells resulting in the formation of discrete cell types. This change is known as differentiation. As the expression of specific genes becomes fixed within certain cell types, it is accompanied by the commitment, to a greater or lesser degree, of the differentiated cell types to specific lineages within the developing organism.

One appraoch to the study of differentiation and cell commitment in the mouse is the use of developmental mutations which reflect the consequences of aberrant gene expression in an otherwise normal environment. Many mutations have been studied so that the gross cause of developmental failure has been well characterized, phenotypically,

357

and is often attributed to defects in certain cell types, for example, A^y, W, Sl, some t alleles and albino deletions (Sherman and Wudl, 1977; Silvers, 1979; Green, 1981). However, although the genetic locus of such mutations is often well defined, the precise molecular details are usually obscure and will remain so until cloning strategies have been successful. Clearly the cloning of many of these loci and a comparison with wild-type alleles is the first objective toward establishing the molecular basis for phenotypes which are associated with the change in gene expression as cell types progressively differentiate. Studies on the transdifferentiation of differentiated cell types in various systems should provide information on the nature and control involved in a switch in gene expression.

The existence of a population of stem cells in the mouse embryo which has the capacity to differentiate into a number of different cell types (pluripotential stem cells) is a useful developmental system for approaching some of the problems outlined above. Teratocarcinoma tumors, or more specifically the undifferentiated embryonal carcinoma (EC) cells within the tumor, have provided a widely used model system because these cells grown in culture will mimic the stem cells of the mammalian embryo (Martin and Evans, 1975; Martin *et al.,* 1977). Compared with preimplantation and postimplantation embryonic material, *in vitro* cultures have the advantage of cellular homogeneity and controlled differentiation combined with the possibility of large-scale culture. This has facilitated biochemical (Dewey *et al.,* 1978; Lovell-Badge and Evans, 1980), immunological (Jacob, 1979; Solter and Knowles, 1979; Gooi *et al.,* 1981; Stinnakre *et al.,* 1981; Hyafil *et al.,* 1983), and molecular analysis (Kurkinen *et al.,* 1983; Buc-Caron *et al.,* 1983; Morello *et al.,* 1983; Stacey and Evans, 1984) of the early differentiation events occurring in the mammalian embryo.

We do not intend to review here how EC cells have been used in the context of developmental studies. We shall, however, emphasize that, although the use of EC cells as a model of early mammalian development has been well justified, their use should not preclude an appreciation of two important limitations of the experimental material. First, although most EC cell lines are phenotypically very similar, they encompass a wide spectrum of differentiation potentials, often associated with a variety of karyotypic changes. Second, EC cells are derived from tumors and they should, therefore, be considered a neoplastic, or at any rate an abnormal, cell type.

The recent development of techniques which have facilitated the direct isolation and culture of pluripotential stem cells from the mouse embryo (Evans and Kaufman, 1981; Martin, 1981) has presented a number of novel approaches for the study of early mammalian development, using some aspects of the EC cell system. The applications which will be considered here focus on the current direction of the authors' research, namely, the differentiation of these recently isolated stem cells in an embryonic environment (Robertson *et al.*, 1983a; Evans *et al.*, 1983, 1985; Bradley *et al.*, 1984). The encouraging results obtained using these stem cells after differentiation *in vitro* or in chimeric association with an adult animal have stimulated us to propose experimental systems for analyzing differentiation, particularly at the molecular level.

II. Embryo-Derived Stem Cells *in Vitro*

A. ISOLATION TECHNIQUES AND PROPERTIES OF EMBRYO-DERIVED STEM CELLS

Since the original reports of the isolation of these stem cell lines (Evans and Kaufman, 1981; Martin, 1981), a number of other workers have reported successful stem cell isolation from preimplantation embryos cultured *in vitro*. In our laboratory, we have concentrated on isolating lines from blastocyst stage embryos (delayed or nondelayed) and have carried out studies on the timings of the various manipulations we consider to be important factors in the eventual isolation of a stem cell line.

The isolation of a line probably depends on a delicate balance between the number of stem cells present in the egg cylinder, the cloning efficiency of these cells, and the stage in their commitment to specific-differentiated lineages at the time the egg cylinder is disaggregated. It is reasonably clear that stem cells are present and can be isolated from the developing egg cylinder during its growth *in vitro* at a well-defined time. This timing depends very markedly on whether or not the embryos are subject to a period of implantational delay (Evans and Kaufman, 1981). In addition, the efficiency of isolation achieved when delayed embryos are disaggregated at the appropriate time is over three times that attained when normal nondelayed blastocysts are used (approximately 25 as opposed to 8%).

There are several possible explanations to account for the variation in the efficiencies of isolation which might exemplify the differences between the inner cell mass (ICM) populations of delayed and nondelayed embryos. One explanation for this difference might be related to cell numbers, believed to be elevated in delayed embryos. It is also possible that the slower growth (and probably differentiation) of delayed embryos following explantation in culture would provide a wider "developmental window" during which maximal numbers of uncommited stem cells are present within the proliferating egg cylinder. Finally, the entry of ICM cells within the blastocyst into the "delayed" state represents an unusual and perhaps "regulatory" event which might elevate the stem cell pool relative to the differentiated cells within the developing egg cylinder. The high efficiencies of isolation clearly indicate that the isolated stem cells are, at least initially, normal embryo cells.

Primary cultures of these lines have a perfectly normal karyotype which may be stably maintained following proliferation in tissue culture. In particular, normal euploid XY and XX lines are available, although there is evidence for the instability of the XX constitution (Robertson et al., 1983b). It is quite possible that normal karyotypes are a reflection of the recent isolation and relatively short periods of time that these cells have been maintained in culture. Some recent isolates of EC cells have also exhibited euploid, karyotypes although in other respects, these lines are disimilar to embryo-derived stem cells.

It is clear from various reports (Axelrod and Lader, 1983; Martin and Lock, 1983; Robertson et al., 1983a) that there is no apparent strain dependence for the isolation of these lines. In particular, stem cell lines are available from outbred strains of mice. This contrasts with the available EC cell lines which can only be isolated from teratocarcinoma tumors formed by the ectopic growth (within histocompatible host strains) of embryos from a few permissive inbred mouse strains (Solter et al., 1975).

B. Stem Cell Lines from Embryos Carrying Developmental Lethal Mutations

One of the immediate applications of the technique of stem cell isolation from any mouse strain is the possibility of isolating lines from mouse strains which carry developmental lethal mutations. The possibility of isolating a stem cell line homozygous for a given defect is clearly an important advance because this will present substantial material not only for biochemical analysis, but will also enable in vitro or in vivo studies of differentiative abilities to be carried out. To date, a single stem cell line homozygous for a developmental lesion has been isolated (Magnuson et al., 1982). The fact that this t^{w5}/t^{w5} line has been isolated indicates that the cause of developmental failure is not due to an inability of the cells to proliferate, and in addition, the development of tumors in hosts have demonstrated an apparently normal differentiation potential (Magnuson et al., 1983).

The study of stem cell lines carrying embryonic lethals is most applicable to those which act early during development because of the possibility of *in vitro* studies; however, the availability of chimera systems should allow the investigation of homozygous lines carrying mutations which have effects later in gestation.

The nature of homozygous lethal mutations requires that parental crosses to produce homozygotes are between heterozygous individuals. As a result, it is to be expected that only a proportion of cell lines established will be derived from homozygous embryos. The detection of possible homozygosity might proceed using karyotypic or linked biochemical markers, but such markers cannot usually confirm homozygosity because of the possibility of recombination in the region of interest. Such studies may then proceed by proving that a significant proportion of isolates have the desired markers. In the absence of closely linked molecular markers, homozygosity is difficult to prove definitively.

As was emphasized previously, one of the fundamental problems is that the majority of developmental lethals are not characterized at the molecular level. One mutation for which closely linked DNA probes are now available is lethal yellow (A^y), an allele at the agouti locus (Copeland *et al.*, 1983). In the absence of a molecular probe and a known gene product, establishing more than the morphological data which has characterized classic descriptions of mutations would be difficult. In Section IV we shall present one possible route around this problem.

C. Expression of Introduced Genes in Culture Systems

Where the products of developmentally significant genes are known and have been cloned, the potential is available to consider their expression in a culture system. It has been demonstrated that cloned genes will express after their introduction into the nuclei of various cell types (Kondoh *et al.*, 1983; see also an article by Kondoh in this volume). This type of study is limited by the range of cell types which can be successfully cultured *in vitro* combined with the necessity of a very sensitive assay system which will distinguish the expression of the introduced and native genes. Accepting the limitations of the general applicability of this technique, this type of study can be used to define sequences necessary for expression of a particular gene. However, it would not be possible to ask the question of whether the integrated gene was developmentally regulated. Is it expressed at the correct level in the appropriate tissue at the right time? What are the developmental switches which take place and are these regulated?

In a number of cases, genes expressed during development have been examined by transferring them into multipotential cells, the progeny of which will eventually contribute to the tissue where the gene of interest should be expressed. One type of multipotential cell to which this has been successfully applied is the fertilized zygote (reviewed by Gordon, 1984). Both appropriate and inappropriate expression have been observed by various workers using a variety of cloned genes (Lacy *et al.*, 1983; McKnight *et al.*, 1983; Storb *et al.*, 1984; Magram *et al.*, 1985). The other type of multipotential system to which this has been applied is the transformation of EC cells with cloned genes (Kondoh *et al.*, 1984; Scott *et al.*, 1984). The latter system has a number of advantages including the availability of an *in vitro* system of differentiation and the possibility of "shot-gun" experiments which facilitates the selection of integration sites prior to an examination of expression. A number of EC cell lines has been used for this purpose, but the differentiation observed *in vitro* or as an *in vivo* tumor is disorganized. Although discrete cell types can be recognized following such differentiation, it is not always reasonable to extrapolate the experimental observations to differentiation occurring during normal embryonic development. Fortunately EC cells have demonstrated an ability to differentiate *in vivo* when in chimeric association with a host embryo (Brinster, 1974; Papaioannou *et al.*, 1975; Mintz and Illmensee, 1975). However, this property has proved to be somewhat unpredictable, the efficiency of colonization is generally low, the extent of contribution is usually limited (and depends on the particular EC cell line used), and chimeric mice formed often developed tumors of an EC cell origin (reviewed by Papaioannou and Rossant, 1983).

Results in our laboratory on the chimeric association of our embryo-derived stem (EK) cells have been very encouraging (Robertson *et al.*, 1983a). The efficiencies of chimera formation were high, contributions to the chimera were extensive, mosaicism was fine grained and included functional contributions to the germ line (Bradley *et al.*, 1984).

III. Cultured Embryo-Derived Stem Cells in Chimeras *in Vivo*

A. High-Efficiency Chimera Formation

One of the most important features of embryo-derived stem cell lines is their ability to proliferate and differentiate following introduction into the early embryo. An interesting feature of this differentiation is that lines derived from parthenogenetically activated embryos (Kaufman *et al.*, 1983) are quite capable of contributing to chimeras, despite the homozygous genome in many of these lines. A summary of efficiencies of chimera formation is presented in Table IA and B.

Our studies have indicated a number of important differences between embryo-derived stem cells and EC cells following their proliferation and differentiation in an embryonic environment.

TABLE IA

Chimera-Forming Efficiencies of a Variety of
Embryo-Derived Stem Cell (EK Cell) Lines

Cell line	Strain	Chromosomes	Number injected	Number born (%)	Number chimeric (%)
B2B2	129	40 XY	66	40 (60.1)	14 (35.0)
CP2	129	40 XY	123	94 (76.4)	18 (19.1)
CP3	129	40 XY	156	111 (71.1)	43 (38.7)
CP4-2	129	40 XX:XX[del]	29	8 (27.6)	5 (62.5)
A13	129	40 XX	160	109 (68.1)	36 (33.0)
X1	Rb163H	38 XY	79	52 (73.5)	4 (7.7)
CL6	LT	40 XY	161	103 (64.0)	21 (20.4)
Rm53	Rm	— XY	63	48 (76.2)	8 (16.7)
CZ4	CFLP	— XY	8	6 (75.0)	2 (33.3)
CP1	129	40 XY	246	161 (65.7)	70 (43.4)
CC1.1	129[cc]	40 XY	83	61 (73.5)	37 (60.6)
CC1.2	129[cc]	40 XY	321	205 (64.0)	127 (61.9)
HD1[a]	129	40 XX[del]	42	30 (71.4)	8 (26.6)
HD10[a]	129	40 —	113	81 (71.1)	10 (12.3)
HD14[a]	129	40 XX	110	68 (61.8)	17 (25.0)
HD15[a]	F_1[b]	— —	46	31 (67.4)	1 (3.2)
DP3[a]	F_1[b]	— —	13	10 (76.9)	4 (40.0)
Total			1819	1218 (67.0)	420 (34.5)

[a] Lines derived from pathenogenetically activated embryos (Kaufman et al., 1983).

[b] F_1 (CBA × C57BL).

1.   The efficiencies of formation (scored at birth) differ from those observed when EC cell lines are used in a similar combination (see Table IA). It is somewhat difficult to assign a direct cause to this, but it might be due in part to the surface membrane determinants on these cells (see Section V) and the prenatal death of a large number of chimeric conceptuses (see below).

2.   The tumorigenicity of EC cells, even after differentiation within an embryo, is reflected in both live-born animals and prenatal conceptuses. Before birth, the development of abnormal individuals is responsible for considerable prenatal loss of chimeras. At or after birth, an average of 9.5% of chimeras are abnormal. By contrast, of over 500 chimeras constructed using our embryo-derived stem cells, none have ever been observed to develop tumors. Additionally, prenatal losses are low, close to the levels of microsurgically operated and transferred control embryos.

TABLE IB

A SUMMARY OF THE CHIMERA-FORMING EFFICIENCIES
OF A VARIETY OF EC CELL LINES[a]

Tumor/ cell line[b]	Strain	Number born	Number chimeric	
			Normal (%)[c]	Abnormal (%)[d]
OTT6050	129	248	29 (11.7)	2 (6.4)
OTT5568	129	168	28 (16.7)	1 (3.4)
METT-1	129	324	46 (14.2)	1 (2.1)
LT72484	LT/Sv	74	8 (10.8)	3 (27.3)
C17	C3H	77	5 (6.5)	8 (61.5)
C86	C3H	71	5 (7.0)	1 (16.7)
C3H145	C3H	201	0 (0.0)	0 (0.0)
P19	C3H/He	62	0 (0.0)	10 (100.0)
P10	C3H/He	58	31 (53.4)	0 (0.0)
F1/9	F_1[e]	18	0 (0.0)	0 (0.0)
1009	129XB	137	0 (0.0)	0 (0.0)
Total		1438	152 (10.5)	16 (9.5)

[a] (Adapted from Papaioannou and Rossant, 1983.)

[b] The different clonal lines (excluding cell hybrids) have been combined under their parental tumor for clarity.

[c] Percentage of number born.

[d] Percentage of number chimeric.

[e] F_1 (C3H × 129).

In summary, it is important to state that general comparisons involving EC cell lines in this context are not easy to make because EC cell lines cover a broad spectrum in their differentiative abilities. It is clear, however, that the very best EC cell lines have shown extensive, fine-grained and tumor-free differentiation in chimeras (Rossant and McBurney, 1983).

B. DISTRIBUTIONS OF CELLS IN CHIMERAS

Another feature which distinguishes the chimeras constructed using EC cells and our embryo-derived stem cells is the consequence of cellular proliferation within the embryo. We have performed an extensive distribution analysis on a number of chimeric animals utilizing electrophoretic variants of the glucose-6-phosphate isomerase (GPI) enzyme. The results of this analysis are presented in Table II.

The most significant features of this table are patterns of distribution within individuals, even when the overall level of contribution is low. This type of contribution is more indicative of that found in an

TABLE II

CONTRIBUTIONS OF EMBRYO-DERIVED STEM CELLS (EK CELLS) TO VARIOUS TISSUES IN CHIMERAS AS ASSESSED BY GPI ISOENZYME ANALYSIS[a]

Number	1	2	3	4	5	6	7	8	9	10	11	12	13	14	15	16	17	18	19	20	21	22	23	24	25	26	27	28	29	30	31	32	33	34
Liver	*	*	*	*	*	*	*	*	*	*	*	*	*	*	*	*	*	O	O	O	O	O	*	*	*	*	*	*	*	*	*	*	*	—
Spleen	*	*	*	*	*	*	*	*	*	*	*	*	*	*	*	*	*	O	O	O	O	O	*	*	*	*	*	*	O	*	*	*	*	—
Heart	*	*	*	*	*	*	*	*	*	*	*	*	*	*	*	*	*	*	O	O	*	O	*	*	*	*	*	*	*	*	*	—	*	*
Lung	*	O	O	*	*	*	*	*	*	O	*	*	*	*	*	*	*	O	O	O	O	*	—	*	*	*	*	*	*	*	*	*	*	*
R. kidney	*	*	*	*	*	*	*	*	*	*	*	*	*	*	*	*	*	O	O	O	*	*	*	*	*	*	*	*	*	*	*	*	*	*
L. kidney	*	*	*	*	*	*	*	*	*	*	*	*	*	*	*	*	*	O	O	O	O	O	*	*	*	*	*	*	*	*	*	*	*	*
Adrenals	*	—	*	*	*	*	*	*	*	*	*	O/*	O/*	*	*	*	*	O	O	O	O	*	*	*	*	*	*	*	*	*	*	*	*	*
Pancreas	*	—	—	—	—	—	*	*	*	—	*	—	—	*	*	*	*	*	O	O	*	O	*	*	*	*	*	*	*	*	*	*	—	—
Stomach	—	—	—	—	*	*	*	—	*	—	*	—	—	O	*	*	*	O	O	O	*	O	*	*	*	*	*	*	O	*	*	*	—	*
R. gonad	*	O	—	*	*	*	*	*	*	—	*	—	*	*	*	*	*	O	O	O	*	O	—	*	*	*	*	*	O	*	*	—	*	*
L. gonad	*	—	*	—	*	*	*	—	*	*	*	—	*	*	—	*	*	O	O	O	*	—	*	*	*	*	*	*	*	*	*	*	—	*
Bladder	—	—	—	*	*	—	—	*	*	—	*	*	—	O	*	*	*	O	O	O	*	O	*	*	*	*	*	*	*	*	*	*	—	—
Uterus	*	—	—	—	*	*	*	*	*	*	*	*	*	*	*	*	*	O	*	O	*	O	*	*	*	*	*	*	*	*	*	*	—	*
Muscle	*	—	—	—	*	*	*	*	*	—	*	*	*	*	*	*	*	*	*	O	*	*	—	*	*	*	*	*	O	*	*	*	—	*
Abd. wall	—	—	—	—	—	—	*	*	*	—	*	—	*	*	*	*	*	O	O	O	O	O	*	*	*	*	*	*	O	*	*	*	O	*
Brain	—	*	*	*	*	*	*	*	*	*	*	O	*	*	*	*	*	*	O	O	*	O	*	*	*	*	*	*	O	*	*	*	*	*
Gut	—	*	—	*	*	*	*	*	*	*	*	*	*	*	*	*	*	O	*	O	*	O	—	*	*	*	*	*	*	*	*	*	*	*
Thymus	—	—	—	—	—	—	—	—	—	—	*	—	*	—	*	*	*	*	O	O	O	O	*	*	*	*	*	*	O	*	*	*	—	—
Tail	—	O	O	*	*	—	*	*	*	—	*	*	*	O	—	*	*	O	O	O	O	O	*	*	*	*	*	*	O	—	*	*	—	*
Tongue	—	—	—	*	*	—	*	*	*	—	*	—	—	O	*	*	*	O	O	O	O	O	*	*	*	*	*	*	*	*	*	—	—	—

[a] *, Chimeric; O, nonchimeric; —, not tested.

[b] 1–3, B2B2 females; 4 and 6, CL6 females; 5, CL6 male; 7, Rm5.3 female; 8 and 9, Rm 5.3 males; 10, 12, and 13, A13 males; 11, A13 female; 14–16, CP1 females; 17, CP1 male (germ line chimera); 18–23, CC1.2 females; 24, WCC1.2 female; 25–28, CC1.1 females; 29, CC1.1 male (germ line chimera); 30, WCC1.1 female; 31, HD10 female; 32, HD14 female; 33 and 34, HD14 males.

ICM-embryo chimera than that in an EC-embryo chimera. In the former instance, contribution is generally fine grained, while in the latter, the distribution is often confined to patches within a tissue. Fine-grained mosaicism probably reflects a similarity in cell-surface phenotypes which permits adhoc mixing of the introduced cells with the host embryo cells. Conversely the "patchy" distribution found within EC-embryo chimeras is probably a result of clonally derived cells remaining clumped in a group during embryogenesis.

The combination of fine-grained mosaicism together with the activity of the Y chromosome within a number of these chimeras has been directly involved in one of the most important characteristics of

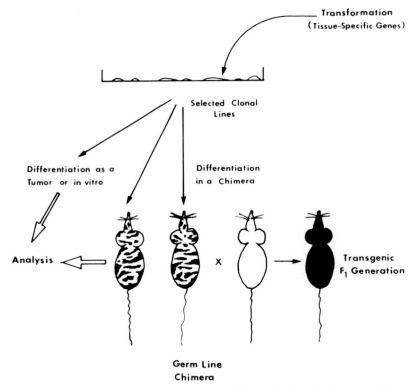

FIG. 1A. An analysis of expression during differentiation of tissue-specific genes transformed into stem cells. Specific genes are stably introduced into stem cells by a variety of transformation techniques. Cloned lines may be selected and the expression examined in the component tissue types resulting from disorganized development as a tumor *in vitro* or following organized differentiation in a chimera. The possibility of germ line chimerism allows analysis in successive generations.

these individuals, namely, the formation of functional sperm from the introduced stem cells (Bradley *et al.*, 1984; Evans *et al.*, 1985). To date we have observed germ cell chimerism in 20 different chimeric males constructed using three different XY cell lines. This high-efficiency event (20 males out of 54 breeding male chimeras derived from these cell lines) reopens the possibility of using cultured stem cells as a route into the genome of mice.

IV. Cultured Stem Cells: A Route into the Mouse Genome

The experimental procedures which summarize the techniques possible with the type of system described are shown in Figs. 1A and B. Principally an examination of introduced tissue-specific genes would be accomplished in a chimera. Although the proportion of cells contributing to specific lineages is variable, if transformed cells behaved like

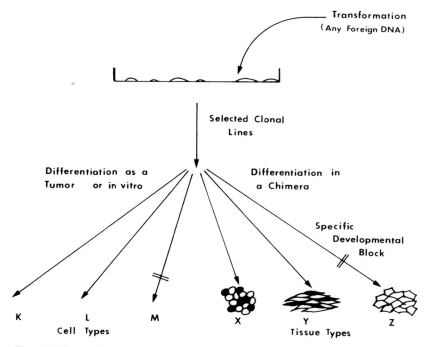

FIG. 1B. The application of insertional mutagenesis as a mechanism for identifying developmental genes and their effects. The stable integration of any DNA fragment will disrupt genes where insertion occurs. If a gene important in development is disrupted, then it is likely that the differentiation of the stem cells will be blocked. This should be detected as an exclusion of contributions from clonal lines of stem cells to specific cell types seen in tumors or tissue types in chimeras.

their nontransformed parents, transmission of the genome would also facilitate studies in an F_1 generation.

It is important to state that transmission is not a necessary prerequisite for examining tissue-specific expression of introduced genes, because of the availability of tissues with large percentage contributions derived from the modified stem cells within chimeras. The application of *in situ* hybridization techniques to the tissues in chimeras may allow the distinction of "modified" cells expressing specific genes in close proximity to normal control host embryo cells.

The other use of the system is to provide a route to isolate genes which are important in development using the mechanism of insertional mutagenesis. This type of approach relies on high-efficiency transformation techniques for which defective retroviral vectors will probably prove most suitable (Mann *et al.*, 1983). The possibility exists that insertion of any exogenous DNA into developmentally important genes will block or alter its function. As a result, it is assumed that mutated cells will be excluded from specific lineages within a chimera.

Although most mutations will probably be recessive and masked by their "wild-type" allele, dominance is to be expected in a number of situations. The simplest method initially would be to use XY cell lines, where insertions into the X chromosome would be hemizygous and therefore expressed. Alternatively homozygosity could be achieved on the autosomes by deriving lines from the haploid class of parthenogenetically activated eggs which have been infected with retroviral vector prior to the recovery of diploid pluripotential stem cell lines.

Both of these techniques could be used in combination with breeding studies which would screen for an absence of individuals which carry a homozygous insertion from heterozygous crosses. It appears quite likely that a range of lethal mutations would be uncovered, since experience with zygote injection and retroviral infection of the fetus has uncovered a number of different lethals (Schnieke *et al.*, 1983; Wagner *et al.*, 1983). Cloning the sequences flanking the integration sites of these inserts is a logical experimental progression following the discovery of a mutation. The development of shuttle vector systems and high-efficiency-cloning vectors will facilitate the isolation of flanking sequences (Cepko *et al.*, 1983; Lobel *et al.*, 1985).

V. Concluding Remarks

Although we have stressed the point that embryo-derived stem cells offer many advantages over the use of EC cells as a tool of the developmental biologist, EC cells will continue to have their uses, especially where specific differentiation steps have been well characterized *in vitro* (Strickland and Mahdavi, 1978; Hogan *et al.*, 1981).

Throughout this chapter, we have tried to contrast the properties of EC cells and embryo-derived stem cells. It is perhaps pertinent to consider the basis of these differences. In many ways, this is a difficult consideration because, excluding spontaneous testicular tumors in 129 mice, both of these cells types are basically derived from the stem cells of the embryonic egg cylinder. It is tempting to assign many of the different features exhibited by EC cells as resulting from the period of *in vivo* growth in the tumor form. It is possible that the "host-related" formation of a teratocarcinoma tumor rather than a benign teratoma is a "transformation" phenomena. Whatever the causal basis for progressive growth, it is clear that the growth of an embryo in an ectopic site is very different to the growth and differentiation undergone when stem cells are isolated directly *in vitro*. In an ectopic site, teratocarcinomas are selected as those tumors in which the stem cells do not differentiate in response to the normal signal. There might be some regulative phenomena in the undifferentiated state which is reflected in a subtle change in cell-surface phenotype. As a result, the stem cells may be unable to respond to the differentiation signals and, therefore, they will proliferate indefinitely despite the developmental signals and the confines of the ectopic environment. By contrast, the technique of stem cell isolation requires the physical disruption of most cell–cell contacts, thereby depriving the cells of differentiation signals which would normally operate during growth in an ectopic site *in vivo*.

It is becoming clear that embryo-derived stem cells offer a unique experimental tool which will be increasingly exploited in conjunction with molecular biological techniques. These studies will help to unravel the complexity of gene expression during development and in neoplasia.

ACKNOWLEDGMENTS

We would like to thank Martin Evans for helpful comments on this manuscript and Pam Fletcher and Lesley Cooke for excellent technical assistance. This work has been supported by grants from the Cancer Research Campaign and the Medical Research Council. A. B. is a Beit Memorial Research Fellow.

REFERENCES

Axelrod, H. R., and Lader, E. (1983). *Cold Spring Harbor Conf. Cell Prolif.* **10,** 646–665.
Bradley, A., Evans, M., Kaufman, M. H., and Robertson, E. (1984). *Nature (London)* **309,** 255–256.
Brinster, R. L. (1974). *J. Exp. Med.* **140,** 1049–1056.

Buc-Caron, M.-H., Darmon, M., Poiret, M., Sellem, C., Sala-Trepat, J.-M., and Erdos, T. (1983). *Cold Spring Harbor Conf. Cell Prolif.* **10**, 411–420.

Cepko, C. L., Roberts, B. E., and Mulligan, R. C. (1983). *Cell* **37**, 1053–1062.

Copeland, N. G., Jenkins, M. A., and Lee, B. K. (1983). *Proc. Natl. Acad. Sci. U.S.A.* **80**, 247–249.

Dewey, M. J., Filler, R., and Mintz, B. (1978). *Dev. Biol.* **65**, 230–244.

Evans, M. J., and Kaufman, M. H. (1981). *Nature (London)* **292**, 154–156.

Evans, M., Robertson, E., Bradley, A., and Kaufman, M. H. (1983). *In* "Current Problems in Germ Cell Differentiation" (A. McLaren and C. C. Wylie, eds.), pp. 139–155. Cambridge Univ. Press, London and New York.

Evans, M., Bradley, A., and Robertson, E. (1985). *In* "Genetic Manipulation of the Mammalian Ovum and Early Embryo." Cold Spring Harbor, New York, in press.

Gooi, H. C., Feizi, T., Kapadia, A., Knowles, B. B., Solter, D., and Evans, M. J. (1981). *Nature (London)* **292**, 156–158.

Gordon, J. W. (1984). *Dev. Genet.* **4**, 1–20.

Green, M. C. (1981). "Genetic Variants and Strains of the Laboratory Mouse." Fischer, Stuttgart.

Hogan, B. L. M., Taylor, A., and Adamson, E. (1981). *Nature (London)* **291**, 235–237.

Hyafil, F., Babinet, C., Huet, C., and Jacob, F. (1983). *Cold Spring Harbor Conf. Cell Prolif.* **10**, 197–207.

Jacob, F. (1979). *Curr. Top. Dev. Biol.* **13**, 117–137.

Kaufman, M. H., Robertson, E. J., Handyside, A. H., and Evans, M. J. (1983). *J. Embryol. Exp. Morphol.* **73**, 249–261.

Kondoh, H., Yasuda, K., and Okada, T. S. (1983). *Nature (London)* **301**, 440–442.

Kondoh, H., Takahashi, Y., and Okada, T. S. (1984). *EMBO J.* **3**, 2009–2014.

Kurkinen, M., Cooper, A. R., Barlow, D. P., Jenkins, J. R., and Hogan, B. L. M. (1983). *Cold Spring Harbor Conf. Cell Prolif.* **10**, 389–401.

Lacy, E., Roberts, S., Evans, E. P., Burtenshaw, M. D., and Costantini, F. D. (1983). *Cell* **34**, 343–358.

Lobel, L. I., Patel, M., King, W., Nguyen-Huv, M. C., and Goff, S. P. (1985). *Science* **222**, 329.

Lovell-Badge, R. H., and Evans, M. J. (1980). *J. Embryol. Exp. Morphol.* **59**, 187–206.

McKnight, G. S., Hammer, R. E., Kuenzel, E. A., and Brinster, R. L. (1983). *Cell* **34**, 335–341.

Magnuson, T., Epstein, C. J., Silver, L. M., and Martin, G. R. (1982). *Nature (London)* **298**, 750–753.

Magnuson, T., Martin, G. R., Silver, L. M., and Epstein, C. J. (1983). *Cold Spring Harbor Conf. Cell Prolif.* **10**, 671–681.

Magram, J., Chada, K., and Constantini, F.-J. (1985). *Nature (London)* **315**, 338–340.

Mann, R., Mulligan, R. C., and Baltimore, D. (1983). *Cell* **33**, 153–159.

Martin, G. R. (1981). *Proc. Natl. Acad. Sci. U.S.A.* **78**, 7634–7636.

Martin, G. R., and Evans, M. J. (1975). *Proc. Natl. Acad. Sci. U.S.A.* **72**, 3585–3588.

Martin, G. R., and Lock, L. F. (1983). *Cold Spring Harbor Conf. Cell Prolif.* **10**, 635–646.

Martin, G. R., Wiley, L. M., and Damjanov, I. (1977). *Dev. Biol.* **61**, 230–244.

Mintz, B., and Illmensee, K. (1975). *Proc. Natl. Acad. Sci. U.S.A.* **72**, 3585–3589.

Morello, D., Gachelin, G., Daniel, F., and Kourilsky, P. (1983). *Cold Spring Harbor Conf. Cell Prolif.* **10**, 421–437.

Papaioannou, V. E., and Rossant, J. (1983). *Cancer Surv.* **2**, 165–183.

Papaioannou, V. E., McBurney, M. W., Gardner, R. L., and Evans, M. J. (1975). *Nature (London)* **258**, 69–73.

Robertson, E. J., Kaufman, M. H., Bradley, A., and Evans, M. J. (1983a). *Cold Spring Harb. - Conf. Cell Prolif.* **10,** 647–663.

Robertson, E. J., Evans, M. J., and Kaufman, M. H. (1983b). *J. Embryol. Exp. Morphol.* **74,** 297–309.

Rossant, J., and McBurney, M. W. (1983). *Cold Spring Harbor Conf. Cell Prolif.* **10,** 625–633.

Schnieke, A., Harbers, K., and Jaenish, R. (1983). *Nature (London)* **304,** 315–320.

Scott, R. W., Vogt, T. F., Croke, M. E., and Tilghman, S. (1984). *Nature (London)* **310,** 562–567.

Sherman, M. I., and Wudl, L. R. (1977). *In* "Concepts in Mammalian Embryogenesis" (M. I. Sherman, ed.), p. 136. MIT Press, Cambridge, Massachusetts.

Silvers, W. K. (1979). "The Coat Colours of Mice." Springer-Verlag, Berlin and New York.

Solter, D., and Knowles, B. B. (1979). *Curr. Top. Dev. Biol.* **13,** 139–165.

Solter, D., Damjanov, I., and Koprowski, H. (1975). *In* "The Early Development of Mammals" (N. Balls and A. Wild, eds.), pp. 243–264. Cambridge Univ. Press, London and New York.

Stacey, A. J., and Evans, M. J. (1984). *EMBO J.* **3,** 2279–2286.

Stinnakre, M.-G., Evans, M. J., Willison, K. R., and Stern, P. L. (1981). *J. Embryol. Exp. Morphol.* **61,** 117–131.

Storb, U., O'Brien, R. L., Mullen, M. D., Gallahon, K. A., and Brinster, R. L. (1984). *Nature (London)* **310,** 238–241.

Strickland, S., and Mahdavi, V. (1978). *Cell* **15,** 393–403.

Wagner, E. F., Covarrubias, L., Stewart, T. A., and Mintz, B. (1983). *Cell* **35,** 647–655.

CHAPTER 26

PHENOTYPIC STABILITY AND VARIATION IN PLANTS

Frederick Meins, Jr.

FRIEDRICH MIESCHER-INSTITUT
BASEL, SWITZERLAND

I. Introduction

During development, cells become progressively determined for specific fates. This process involves stable changes in phenotype that persist in the absence of the external stimuli that brought about the change (Heslop-Harrison, 1967; Meins and Binns, 1979). Totipotent plant cells provide an experimental system well suited for studying the nature of these stable changes. The reason for this is that the developmental potential and genetic integrity of cells exhibiting different phenotypes in culture can be tested by regenerating plants and then performing genetic crosses.

This chapter deals with phenotypic stability and variation in plants. I will summarize the evidence for stable changes in plant development and then show that some of these changes result from potentially reversible alterations in cellular heredity.

II. Phenotypic Stability

It has long been recognized that stable changes occur in the development of plants (Lang, 1965a). Induction of flowering in certain species depends on day length (Lang, 1965b). The flowering state, once established, can persist for long periods of time in the absence of the inductive stimulus. Some species also require a cold treatment, vernalization, for flowering (Purvis, 1961). This treatment induces stable

373

changes to a competent state in which the plant can react to a particular day length.

Woody periennial plants exhibit morphologically distinct juvenile and adult phases of development. Only the adult phase normally flowers (Goebel, 1900). In the case of English ivy, *Hedera helix,* grafting experiments have shown that the adult and juvenile phases are stable, but not permanent. When cuttings from shoots exhibiting one phase are grafted onto host plants exhibiting the alternative phase, they continue to grow and to express the phase of the shoot from which they were derived (Brink, 1962). Adult shoots treated with gibberellin revert in a stable fashion to the juvenile state (Robbins, 1960; Rogler and Hackett, 1975).

Stable changes also play an important role in the determination of meristems. The plant body is derived from the shoot and root meristems established during embryogenesis. When cultured in isolation, root meristems continue to form roots and shoot meristems continue to form shoots (Reinhard, 1954; Torrey, 1954; Smith and Murashige, 1970). Thus, the two types of meristems exhibit a stable restriction of developmental potency which persists in populations of dividing cells. The formation of leaves from the shoot meristem is accompanied by a further restriction in potency. Primordia of leaves excised from the fern *Osmunda cinnamomea* develop in culture into either complete shoots or individual leaves (Steeves, 1966). Very early primordia, although morphologically distinct, only form complete shoots. With increasing maturity, the incidence of shoot formation decreases; older primordia only form individual leaves. A similar stable and progressive restriction in developmental potency can be inferred from the experimental intervention of flower formation (e.g., Cusick, 1956; Tepfer *et al.,* 1963; Soetiarto and Ball, 1969; Hicks and Sussex, 1971).

When and where in the plant stable changes occur depends on the particular organ and the plant species (Meins, 1985). There are even cases in which gradients of determination are discernible within individual organs. Explants from different regions of the *Streptocarpus wendlandii* leaf form plants with different proportions of leaf and flower stalk (Oehlkers, 1955, 1956). The capacity to form proliferating culture is restricted to explants from a small region near the base of young leaves of *Sorghum bicolor* (Wernicke and Brettell, 1980). Within this region, there is a zone that gives rise to root-type cultures and a separate, more restricted zone that gives rise to cultures capable of forming roots and adventitious embryos.

The studies with *Sorghum* illustrate the fact that tissues cultured from different parts of the same plant can exhibit persistent differences in phenotype. In a few rare cases, the tissues retain certain differentiated traits characteristic of the tissue from which they were derived: cultures established from maize endosperm can produce the endosperm-specific protein, zein (Shimamoto *et al.*, 1983); organ-specific essential oil components are produced by cultures of *Ruta graveolens* (Nagel and Reinhard, 1975); and certain cultures of tobacco (Boutenko and Volodarsky, 1968), maize (Khavin *et al.*, 1979), and cherry (Raff *et al.*, 1979) contain organ-specific antigens.

More commonly, tissues from different parts of the same plant exhibit different properties, but not necessarily the properties of the parent organ. Cultures from juvenile tissues of English ivy and *Citrus grandis* can be induced to form shoots, whereas cultures from the adult tissues form adventitious embryos (Banks, 1979; Chaturvedi and Mitra, 1975). With few exceptions, tissue cultures established from cereal species are highly organized and heterogeneous in appearance (e.g., Cure and Mott, 1978; Mott and Cure, 1978; King *et al.*, 1978; Springer *et al.*, 1979). In the case of maize, cultures initiated from most parts of the plant, e.g., roots, bases of young leaves, and intercalary meristems, consist of root primordia arrested at different stages of development. Cultures initiated from the scutellum of immature embryos contain, in addition, embryogenetic tissue and primordia for shoots and leaves (Green and Rhodes, 1982; Green *et al.*, 1983).

III. Phenotypic Variation

The few examples cited above provide good evidence that stable states arise during development and that some of these states persist in populations of dividing cells. Studies of cells and tissues in culture show that these states are not necessarily permanent. Plant cells exhibit considerable developmental plasticity.

In their classical experiments, Skoog and Miller (1957) showed that organ formation in explants of tobacco stem can be regulated by adding the plant hormones auxin and cytokinin to the culture medium. When the concentration of cytokinin is high, relative to auxin, shoots form; when the concentration of cytokinin is low, relative to auxin, roots form; and, at intermediate concentrations, the tissues grow as an unorganized callus. Essentially the same results can be obtained with cloned lines of tobacco established from single cells (Vasil and Hildebrandt, 1965). This indicates that the tobacco cells have the potential to form both root meristems and shoot meristems. Since these meristems are alternative stable states, it appears that the hormones induce cells or groups of cells to become committed to specific developmental fates.

Cloning experiments show that most, if not all, cells in recently initiated cultures of tobacco pith tissue are totipotent. Nevertheless, within an individual explant, only a very low proportion of the cells, approximately 10^{-3}, exhibits this potentiality (Meins et al., 1982). These results illustrate the important point that totipotent cells can vary in their developmental competence and responsiveness to inducers. It is now generally recognized that there are distinct stable states of competence for adventitious embryogenesis, root formation, and shoot formation (Gresshof, 1978; Henshaw et al., 1982). In some cases, these states are already established in the plant and result in the differences in behavior of tissues cultured from different parts of the same plants cited earlier. In other cases, it is possible to manipulate these states in culture.

Pretreatment of tissues with different combinations of hormones can change the types of organs formed when tissues are subsequently cultured on an organ-inducing medium (Walker et al., 1979; Christianson and Warnick, 1983, 1984). This suggests that different stages of organogenetic competence can be established without overt organogenesis. There are also examples of stable changes akin to transdetermination in animals (Hadorn, 1967). Callus cultures initiated from leaves and stems of C. grandis form shoots under inductive conditions, whereas cultures initiated from reproductive organs form adventitious embryos. On prolonged culture, stem-derived tissues sometimes give rise to variants which form adventitious embryos (Chaturvedi and Mitra, 1975).

So far the discussion has focused on the variation in the competence of tissues to form organs. Plant tissues in culture also undergo extensive variation in differentiated functions, such as their capacity to produce specific pigments and alkaloids (Butcher, 1977), their resistance to metabolic inhibitors (Maliga, 1984), and in their requirement for specific growth factors (Meins, 1983). Although some of this variation results from selection of subpopulations of cells within the original explant, most is thought to arise from stable cellular changes in culture. There is good evidence that simply culturing plant cells and tissues generates genetic as well as epigenetic instability (Meins, 1983).

IV. Cellular Basis for Stable Variation

A. SUPRACELLULAR MECHANISMS

The important question that arises is whether stable changes occur at the supracellular level or result from alteration in individual cells.

If we assume that stability is a supracellular phenomenon, what types of mechanisms could account for both flexibility and stability? One possibility is that particular states are an inherent property of tissue organization. It could be argued that certain cellular patterns are self-perpetuating, just as crystals are able to grow and sometimes regenerate after breaking (Bonner, 1963). This is well documented in the case of protozoa where patterns of cilia and the formation of an extracellular shell are transmitted to daughter cells (Nanney, 1968). Direct evidence for this type of stability in plants comes from studies of the water fern *Azolla* (Gunning *et al.*, 1978). The root forms from a single apical cell by a fixed pattern of divisions. Depending upon the plane of the division two self-perpetuating, enantiomorphic forms of root can develop. It is unlikely, however, that this provides a general explanation for stability. Certain stable changes in morphogenetic potential can persist for many cell generations in unorganized tissues maintained in culture (Banks, 1979; Chaturvedi and Mitra, 1975).

Another possibility is that stability results from the kinetic structure of tissues. Certain forms of cell–cell interactions and gradients of morphogens are self-sustaining (Harrison, 1982; Meinhardt, 1982). Lang (1965a) has pointed out the "contagious" nature of differentiation whereby plant cells can induce their own type of differentiation in neighboring cells. There is evidence that this is involved in the progressive differentiation of vascular tissues (Sachs, 1969; Gersani and Sachs, 1984).

B. EPIGENETIC CHANGES

It is widely believed that determination in plants is primarily at the tissue and organ level (Henshaw *et al.*, 1982). The argument is roughly the following: determination persists in populations of dividing cells implying that determined cells are altered in their cellular heredity, i.e., they transmit the committed state to daughter cells. Nevertheless, complete plants can be regenerated from single cells of determined organs (Narayanaswamy, 1977). Superficially, this is a paradox. How can cells remain totipotent and inherit different potentialities? The crux of the problem is the distinction between stability and permanence. Part of an organism is said to be determined to form a specific structure if it continues to form the structure when placed in a new environment (Spemann, 1938). Thus determination is relative: it depends on the properties of the tissue or organ and the conditions of the experiment. Stability under one set of conditions does not imply that a developmental state is permanent (Meins and Binns, 1979).

The same argument may be extended to the cellular level. In animals, it has long been recognized that certain developmental states are inherited by individual cells (e.g., Cahn and Cahn, 1966; Coon, 1966; Gehring, 1968; Okada *et al.,* 1973). Studies of ciliate protozoa provide direct evidence that genetically equivalent cells can inherit different phenotypes (Beale, 1958). Recognition of this fact has led to the proposal that there is an *epigenetic* (i.e., developmental) system of inheritance concerned with somatic transmission of patterns of gene expression (Nanney, 1958). Thus, alterations in cellular heredity that do not result from permanent genetic changes are known as *epigenetic changes* (Harris, 1964). Because their underlying mechanism is still unknown, epigenetic changes are defined operationally (Meins, 1983): they are alterations in cellular heredity that are directed, i.e., the changes in phenotype occur at high rates under inductive conditions and at low rates otherwise. Since the genome is not permanently altered, the range of phenotypes is limited, and the changes are potentially reversible. Finally, by definition, epigenetic changes are not transmitted through meiosis.

In the case of totipotent plant cells, epigenetic changes can be distinguished from classical mutations by a plant regeneration test. If a variant phenotype is a mutant, then the new phenotype should persist in tissues cultured from the regenerated plants, and it should be possible to transmit the phenotype in a regular fashion when the plants are crossed (Meins, 1983). On the other hand, epigenetic changes should be lost at some stage of the plant regeneration process or during meiosis.

C. HABITUATION

Direct evidence for epigenetic changes in plant development has come from studies of a type cellular alteration that arises in culture known as *habituation*. Pith parenchyma tissue excised from tobacco plants requires both auxin and a cell-division factor, usually provided as a cytokinin, for continued proliferation in culture (Jablonski and Skoog, 1954). After a variable number of passages, these tissues sometimes habituate, i.e., they lose their exogenous requirement for cytokinin or for both cytokinin and auxin (Meins, 1982). Thereafter, the habituated tissues and clones derived from these tissues can be subcultured indefinitely without the added growth factor (Lutz, 1971; Tandeau de Marsac and Jouanneau, 1972; Binns and Meins, 1973). The habituated state is inherited at the cellular level.

Cultured plant cells undergo extensive genetic variation during prolonged culture (Meins, 1983; Maliga, 1984). Therefore, it was important to find out whether habituation was the result of rare, random mutations. In the case of cytokinin habituation of tobacco pith cells, which has been studied in detail, the heritable change appears to have an epigenetic basis. The induction of habituation is a directed process that occurs under specific culture conditions (Meins, 1974; Meins and Lutz, 1980). Under these conditions, cells shift to the habituated state at rates greater than 4×10^{-3} per cell generation, which is 100–1000 times faster than the rate expected for somatic mutation (Meins and Lutz, 1980). Moreover, unlike classical mutation, the rate of conversion depends on both the developmental and physiological state of the pith cell (Meins *et al.*, 1980).

Although the cytokinin-habituated state is extremely stable and persists for many cell generations in culture, it is not permanent. Cloned lines of habituated and cytokinin-requiring pith cells regularly give rise to fertile plants when incubated on a shoot-inducing medium (Binns and Meins, 1973). Pith tissues from these plants, like pith tissue from seed-grown plants, require cytokinin for proliferation in culture. Therefore, at least some progeny of habituated cells can revert to the cytokinin-requiring state.

Because plants are regenerated from clones containing many cells, it may be argued that reversion results from rare back mutation followed by selection. This is unlikely. Measurements show that plantlets with cytokinin-requiring tissues arise from the clones at rates greater than 3×10^{-3} per cell generation (Meins and Binns, 1982). Therefore, the plants either develop directly from habituated cells or from revertant cells that arise in the cloned lines at very high rates. It appears that habituation involves epigenetic changes in the expression of a normally silent genetic potentiality of the tobacco pith cell.

There is also evidence that epigenetic changes of this type occur during the development of the tobacco plant. Tissues from different organs differ in their competence for cytokinin habituation in culture (Meins *et al.*, 1980; Meins and Lutz, 1979). Pith tissue consists of two types of cells: inducible cells, which habituate rapidly in culture, and noninducible cells, which do not habituate under standard inductive conditions. Tissues cultured from explants of leaf lamina, on the other hand, appear to consist exclusively of noninducible cells. Finally, cells from the cortex of the stem exhibit a constitutive phenotype; they are initially cytokinin habituated when placed in culture and remain so when subcultured. The constitutive and noninducible phenotypes persist when cells are cloned indicating that they are inherited at the cellular level.

Plants can be regenerated from noninducible, inducible, and constitutive cells at comparable high rates on a per cell generation basis (Meins and Binns, 1982). Tissues cultured from the pith, leaf, and stem cortex of the regenerated plants exhibit the same state of habituation as comparable tissues from seed grown plants (Meins *et al.,* 1983; Meins and Foster, 1985) indicating that these states, although inherited by individual cells, are not permanent.

The experiments outlined above show that in some cases, at least, stable changes in plant development result from alterations in cellular heredity. These alterations are directed and are potentially reversible indicating that they have an epigenetic basis. Tissues from regenerated plants exhibit the proper cytokinin requirement regardless of the epigenetic state of the cloned cells from which they were derived. This suggests that during plant regeneration a "resetting" mechanism operates which switches cells back to the appropriate developmental state (Meins and Foster, 1985).

V. Concluding Remarks

Stable changes in phenotype are a fundamental property of developing systems. They allow cells to "remember" their past and provide the basis for the progressive nature of development. Although botanists have tended to emphasize the plasticity of plant development, there is ample evidence that plants undergo determination and that certain determined states persists when tissues and organs are serially propagated in culture.

At present, there are too few well-documented studies to assess the relative importance of supracellular mechanisms and cellular heredity in stabilizing developmental states. The studies of cytokinin habituation show that alterations in cellular heredity can occur in plant development. These studies emphasize the distinction between stability and permanence. Cells that are committed under one set of conditions and transmit the committed state to their mitotic progeny can remain totipotent.

REFERENCES

Banks, M. S. (1979). *Z. Pflanzenphysiol.* **92,** 349–353.
Beale, G. H. (1958). *Proc. R. Soc. London Ser. B* **148,** 308–314.
Binns, A., and Meins, F., Jr. (1973). *Proc. Natl. Acad. Sci. U.S.A.* **70,** 2660–2662.
Bonner, J. T. (1963). "Morphogenesis," pp. 25–54. Atheneum, New York.
Boutenko, R. G., and Volodarsky, A. D. (1968). *Physiol. Veg.* **6,** 299–309.
Brink, A. (1962). *Q. Rev. Biol.* **37,** 1–22.
Butcher, D. N. (1977). *In* "Plant Cell, Tissue and Organ Culture" (J. Reinert and Y. P. S. Bajaj, eds.), pp. 344–357. Springer-Verlag, Berlin and New York.
Cahn, R. D., and Cahn, M. B. (1966). *Proc. Natl. Acad. Sci. U.S.A.* **55,** 106–114.
Chaturvedi, H. C., and Mitra, G. C. (1975). *Ann. Bot.* **39,** 683–687.
Christianson, M. L., and Warnick, D. A. (1983). *Dev. Biol.* **95,** 288–293.

Christianson, M. L., and Warnick, D. A. (1984). *Dev. Biol.* **101**, 382–390.

Coon, H. G. (1966). *Proc. Natl. Acad. Sci. U.S.A.* **55**, 66–73.

Cure, W. W., and Mott, R. L. (1978). *Physiol. Plant.* **42**, 91–96.

Cusick, F. (1956). *Trans. R. Soc. Edinburgh* **63**, 153–166.

Gehring, W. (1968). *In* "The Stability of the Differentiated State" (H. Ursprung, ed.), pp. 134–154. Springer-Verlag, Berlin and New York.

Gersano, M., and Sachs, T. (1984). *Differentiation* **25**, 205–208.

Goebel, K. (1900). "Organography of Plants" (I. B. Balfour, transl.), Part I. Clarendon, Oxford.

Green, C. E., and Rhodes, C. A. (1982). *In* "Maize for Biological Research" (W. F. Sheridan, ed.), pp. 367–372. Univ. of North Dakota, Grand Forks.

Green, C. E., Armstrong, C. L., and Anderson, P. C. (1983). *Adv. Genet. Plants Anim.* **20**.

Gresshof, P. M. (1978). *In* "Phytohormones and Related Compounds—A Comprehensive Treatise" (D. S. Letham, P. B. Goodwin, and T. J. V. Higgins, eds.), Vol. II, pp. 1–29. Elsevier, Amsterdam.

Gunning, B. E. S., Hughes, J. E., and Hardham, A. R. (1978). *Planta* **143**, 121–144.

Hadorn, E. (1967). *In* "Major Problems in Developmental Biology" (M. Locke, ed.), pp. 85–104. Academic Press, New York.

Harris, M. (1964). "Cell Culture and Somatic Variation." Holt, New York.

Harrison, L. G. (1982). *Symp. Soc. Dev. Biol.* **40**, 3–33.

Henshaw, G. G., O'Hara, J. F., and Webb, K. J. (1982). *Symp. Br. Soc. Cell Biol.* **4**, 231–251.

Heslop-Harrison, J. (1967). *Annu. Rev. Plant Physiol.* **18**, 325–348.

Hicks, G. S., and Sussex, I. M. (1971). *Bot. Gaz.* **132**, 350–363.

Jablonski, J. R., and Skoog, F. (1954). *Physiol. Plant.* **7**, 16–24.

Khavkin, E. E., Misharin, S. I., and Polikarpochkina, R. T. (1979). *Planta* **145**, 245–251.

King, P. J., Potrykus, I., and Thomas, E. (1978). *Physiol. Vég.* **16**, 381–399.

Lang, A. (1965a). *Encycl. Plant Physiol.* **15**, 409–423.

Lang, A. (1965b). *Encycl. Plant Physiol.* **15**, 1380–1536.

Lutz, A. (1971). *Colloq. Int. C.N.R.S.* **193**, 163–168.

Maliga, P. (1984). *Annu. Rev. Plant Physiol.* **35**, 519–542.

Meinhardt, H. (1982). *Soc. Dev. Biol. Symp.* **40**, 439–461.

Meins, F., Jr. (1974). *In* "Tissue Culture and Plant Science 1974" (H. E. Street, ed.), pp. 233–264. Academic Press, New York.

Meins, F., Jr. (1982). *In* "Molecular Biology of Plant Tumors" (J. Schell and G. Kahl, eds.), pp. 3–31. Academic Press, New York.

Meins, F., Jr. (1983). *Annu. Rev. Plant Physiol.* **34**, 327–346.

Meins, F., Jr. (1985). *In* "Plant Cell Culture Technology" (M. M. Yeoman, ed.). Blackwell, Oxford, in press.

Meins, F., Jr., and Binns, A. N. (1979). *BioScience* **29**, 221–225.

Meins, F., Jr., and Binns, A. N. (1982). *Differentiation* **23**, 10–12.

Meins, F., Jr., and Foster, R. (1985). *Dev. Biol.* **107**, 1–5.

Meins, F., Jr., and Lutz, J. (1979). *Differentiation* **15**, 1–6.

Meins, F., Jr., and Lutz, J. (1980). *Planta* **149**, 402–407.

Meins, F., Jr., Lutz, J., and Foster, R. (1980). *Planta* **150**, 264–268.

Meins, F., Jr., Foster, R., and Lutz, J. (1982). *Planta* **155**, 473–477.

Meins, F., Jr., Foster, R., and Lutz, J. D. (1983). *Dev. Genet.* **4**, 129–141.

Mott, R. L., and Cure, W. W. (1978). *Physiol. Plant.* **43**, 139–145.

Nagel, M., and Reinhard, E. (1975). *Planta Med.* **27**, 151–158.

Nanney, D. L. (1958). *Proc. Natl. Acad. Sci. U.S.A.* **44**, 712–717.

Nanney, D. L. (1968). *Science* **160**, 496–502.

Narayanaswamy, S. (1977). *In* "Plant Cell, Tissue and Organ Culture" (J. Reinart and Y. P. S. Bajaj, eds.), pp. 179–248. Springer-Verlag, Berlin and New York.

Oehlkers, F. (1955). *Z. Naturforsch.* **10**, 158–160.

Oehlkers, F. (1956). *Z. Naturforsch.* **11**, 471–480.

Okada, T. S., Eguchi, G., and Takeichi, M. (1973). *Dev. Biol.* **34**, 321–333.

Purvis, O. N. (1961). *Encycl. Plant Physiol.* **16**, 76–112.

Raff, J. W., Hutchinson, J. F., Knox, R. B., and Clarke, A. E. (1979). *Differentiation* **12**, 179–186.

Reinhard, E. (1954). *Z. Bot.* **42**, 353–376.

Robbins, W. J. (1960). *Am. J. Bot.* **47**, 485–491.

Rogler, C. E., and Hackett, W. P. (1975). *Physiol. Plant.* **34**, 141–147.

Sachs, T. (1969). *Ann. Bot.* **33**, 263–275.

Shimamoto, K., Ackermann, M., and Dierks-Ventling, C. (1983). *Plant Physiol.* **73**, 915–920.

Skoog, F., and Miller, C. O. (1957). *Symp. Soc. Exp. Biol.* **11**, 118–131.

Smith, R. H., and Murashige, T. (1970). *Am. J. Bot.* **57**, 562–568.

Soetiarto, S. R., and Ball, E. (1969). *Can. J. Bot.* **47**, 1067–1076.

Spemann, H. (1938). "Embryonic Development and Induction." Yale Univ. Press, New Haven, Connecticut.

Springer, W. D., Green, C. E., and Kohn, K. A. (1979). *Protoplasma* **101**, 269–281.

Steeves, T. A. (1966). *In* "Trends in Plant Morphogenesis" (E. G. Cutter, ed.), pp. 200–219. Longmans, London.

Tandeau de Marsac, N., and Jouanneau, J.-P. (1972). *Physiol. Vég.* **10**, 369–380.

Tepfer, S. S., Greyson, R. I., Craig, W. R., and Hindman, J. L. (1963). *Am. J. Bot.* **50**, 1035–1045.

Torrey, J. G. (1954). *Plant Physiol.* **29**, 279–287.

Vasil, V., and Hildebrandt, A. C. (1965). *Science* **150**, 889–892.

Walker, K. A., Wendeln, M. L., and Jaworski, E. G. (1979). *Plant Sci. Lett.* **16**, 23–30.

Wernicke, W., and Brettell, R. (1980). *Nature (London)* **287**, 138–139.

CHAPTER 27

FLEXIBILITY AND COMMITMENT IN PLANT CELLS DURING DEVELOPMENT

Daphne J. Osborne*

DEPARTMENT OF PLANT SCIENCES
THE SCHOOL OF BOTANY
UNIVERSITY OF OXFORD
OXFORD, ENGLAND

and

Michael T. McManus

DEPARTMENT OF BIOCHEMISTRY
UNIVERSITY OF OXFORD
OXFORD, ENGLAND

I. Introduction

Developmental biologists speculate on similarities and differences in the mechanisms regulating differentiation in plants and animals. Because cells in tissues from differentiated parts of plants can be induced to undergo cell division and then express a complete spectrum of morphogenetic characters (and hence gene products) not expressed by the original parent tissue, plants have been thought to be more plastic in their development than animals. Because tissue cultures derived from leaves or root cells can, for example, be induced to produce whole new individuals, plant cells have been considered to be totipotent. New

383

individuals cannot be regenerated in this way in animals. It is, therefore, frequently assumed that there is some basic underlying difference between these two life forms.

In this presentation, we take the view that the molecular mechanisms underlying differentiation in plants and animals are intrinsically similar, but develop the concept that, in plant tissues, the forward progression of different cells along their differentiation pathways is not closely synchronized in time. As a result, we can reconcile the considerable heterogeneity of cell commitment which persists in mature plant tissues, with some cells remaining essentially uncommitted while others are arrested at stages permitting a fewer variety of options (flexibility) for further change in cell determination. Unlike most animal cells, few plant cells, if any, appear to reach a phase of commitment in which, in culture, they will faithfully reproduce daughter cells of only their own committed phenotype.

II. Definitions of Commitment and Flexibility

A. COMMITMENT

We can best describe cell commitment as a state of specific differentiation in which certain environmental or hormonal signals are recognized and evoke specific morphological or biochemical responses. Each state of commitment results from earlier recognition and response events directing development along increasingly more specialized pathways. The initial options for further developmental change may be many (i.e., the meristematic cell recognizes and responds to many signals), but as each recognition and response is achieved and sequential stages of competence are passed, the options for further differentiation and development become progressively restricted. Theoretically, a condition of terminal differentiation must eventually be reached. For those cells that die after forming the structural and conducting elements of the plant (i.e., phloem, xylem, and lignified schlerenchyma), the terminal state of differentiation is apparent. However, the question of whether a cell that retains its nucleus and cytoplasmic organization will also always possess an option for further developmental change must remain open (Wareing, 1982). Examples of such cells are those of the flesh of ripening fruits and those cells of the root cap that are sloughed off as the cap progressively differentiates and grows forward from the root meristem. But as long as the information encoded in the nuclear genome remains essentially intact, we cannot assume that the expression of genetic information may not be further modified, derepressed, or changed by the cells' recognition and response to specific internal or external cues.

Commitment of cells to a particular differentiation pathway can normally be identified by the physiological response that is evoked by specific environmental or chemical signals. Additionally, this determinant condition can be marked biochemically by the presence of cell-specific molecular or antigenic components, usually proteins (Osborne, 1984).

B. FLEXIBILITY

In contrast to commitment, flexibility implies a condition in which many options for signal recognition and response are retained so that the potential to switch to other differentiation pathways remains.

In developmental terms, cell flexibility *in vivo* is considerably more limited than cell flexibility *in vitro*. All mature plant tissues *in vivo* contain a complex of cells in a variety of different states of competence, all of which are held in status quo under the environmental constraints of surrounding cell inputs (i.e., chemical or physical gradients). Only if such constraints are withdrawn or altered can these cells exhibit flexibility as quantitatively or qualitatively modulated gene expression. If, however, cell division is induced *in vivo,* a new population of cells can be generated with extended options for altered gene expression, but the differentiation of these cells is still in part directed by signals from their neighbors. For example, when a shoot is excised and maintained upon a suitable synthetic medium, new roots will usually develop only at the physiologically basal end. It is only when small parts of plants are isolated and cultured upon a synthetic medium that the cells have the conditions in which they will express the greatest potential flexibility for further differentiation.

C. EXAMPLES IN ANIMALS

Before considering further aspects of commitment and flexibility in plants, it is useful to deal briefly with two well-characterized examples during differentiation in animals, one of commitment and one of flexibility.

During blood cell formation (hematopoiesis), each single pluripotent stem cell can give rise to a number of different types of "committed" progenitor cells. Each of these progenitors can progress along only one specific differentiation pathway eventually producing a specific type of blood cell (e.g., an erythrocyte). The differentiation commitment of one specific progenitor type within a progenitor mixture can be directed by specific glycoproteins to produce cells that become granulocyte–macrophage. This glycoprotein factor is ineffective in inducing granulocyte–macrophage differentiation from any of the other progenitor cells, or in regulating the development of other blood cell types from their appropriate progenitors (Metcalf, 1981). In other words, only granulocyte–macrophage progenitors are committed to granulocyte–macrophage pathways, and only they have the competence to recognize and respond to the specific differentiation signals.

Flexibility can be demonstrated in stem cells of the optic nerve. When these cells are taken from 7-day-old rats and cultured for 3 days in Dulbecco's modified Eagles medium in the presence of 20% fetal calf serum, 8–20% of them can be identified immunologically (using cell-specific monoclonal antibodies) as fibrous astrocytes and 0.5–3.0% as oligodendrocytes. In the absence of serum, only 0.1–2.0% can be detected as astrocytes, but 20–40% then develop as oligodendrocytes (Raff et al., 1983). The same trend is repeated with cultures of optic tissue from older rats (14 days), but then the number of cells that can differentiate into either cell type is reduced. More of the cells in older rats can be presumed to have become committed to other differentiation pathways and can no longer be induced to follow the option of either astrocyte or oligodendrocyte pathways. In other words, part of their developmental flexibility has been lost.

How far is the interpretation of these examples applicable to developmental studies in plants?

III. Flexibility and Determination in Cultured Plant Cells

Traditionally, the transfer of mature plant tissues to an *in vitro* environment (so removing them from *in vivo* constraints), provides the opportunity for new gene expression. In tobacco (*Nicotiana* sp.) for example, complete regeneration from cells in culture has been achieved with relative ease by manipulation of the hormonal (auxin and cytokinin) and nutrient medium (Skoog and Miller, 1957). Antigenic components whose expression is restricted to particular tissues in the parent plant may, however, continue to be produced by their progeny in tissue culture: this is so for antigenic proteins specific for vascular tissue in maize (*Zea mays*) seedlings, the cells of which have been maintained as a nonvascular suspension culture for 19 transfers (Khavkin et al., 1980). Thus, although loss of phenotypic vascular expression may occur, some commitment to the original specific gene expression is retained. Antigens produced only in one specific organ of the intact plant may sometimes also be expressed by cells of callus cultures derived from other (nonproducing) tissue types. An antigenic protein produced only by the styles of a particular self-incompatibility (S) genotype of *Prunus avium* is, for example, produced by callus cells from the leaves—an expression that does not occur in leaves *in vivo* (Raff and Clarke, 1981). Changes in commitment that follow upon removal of a tissue from the *in vivo* environment are, therefore, not necessarily comparable with the developmental processes that take place in the whole plant.

Complete dedifferentiation from a determinant condition *in vivo* to that of totipotent competence *in vitro* may, therefore, require multiple, complex, and sequential gene interactions. This could account for the difficulties encountered in achieving high-percentage regeneration of new plantlets in certain tissue culture propagation systems (e.g., oil palm *Elaeis guineensis*), where the yield of new individuals is low (Jones, 1983). It may well be that fewer cells of the parent oil palm tissue retain the degree of flexibility (or developmental options) present in tobacco, and in this respect, the situation in oil palm could be more akin to that observed in animals.

IV. Commitment and Flexibility in the Whole Plant

Newly divided cells in the shoot or root meristem can be considered as totipotent stem cells providing the progenitors from which all other cell types are derived. Studies of shoot apices during conversion from the vegetative to the reproductive (flowering) condition provide evidence for the direct transition of the meristematic cells from one existing committed condition to another. In mustard (*Sinapis alba*), plants remain vegetative when grown in 8 hours of light per day, but are induced to flower by exposing them to one long day of 20 hours. Using immunohistochemical techniques, Pierard *et al.* (1980) showed that within 48 hours from the commencement of the long day treatment, two new antigenic proteins (not detectable in vegetative apices) became uniformly distributed in the meristem and continued to increase thereafter. These proteins were discerned 8 hours before the meristem was irreversibly committed to the reproductive condition and were considered as either causal or accompanying "markers" for a new condition of cell determination. The appearance of these proteins is evidence for an altered gene expression being a very early event in the changed competence of meristematic cells as they pass from the vegetative to the reproductive condition.

An environmental cue such as light must be transduced to a chemical signal within the cell in order for cell development to be programmed at the biochemical level. It is known that the switch from the vegetative to the flowering condition is mediated by a light-directed conversion of the photosensitive protein-containing pigment, phytochrome. This possesses an active form, absorbing at 660 nm, which is reversibly converted to an inactive form, absorbing maximally at 730 nm. Developmental events following these conversions have been linked to altered levels of hormones within the perceiving tissue, and many studies have shown that externally supplied hormones will substitute for the specific phytochrome-mediated light involvement (Smith, 1975). Of special interest, therefore, is the demonstration that either light or a cytokinin hormone in the dark can induce the production of a particular protein in cotyledons of germinating cucumber seedlings (Lerbs *et al.*, 1984). Normally, these fleshy cotyledons become green and fix CO_2 on exposure to light. Light induces the respective transcription of nuclear- and chloroplast-coded mRNAs (and their translation) for the small and large subunits of ribulose-1,5-biphosphate carboxylase (Rubisco), an essential enzyme in photosynthesis. In the absence of light, the synthetic cytokinin 6-benzylaminopurine will also induce the production of mRNAs that are translated into immunoprecipitable Rubisco protein. No qualitative or major quantitative differences in Rubisco gene expression were detected between the light and the cytokinin treatment. Although both could effect different causal events leading to Rubisco synthesis, the inference that *in vivo* the control of genetic expression by light might be linked through that of endogenous cytokinin is agreeably compelling.

Following upon studies aimed at identifying which cells are committed to stelar differentiation in root meristems, the cytochemical localization of a naphthol AS-D-positive esterase activity has been used as an early biochemical marker (Rana and Gahan, 1982). The presence of esterase provides a method for predicting specific commitment (and hence specific progenitor status) for the vascular condition before morphological identification is possible. This commitment pathway to xylem does not necessarily arise only from progenitors in the meristem, but can be induced from cells that are already differentiated as components of the root cortex. When the central vascular connection in roots of pea seedlings (*Pisum sativum*) is severed, new vascular elements are induced in the parenchyma cortex to bridge the cut, and those cells that are eventually converted to xylem show the early localization of esterase activity. Removal of the treated (cut) region and its transfer onto a basal culture medium at different times after the cut has been made has shown that the esterase activity and general determination to produce vascular tissue is irreversibly developed within some 8–10 hours (Rana and Gahan, 1983). Since cell division was not observed prior to commitment to the vascular differentiation pathway, it would appear that cortical parenchyma is an example of cells that retain the capacity (flexibility) to pass directly from one committed state of differentiation to another.

V. Competence for Cell Separation—Committed Cells in Abscission Zones

In order to study the degree of cell flexibility and commitment during plant development *in vivo,* we believe that three important criteria should be met. First, differentiation should be followed in a single cell type in which a specific physiological response to a specific signal can be readily distinguished. Second, the cells should possess some specific antigenic markers to permit their biochemical identification. Third, it should be possible to ascertain the extent of retention of both physiological and biochemical competence through a cycle of induction and repression events. Toward fulfilling these requirements, the positionally differentiated cells that separate from each other at organ shedding (abscission) have offered us an attractive model system for studying commitment and flexibility in the mature plant.

In the dwarf bean (*Phaseolus vulgaris*), cells possessing this potential to separate one from another are present at the junction of the leaf blade and the petiole (Fig. 1a). In response to the hormone ethylene produced when the blade and pulvinus become senescent (or in response to applied ethylene), one to two rows of cells at the position of shedding (zone cells, Fig. 1b) are induced to enlarge and secrete enzymes which include a zone-specific isozyme of β-1,4-glucanhydrolase (Koehler *et al.,* 1981). As a result, the middle lamellae between the

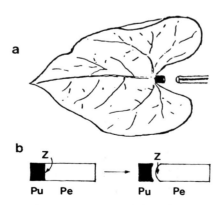

FIG. 1. (a) Separation position of the abscission zone between leaf blade pulvinus and petiole in *P. vulgaris.* (b) Excised segments (explants) containing abscission zones showing position of pulvinus (Pu), petiole (Pe), and zone cells (Z).

zone cells (but not between other cells) are degraded, the cells lose cohesion, and the organ is shed. Since none of these changes occurs in the neighboring nonzone tissue, the first of the three criteria is fully met.

With respect to our second criterion, polyacrylamide fractionation of proteins from ethylene-treated cells of the abscission zone of *P. vulgaris* have been shown to possess specific antigenic proteins that are absent from similarly treated nonzone cells (Osborne *et al.*, 1984). In another plant, *Sambucus nigra,* where the abscission zone is larger (15–30 rows of cells), it is easier to prepare extracts of predominantly abscission zone cells for immunological identification of zone-specific proteins. Antiserum has been raised against these zone cells and competed with extracts of nonzone tissue. The competed IgG (zone-enriched IgG) is immobilized on CN–Br Sepharose columns and used to selectively adsorb zone-specific proteins. After desorption and fractionation on polyacrylamide gels, zone-specific proteins can be readily visualized by an immunoblotting recognition using the zone-enriched IgG. These specific proteins are not detectable in similarly competed and fractionated extracts of nonzone material (McManus *et al.,* unpublished data). Since zone cells possess antigenic determinants that "mark" their competence to recognize and respond in a specific way to the ethylene signal, the system fulfills the requirements of our second criterion. We have no evidence to date, however, to suggest that these proteins are causal to the cells' ethylene recognition and response.

In the whole plant, the presence of a green leaf blade prevents cell separation occurring at the abscission zone even though competent zone cells are already differentiated there. The hormone auxin, indole-3-acetic acid (IAA), will substitute for the blade in maintaining the zone cells in a repressed condition. The commitment of zone cells to retain their competence for ethylene recognition and response (our third criterion) is illustrated by their continued competence for β-1,4-glucanhydrolase production through a cycle of enzyme repression by auxin and a subsequent ethylene reinducement.

The results in Table IA show the extent of ethylene (4 μl/liter) induction of β-1,4-glucanhydrolase in the zone cells at the time of separation of the pulvinus from the dome (day 3); further enzyme production by the cells of the dome can be "turned off" (days 3–6) by the addition of IAA (still in the presence of ethylene 4 μl/liter). Evidence that competence for enzyme induction by ethylene is retained by the IAA-treated dome cells is seen when an enhanced level of ethylene (500 μl/liter) is provided, still in the presence of IAA (days 4–6); then enzyme activity again increases (day 6).

TABLE I

COMMITMENT IN ETHYLENE RESPONSIVE ZONE CELLS OF ABSCISSION EXPLANTS OF *P. vulgaris*[a,b]

Day 3 ⟶ Day 4 ⟶ Day 6

(A) **Relative β-1,4-glucanhydrolase activity[c] (day 0 activity <1)**

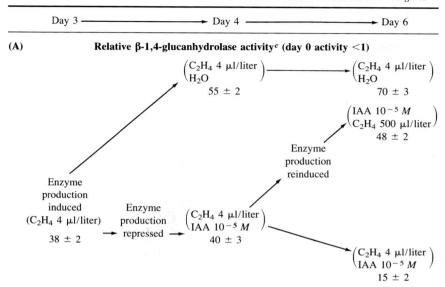

(B) **Ratios of dictyosome stack width to number of membranes per stack[d] (day 0 dictyosomes are suppressed)**

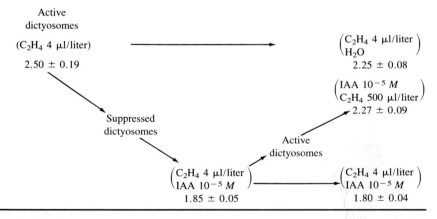

[a] See Fig. 1.

[b] Relationship between production of β-1,4-glucanhydrolase in the zone cells and the conformation of dictyosome stacks. Results for (A) enzyme activity and (B) stack conformation after an induction period in ethylene 4 μl/liter (days 0–3) treatment with IAA 10^{-5} M (day 3–4) and a second induction period in an increased concentration of ethylene 500 μl/liter (days 4–6). IAA applied as 2×1 μl to the dome on day 3. Controls received H_2O.

[c] Relative β-1,4-glucanhydrolase activity measured by the percentage reduction in viscosity of a synthetic carboxymethyl cellulose substrate (details in Osborne and McManus, 1984).

[d] Dictyosomes analyzed from electron micrographs. Examples are in Fig. 2 (details in Osborne *et al.*, 1984).

Confirmation that competence to respond to ethylene is retained by zone cells during an ethylene induction, auxin repression, and ethylene reinduction cycle (our third criterion) is provided by electron micrographic analysis of the conformation of dictyosomes in zone compared with nonzone cells (Table IB). Nonseparating or auxin-repressed abscission zones show dictyosome conformations that are quiescent (Fig. 2a). Three days after exposure to ethylene, dictyosomes with the conformation of actively secreting cells (Fig. 2b) are discernible in the separating zone cells (but not in other cells). On addition of auxin, dictyosome conformations return to the quiescent form as the increase in enzyme activity is arrested (i.e., the induction processes are "turned off"); this conformation is then retained (days 4–6). If, however, the concentration of ethylene is increased, production of enzyme is resumed (Table IA), and the dictyosomes again assume the conformation typical of active secretion (Table IB). Abscission zone cells are, therefore, a good example of a committed plant cell type with an ability to respond uniquely to hormones.

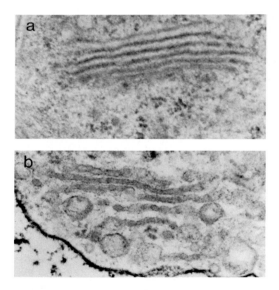

FIG. 2. Dictyosomes from abscission zone cells of *P. vulgaris* in which (a) β-1,4-glucanhydrolase production and secretion is repressed by auxin and (b) enzyme production and secretion is induced by ethylene (see text and Osborne *et al.,* 1984). The dictyosome ratios are (a) 1 : 83 and (b) 2 : 40. Nonzone cells do not show these changes.

VI. Cell Flexibility in Mature Plant Tissues

Whereas abscission zone cells can immediately recognize and respond to an ethylene signal (evidence of their committed state), it has been shown that certain cells in the petiole of *P. vulgaris* can be modified hormonally so that they too will exhibit the competence typical of normal zone cells. This flexibility is exhibited by the cortical parenchyma, and conversion to the new competence occurs in the absence of cell division (J. Roberts, unpublished data). The change to an abscission zone commitment is brought about when the domes of separated explants (Fig. 1b) are treated with 1-μl drops of a solution of IAA (10^{-5}–10^{-3} M) and then maintained continuously in ethylene. The new zone-cell competence is expressed in the formation of a secondary abscission zone at some position along the petiole distant from the initial separation zone (Fig. 3). The precise location of the secondary zone is directed by the concentration of IAA that is applied, as has been shown before for induced abscission zones in the stalks of pear fruits (Pierik, 1980).

The newly determined zone cells of *P. vulgaris* exhibit enzyme production, cell enlargement, and cell separation responses that are similar to those of normal abscission zones suggesting that this competence can be conferred to cortical cells of the petiole provided the appropriate ethylene and auxin gradients are established. Such results also establish the flexibility of the cortical cells to further developmental commitment without prior cell division. It may be, therefore, that the differentiation of abscission zone cells *in vivo* and the development of specific antigenic markers for their specificity in signal recognition and response is normally regulated by local gradients of ethylene and auxin.

VII. Concluding Remarks

In considering the nature of flexibility and commitment in plants, there appears no reason to propose processes that are essentially different from those that operate in animals. Rather it would seem that in plants, only a small percentage of the cells progresses to an advanced state of differentiation so that overall the living tissue retains a greater capability for future developmental change. Removal of cells from the constraints that operate *in vivo* allows them to express their residual flexibility; the comparative ease of regeneration of whole plants *in vitro* may simply be dependent upon the fewness of the cells that attain the irreversibly committed state. (As to the maximum capacity

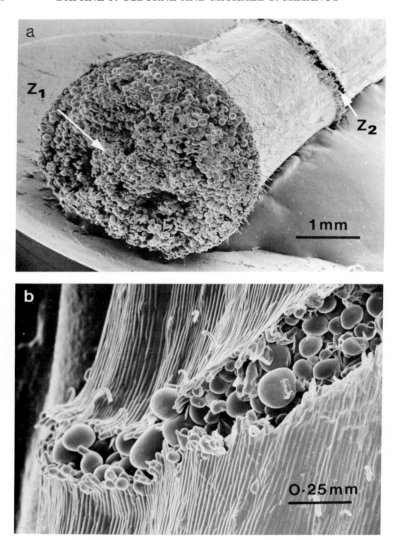

FIG. 3. Differentiation of secondary abscission zones in the cortical parenchyma in petioles of *P. vulgaris*. (a) Dome of zone cells of original abscission zone (Z_1) and the newly committed secondary abscission zone (Z_2). (b) Higher magnification of secondary zone (Z_2) to show cell enlargement and separation. For details, see text. (Scanning electron micrographs prepared by Dr. J. A. Sargent, Developmental Botany, AFRC Weed Research Organization.)

of regeneration from a single tissue in animals, see the article by Schmid and Alder on medusa in this volume.)

Most cells in a plant tissue probably retain this considerable flexibility, but it is evident that certain specific forms of irreversible commitment do commonly occur, for example, conversion of a vegetative meristem to one that expresses the reproductive condition. Furthermore, as in animals, some plant cells do proceed to a (possibly) terminal commitment in which they continue to show a highly specific response to a specific signal and are not known to differentiate further (e.g., abscission zone cells in response to ethylene).

Differentiation has been considered here as an one-way progressive program of commitment from which there is no backward turn except by the operation of complex switching programs involving cell division. In plants, it is clear that many cells retain the facility to operate such options at cell division, particularly *in vitro,* so that every potential cell type can be regenerated. Therein would appear to lie the basis of the difference between plants and animals, for although cells of many animal tissues will divide *in vitro,* to date there has been limited success only in persuading any but embryonic cells to operate an appreciable part of a backward-switching program. Instead, the state of commitment of the parent cell type is reiterated in the progeny (but see other chapters in this volume) so that the prospect for regenerating whole new organs from them is precluded.

It remains to be discovered whether the relatively few highly advanced differentiated plant cells and the many such differentiated cells in animal tissues can one day be released from what presently appear to be unswitchable committed states.

ACKNOWLEDGMENTS

The authors wish to thank Mrs. Rachael Daubney and Mrs. Shiela Dunford for their expert assistance in the preparation of this manuscript.

REFERENCES

Jones, L. H. (1983). *Biologist* **30,** 181–188.
Khavkin, E. E., Markov, E. Yu, and Misharin, S. I. (1980). *Planta* **148,** 116–123.
Koehler, D. E., Lewis, L. N., Shannon, L. M., and Durbin, M. L. (1981). *Phytochemistry* **20,** 409–412.
Lerbs, S., Lerbs, W., Klyachko, N. L., Romanko, E. G., Kulaeva, O. N., Wollgiehn, R., and Parthier, B. (1984). *Planta* **162,** 289–298.
Metcalf, D. (1981). *In* "Cellular Controls in Differentiation" (C. W. Lloyd and D. A. Rees, eds.), pp. 125–144. Academic Press, London.
Osborne, D. J. (1984). *Cell Differ.* **14,** 161–169.
Osborne, D. J., and McManus, M. T. (1984). *In* "Ethylene: Biochemical, Physiological

and Applied Aspects" (Y. Fuchs and E. Chalutz, eds.), pp. 221–230. Nijhoff/Dr. W. Junk, The Hague.

Osborne, D. J., McManus, M. T., and Webb, J. (1984). *In* "Ethylene and Plant Development" (J. A. Roberts and G. A. Tucker, eds.), pp. 197–212. Butterworths, London.

Pierard, D., Jacqmard, A., Bernier, G., and Salmon, J. (1980). *Planta* **150**, 397–405.

Pierik, R. L. M. (1980). *Physiol. Plant.* **48**, 5–8.

Raff, J. W., and Clarke, A. E. (1981). *Planta* **153**, 115–124.

Raff, M. C., Miller, R. H., and Noble, M. (1983). *Nature (London)* **303**, 390–396.

Rana, M. A., and Gahan, P. B. (1982). *Ann. Bot.* **50**, 757–762.

Rana, M. A., and Gahan, P. B. (1983). *Planta* **157**, 307–316.

Skoog, F., and Miller, C. O. (1957). *Soc. Exp. Biol.* **11**, 118–131.

Smith, H. (1975). "Phytochrome and Photomorphogenesis." McGraw Hill, New York.

Wareing, P. F. (1982). *In* "The Molecular Biology of Plant Development" (H. Smith and D. Grierson, eds.), pp. 517–541. Blackwell, Oxford.

CHAPTER 28

INDUCTION OF EMBRYOGENESIS AND REGULATION OF THE DEVELOPMENTAL PATHWAY IN IMMATURE POLLEN OF *NICOTIANA* SPECIES

Hiroshi Harada, Masaharu Kyo, and Jun Imamura

INSTITUTE OF BIOLOGICAL SCIENCES
THE UNIVERSITY OF TSUKUBA
SAKURA-MURA, IBARAKI, JAPAN

I. Introduction

Pollen embryogenesis, a typical androgenic phenomenon, has raised a number of intriguing questions in developmental physiology and has aroused much interest on the part of plant physiologists due to its ability to function as an experimental system in providing useful tools for plant breeders and geneticists. Moreover, pollen grains are quite uniform in size and shape as long as they belong to the same species or cultiver, and, in many cases, they can easily be isolated by gently pressing anthers in a liquid medium, giving a homogeneous cell suspension. Thus, they constitute unique experimental materials for the study of cell commitment and instability in the course of plant cell differentiation.

It was reported for the first time in 1966 that pollen grains, which were thought to be well-differentiated cells in terms of their morphology and function, could develop to haploid plants through an analogic pathway of zygotic embryogenesis (Guha and Maheshwari, 1966). Since then, much attention has been paid to this phenomenon

397

by plant physiologists, embryologists, and breeders. Now, more than 500 papers have been published on this subject and androgenesis has been reported in 171 species, including some hybrids, belonging to about 60 genera and 26 families of angiosperm (Maheshwari *et al.*, 1982). However, information about the physiological and biochemical changes taking place in the early period of pollen embryogenesis is limited, and the mechanism related to the acquisition of embryogenic capacity is still a mystery.

In this chapter we shall (1) review several factors that enhance embryogenic response of immature pollen in an early period of anther culture and discuss their significance, and (2) summarize the results of another of our attempts to establish a workable culture system with isolated pollen for research on the developmental potentiality and flexibility of pollen cells.

II. Factors That Enhance the Embryogenic Response in an Early Period of Anther Culture

Several factors that affect the rate of pollen embryogenesis have been reported by a certain number of investigators (Maheshwari *et al.*, 1982). The genotype, age, and culture conditions (e.g., temperature, photoperiod) of mother plants; pollen stage; components of culture medium (e.g., charcoal, glutamine); and physical factors (e.g., temperature, light intensity) during culture are important parameters involved in the embryogenic response. Cold treatments of anthers or buds have also been reported to stimulate pollen embryogenesis (Nitsch and Norreel, 1973; Tyagi *et al.*, 1979; Rashid and Reinert, 1980). This treatment of anthers is effective when given right after their excision from mother plants. As will be reported in the following, we have examined a wide range of factors stimulating pollen embryogenesis in *Nicotiana tabacum* anthers.

In our experiments, flower buds of *N. tabacum* cv. Samsun and cv. MC were collected, and the anthers that had pollen at mitosis or the early binucleate stage were carefully taken out immediately after harvest of the flower buds. Sterilized anthers were placed on 0.8% agar medium containing Nitsch's medium (Nitsch, 1969) elements, 2% sucrose, and 0.4% activated charcoal. Generally, 30 anthers were put in a 6-cm petri dish and subjected to different treatments.

A. EFFECTS OF REDUCED ATMOSPHERIC PRESSURE AND ANAEROBIC CONDITIONS

Treatments with reduced atmospheric pressure definitely augmented the number of anthers producing plantlets as well as the aver-

TABLE I

EFFECTS OF REDUCED ATMOSPHERIC PRESSURE ON POLLEN EMBRYOGENESIS ON *N. tabucum* CV. SAMSUN
40 DAYS AFTER THE BEGINNING OF CULTURE

		Duration of treatment				
	Control	10 Minutes	20 Minutes	60 Minutes	90 Minutes	24 Hours
Number of anthers cultured	33	37	37	36	33	37
Anthers with plantlets (%)	6	65	54	50	49	0
Total number of plantlets produced	57	447	640	603	343	0
Average number of plantlets per anther	2	12	17	17	10	0

age number of plantlets per anther. Significant results were obtained with 10-, 20-, and 60-minute treatments, while 24-hour treatments completely inhibited plantlet formation (Table I) (Imamura and Harada, 1980a).

To investigate whether the stimulatory effect of reduced atmospheric pressure on pollen embryogenesis was due mainly to reduced air pressure itself or to decreased partial oxygen pressure, anthers were held under a 100% N_2 stream at 1 bar for 15, 30, and 60 minutes. The 30- and 60-minute treatments increased the number of plantlets per anther (Table II). Treatments with N_2 containing 0, 2.5, and 5% O_2 increased the percentage of anthers with plantlets and the average number of plantlets per anther (Fig. 1). Our results indicated that the quasi-anaerobic treatment (2.5 or 5% O_2) was more effective than the purely anaerobic treatment (100% N_2) (Imamura and Harada, 1981).

As reported above, significant stimulation of tobacco pollen embryo formation was obtained with the short-term quasi-anaerobic treatment. Our work on tobacco pollen indicates that a similar treatment might also increase the rate of embryo formation in other plants in

TABLE II

EFFECTS OF N_2 ON POLLEN EMBRYOGENESIS OF *N. tabacum* CV. SAMSUN
40 DAYS AFTER THE BEGINNING OF CULTURE

		Duration of treatment		
	Control	15 Minutes	30 Minutes	60 Minutes
Number of anthers cultured	32	32	32	31
Total number of plantlets produced	290	225	807	770
Average number of plantlets per anther	9	7	25	25

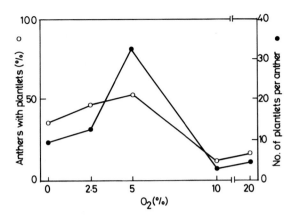

FIG. 1. The effects of varying concentrations of O_2 on pollen embryogenesis of *Nicotiana tabacum* cv. Samsun. One-hour treatments, 34 anthers each. Zero on the abscissa indicates 100% N_2.

which it is considered difficult or nearly impossible to induce pollen embryogenesis.

By lowering the level of dissolved oxygen in a cell suspension culture of carrot tissue below a critical level, Kessell *et al.* (1977) obtained an increase in the number of adventive embryos formed. It has been reported that decreasing dissolved oxygen in the cell suspension culture of carrot led to an increase in both the CN^- sensitivity and cellular concentration of ATP. Respiration of tobacco pollen measured after the reduced atmospheric pressure treatment indicated a clear shift from CN^- insensitive to CN^- sensitive respiration (Imamura and Harada, 1981). However, no increase in cellular ATP concentration was observed. In order to increase the number of embryos in carrot suspension culture, it was necessary to hold the culture in an anaerobic condition for several days. In the case of tobacco anther culture, 30 minutes of reduced atmospheric pressure is appropriate for stimulating pollen embryogenesis. This difference may indicate that a new but different metabolic pathway can be induced in carrot cells and tobacco pollen, respectively, by anaerobic treatments. In seeds of several species, the sensitivity to cyanide increases markedly during the germination period. It has been suggested that the early respiratory activity was through the pentose–phosphate pathway, and that it shifted at some later stage to the Embden–Meyerhof pathway. During the initial phase of pollen embryo induction, the developmental pathway of pollen is supposed to shift from normal pollen maturation to embryogenesis. At this very critical stage which can, in some sense, be compared to the stage of seed germination, there may be some changes in metabolic pathways that favor pollen embryogenesis.

B. Effects of Abscisic Acid and Water Stress

We also studied the effects of abscisic acid (ABA) and water stress on the embryo formation of tobacco pollen. Excised anthers were floated on liquid Nitsch's media with or without ABA (10^{-5} M) and/or mannitol for 1–4 days. After these treatments, the anthers were cultured on agar-solidified medium. A 3-day treatment with ABA had a stimulatory effect on the rate of anther producing plantlets as well as on the number of plantlets per anther (Table III). Water stress given to the anthers for 2 days by adding mannitol (0.5 M) in the medium, showed a clear stimulatory effect on plantlet formation (Table IV). ABA level in the pollen within the anthers increased about two times after 24 hours in a culture with the mannitol treatment (Imamura and Harada, 1980b).

It is known that the endogenous ABA level in certain plants increases when they are exposed to water stress (Milborrow, 1974). Therefore, the observed increase in the frequency of pollen embryogenesis with the water-stress treatment might be caused by the increased endogenous level of ABA in pollen. One of the modes of action of ABA is said to be the inhibition of mRNA synthesis. Therefore, it may be possible that the treatment with ABA inhibits synthesis of certain RNAs necessary for gametophyte development of pollen and may trigger the switching of developmental direction from gametophyte to sporophyte.

C. Effects of Actinomycin D

Clear enhancement of tobacco pollen embryogenesis was also obtained when the anthers were treated with actinomycin D after a pre-

TABLE III

Effects of Short-Term ABA (10^{-5} M) Treatment on Pollen Embryogenesis of *N. tabacum* cv. Samsun

	Duration of treatment (days)			
	Control	1	2	4
Number of anthers cultured	32	30	29	22
Anthers with plantlets (%)	63	67	93	36
Total number of plantlets produced	117	398	199	102
Average number of plantlets per anther	4	13	7	5

TABLE IV

EFFECTS OF WATER STRESS ON POLLEN EMBRYOGENESIS
OF *N. tabacum* CV. SAMSUN

	Control	2-Day treatment
Number of anthers cultured	25	25
Anthers with plantlets (%)	28	80
Total number of plantlets produced	103	165
Average number of plantlets per anther	4	8

culture of 0, 2, 4, 6, or 12 hours, but not after a 24-hour preculture. The stimulation of pollen embryogenesis by the inhibitor of mRNA synthesis may support the supposition that the stimulation of pollen embryogenesis with ABA treatment is caused by the inhibition of mRNA synthesis. The treatment with actinomycin D also shortened the time required for the formation of embryogenic pollen grains and the appearance of plantlets.

III. Conditions That Can Modify the Commitment of Isolated Immature Pollen Grains in the Developmental Program

Basic information about the changes taking place during the early period of androgenesis, especially at the subcellular level, is limited, and the precise mechanism of induction of pollen embryogenesis is entirely unknown. Due to the unavoidable inadequacy of methods previously reported by different workers for inducing pollen embryogenesis, it has been difficult to conduct appropriate physiological and biochemical research on pollen embryogenesis. A major disadvantage of anther culture is that one has to work with a very heterogeneous cell population confined within anther walls. During anther culture, the majority of pollen grains lose their viability, some undergo a maturation process, and only a limited number of pollen grains show embryogenic response. Under these conditions, it is quite difficult to undertake meaningful studies on physiological and biochemical changes occurring in the pollen.

Many investigators induced pollen embryogenesis by partial use of isolated pollen culture techniques combined with pretreatment of anthers or buds (e.g., Nitsch and Norreel, 1973; Weatherhead and Henshow, 1979; Tyagi et al., 1979). Percoll density gradient centrifugation was also tried in selecting embryogenic pollen grains (Wernicke et al., 1978; Rashid and Reinert, 1980). In these previously reported pollen culture methods, the preculture of anthers or cold treatment of flower buds, prior to isolation of pollen, are indispensable in obtaining proper embryogenic response. This means that some essential process in initiating embryogenic development takes place in the pollen within anthers during the preculture or cold treatment. The study of this essential but ill-defined process(es) occurring in the initial period of culture may allow us to gain insight into the induction mechanism of pollen embryogenesis. A series of investigations done by Imamura and Harada on the stimulating factors during the early period of anther culture, mentioned above, was conducted from such a point of view. It is now necessary to work out an experimental system that allows induction of pollen embryogenesis at a high rate through isolated pollen culture without any pretreatment of anthers or buds. At the same time, *in vitro* pollen maturation should be realized as a control of pollen embryogenesis.

Recently, successful induction of pollen embryogenesis and plantlet formation through isolated pollen culture without any pretreatment in *N. tabacum* and *Nicotiana rustica* has been reported (Imamura *et al.*, 1982; Ghandimathi, 1982). It was shown in these reports that culturing isolated pollen for an initial several days in plain water or in a medium lacking sucrose was essential for inducing embryogenesis. The frequency of embryogenic response, however, was still low (6% or so). Therefore, we have tried to develop an improved method that allows pollen to follow an embryogenic pathway at high frequency by selecting a homogeneous population of suitable pollen and examining the duration of the starvation period, concentration of osmoticum, various carbon sources, and other components of the culture medium (Kyo and Harada, 1985).

A. Experimental Methods and Results

In our experiments, the binucleate pollen grains of *N. rustica* were isolated in 0.4 M mannitol from the anthers of freshly collected flower buds with a corolla length of 6–9 mm. They were then filtered through nylon mesh (pore size 53 μm), and fractionated by Percoll discontinuous density gradient centrifugation (450 $\times$ g, 5 minutes). The pollen that remained in an interphase between 35 and 45% Percoll solution was collected and cultured in 0.4 M mannitol at a density of 3 $\times$ 10^4 cells/ml for 3 days at 30°C in the dark (hereafter referred to as the first culture or starvation period; mannitol was added as an osmoticum). After the first culture, the embryogenic pollen was selected by the second Percoll fractionation. The pollen floated on 30% Percoll solution was collected and transferred into 0.4 M mannitol solution containing the macroelements of MS medium (Murashige and Skoog, 1962), 40 mM galactose, 3 mM glutamine, and 5 μM ABA (pH 6.8) (the second culture). Ten days after the beginning of the second culture, divided pollen was observed at a high frequency (maximum 78%, Fig. 2). When the divided pollen was transferred into MS medium with 0.4 M mannitol, 3 mM glutamine, and no phytohormones, about 40% of the divided pollen grains developed to embryos or embryogenic calli larger than 100 μm in diameter (the third culture). They continued to grow when transferred onto a solid medium containing MS medium components (without phytohormones) and activated charcoal (0.5%). As mentioned above, a reasonably high frequency of divided pollen, applicable to physiological or biochemical analysis, could be obtained, though some difficulty caused by seasonal fluctuation still needs to be overcome.

We could also induce pollen embryogenesis in *N. tabacum* cv. Samsun with minor modification of culture medium (galactose replaced by sucrose 1 mM, without ABA) (Fig. 3). In this species the frequency of embryo or callus formation from divided pollen was lower than that in *N. rustica*, but the induction rate of divided pollen was equally high (40–60%), even without the second Percoll fractionation.

Using pollen at the same developmental stage as that used for the induction of embryogenic cell division, the pollen maturation process in *N. tabacum* cv. Samsun could also be achieved *in vitro* by giving no starvation period. The rate of cultured immature pollen that attained *in vitro* the state of full maturation was 80–90%, of which 30% (maximum) germinated when transferred into 0.3 M sucrose solution containing 2 mM H_3BO_4.

B. DISCUSSION AND CONCLUSIONS

The density of intact pollen increases as the maturation process proceeds due to an accumulation of starch grains in the pollen. The majority of pollen, which can follow an embryogenic pathway, was that that remained in the Percoll interphase between 35 and 45% and between 50 and 60%, respectively, when pollen isolated from flower buds with a corolla length of 6–9 mm in *N. rustica* and 19–24 mm in *N. tabacum* cv. Samsun was applied. These pollen grains were at the mid-binucleate stage, having no vacuole but a few starch grains. The pollen in this stage is said to be unsuitable for inducing embryogenesis through anther culture (Sunderland and Dunwell, 1977). We have tried to induce embryogenesis in younger (uninucleate to early binucleate) or older (late binucleate) pollen by our method, but have been without success. The pollen at the uninucleate or mitotic stage occasionally showed an equal cell division (not producing clearly dis-

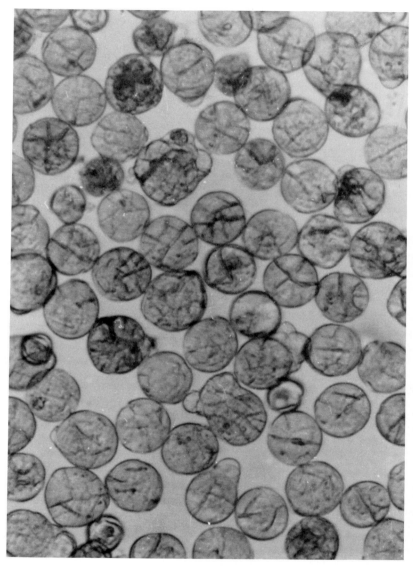

FIG. 2. A view of divided pollen 10 days after the beginning of the second culture.

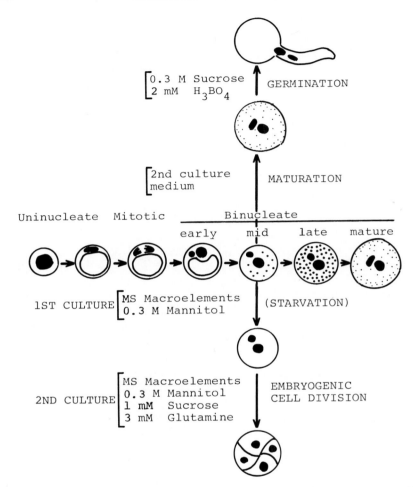

FIG. 3. The developmental processes of *N. tabacum* cv. Samsun pollen *in vivo* and *in vitro*. In intact anthers, pollen develops following normal maturation process (left to right), and after the mitotic stage, the accumulation of starch grains occurs making the density of pollen greater. By selecting flower buds with a certain corolla length and then applying Percoll fractionation to isolated pollen, it was possible to obtain fairly homogeneous pollen population in each developmental stage. Using mid-binucleate pollen population which was characterized by the absence of vacuole and the presence of a few starch grains, we were able to induce dichotomously at a high frequency both maturation (above) and embryogenic cell division (below) *in vitro*.

tinguishable vegetative and generative cells), but so far no further cell division has been observed.

Up to the present, the most effective treatment for inducing pollen embryogenesis is to first culture the pollen in a medium lacking amino acid and carbon sources for a certain period, then culture it in a standard medium with carbon and nitrogen sources. The initial starvation period seems prerequisite to triggering pollen embryogenesis. When no starvation period was given, most of the pollen accumulated starch grains in a manner similar to normal pollen development. Parallel with the induction of pollen embryogenesis, we could obtain completely matured pollen with germinating and fertilizing capacities in an *in vitro* culture of mid-binucleate pollen of *N. tabacum* cv. Samsun. We presume that the deviation from the normal maturation pathway and the acquisition of embryogenic capacity in cultured pollen take place during the starvation period. The optimum duration of the first culture (starvation period) was 3 days at 30°C in *N. rustica*. A shorter duration of the first culture may not be sufficient to trigger the process required for transforming immature pollen into embryogenic cells, and a longer duration may give irreversible physiological damage to the cell.

In the anther culture also, the starvation-induced acquisition of embryogenic capacity seems to occur during the pretreatment made by Dunwell (1981) or in the early period of anther culture. If the nutrients in the culture medium do not reach the pollen inside of the anther wall at the onset of culture, the pollen within the anthers is often in a state of starvation, after consuming the nutrients stored in anther tissues, until the nutrients in the culture medium reach the pollen by penetrating through the anther wall. Aruga and Nakajima (1983) reported that the sugar content in cultured anthers decreases during the initial period of culture. During the cold treatment of flower buds (Rashid and Reinert, 1980, 1983), a similar event seems to take place in anthers for which the supply of nutrients from mother plants is cut. The necessity of pretreatment of anthers or flower buds that has been reported can now be interpreted as an alternative treatment to starvation, at least in the case of *N. tabacum* and *N. rustica*.

Horner and Street (1978) proposed that the origin of pollen embryos in anther culture in *N. tabacum* is the deformed pollen, the so-called S grain, that exists at a low frequency in intact anthers and is characterized by its weakly staining cytoplasm with aetocarmine, distinguishable from normal pollen grains. S grains are said to have a predetermined capacity for embryogenesis prior to the start of the culture and no capacity for maturation. Therefore, they are supposed to be found even in mature anthers. In our isolated pollen culture of *N. rustica* and *N. tabacum* cv. Samsun, however, the origin of the embryogenic pollen was the pollen at the mid-binucleate stage with a few starch grains and no vacuole, which can acquire embryogenic capacity when starved and which certainly have a capacity for maturation as well. The hypothesis proposed by Horner and Street cannot be generalized and further clarification is required.

It is hoped that the experimental system of isolated pollen will continue to serve as a valuable tool in studies of cell commitment and instability in the developmental process of higher plants.

ACKNOWLEDGMENT

This study was supported by a Grant-in-Aid for Special Project Research, Multicellular Organization, from the Ministry of Education, Science and Culture (Project No. 59480023).

REFERENCES

Aruga, K., and Nakajima, T. (1983). *Proc. Congr. Jpn. Assoc. Plant Tissue Cult., 8th, Toyama* Abstract in Japanese.
Dunwell, J. M. (1981). *Plant Sci. Lett.* **21**, 9–13.
Ghandimathi, H. (1982). *Proc. Int. Congr. Plant Tissue Cell Cult., 5th, Tokyo* pp. 527–528.
Guha, S., and Maheshwari, S. C. (1966). *Nature (London)* **212**, 97–98.
Horner, M., and Street, H. E. (1978). *Ann. Bot.* **42**, 763–771.
Imamura, J. (1981). Ph.D. thesis, University of Tsukuba.
Imamura, J., and Harada, H. (1980a). *Naturwissenschaften* **65**, 540–541.
Imamura, J., and Harada, H. (1980b). *Z. Pflanzenphysiol.* **100**, 285–289.
Imamura, J., and Harada, H. (1981). *Z. Pflanzenphysiol.* **103**, 259–263.
Imamura, J., Okabe, E., Kyo, M., and Harada, H. (1982). *Plant Cell Physiol.* **23**, 713–716.
Kessell, R. H. J., Goodwin, C., and Philip, J. (1977). *Plant Sci. Lett.* **10**, 265–274.
Kyo, M., and Harada, H. (1985). *Plant Physiol.*, in press.
Maheshwari, S. C., Rashid, A., and Tyagi, A. K. (1982). *Am. J. Bot.* **69**, 865–879.
Milborrow, B. V. (1974). *Annu. Rev. Plant Physiol.* **25**, 259–307.
Murashige, C., and Skoog, F. (1962). *Physiol. Plant.* **15**, 473–497.
Nitsch, C., and Norreel, B. (1973). *C.R. Acad. Sci. (D) (Paris)* **276**, 303–306.
Nitsch, J. P. (1969). *Phytomorphology* **19**, 389–404.
Rashid, A., and Reinert, J. (1980). *Protoplasma* **105**, 161–167.
Rashid, A., and Reinert, J. (1983). *Protoplasma* **116**, 155–160.
Sunderland, N., and Dunwell, J. M. (1977). *In* "Plant Tissue and Cell Culture" (H. E. Street, ed.), pp. 223–265. Blackwell, Oxford.
Tyagi, A. K., Rashid, A., and Maheshwari, S. C. (1979). *Protoplasma* **99**, 11–17.
Weatherhead, M. A., and Henshow, G. G. (1979). *Z. Pflanzenphysiol.* **89**, 141–147.
Wernicke, W., Harms, C. T., Lorz, H., and Thomas, E. (1978). *Naturwissenschaften* **65**, 540–541.

CHAPTER 29

INSTABILITY OF CHROMOSOMES AND ALKALOID CONTENT IN CELL LINES DERIVED FROM SINGLE PROTOPLASTS OF CULTURED *COPTIS JAPONICA* CELLS

Yasuyuki Yamada and Masanobu Mino*

RESEARCH CENTER FOR CELL AND TISSUE CULTURE
FACULTY OF AGRICULTURE
KYOTO UNIVERSITY
SAKYO-KU, KYOTO, JAPAN

I. Introduction

Genetic homogeneity in cultured cell systems is essential for research on plant cells. Such uniformity is particularly important when studying regeneration of plants with constant chromosome numbers and for stabilizing the production of useful compounds in cultured cells (Yamamoto *et al.*, 1982). Deus and Zenk (1982) suggested that the different rates of alkaloid production which they found in cell lines of *Catharanthus roseus* were related to a high frequency of chromosome aberrations. Chromosome instability as an aspect of genetic variation in cultured cells, however, is a common phenomenon in many plant species (Sunderland, 1977; D'Amato, 1978; Orton, 1980; Inomata, 1982; Ogihara, 1982; Ogura, 1982). Yet there have been only a few reports that describe chromosome stabilization in plant cell cultures (Singh *et al.*, 1972; Evans and Gamborg, 1982).

Cloning of single protoplasts or single cells can be used to reduce the genetic heterogeneity of cell populations and to obtain cell lines that produce useful compounds. Studies of chromosome stability in the root tips of some plant species derived from protoplasts have been reported. Both normal and abnormal chromosome numbers were found

---

* Present address: ZEN-NOH Agricultural Technical Center, Kanagawa, Japan.

CURRENT TOPICS IN
DEVELOPMENTAL BIOLOGY, VOL. 20

$R_1, R_2 = -OCH_2O-, \quad R_3, R_4 = -OCH_3 \quad$ Berberine

$R_1, R_2, R_3, R_4 = -OCH_3 \quad$ Palmatine

$R_1, R_2 = -OCH_2O-, \quad R_3, R_4 = -OCH_2O- \quad$ Coptisine

$R_1 = -OH, \quad R_2, R_3, R_4 = -OCH_3 \quad$ Jateorrhizine

FIG. 1. Berberine and its analogs obtained from cultured *Coptis* cells.

in these regenerated plants (Shepard *et al.*, 1980; Bright *et al.*, 1982). But little information is available on chromosome numbers in callus tissue derived from single protoplasts.

Recently, we obtained 27 single protoplast-derived cell lines of *Coptis japonica* Makino and characterized them by their various chromosomal traits. We present here cytogenetic evidence that a variation in chromosome number occurred in one of our cell clones after 6 months of subculture. The fact that there seems to be no correlation between this variation in chromosome number and the production of berberine (Fig. 1) in the cell also is discussed.

II. Chromosome Instability

Invariably, we found 18 chromosomes in root tips from an intact *C. japonica* plant (Fig. 2A), which agrees with a previous report (Langlet, 1932). In contrast, in all the calluses that we examined, the chromosome number varied considerably from cell to cell. The ninth and tenth generations from our abscisic acid (ABA) cell culture (Sato and Yamada, 1984), which has been routinely subcultured every 3 weeks for about 35 and 36 months, were examined. (These are the forty-sixth and forty-seventh generations since isolation.) The number of chromosomes in most cells of both generations varied from 27 (triploidy) to 36 (tetraploidy), and the distribution patterns were very similar (Fig. 2B and C). The frequency of variant cells increased as the number of chromosomes rose, about 25% of the cells counted being tetraploid. We noted only a few cells that carried 35 chromosomes ($4x - 1$).

Some culture conditions have been shown to destroy specific cell types in a cell population (Bayliss, 1975; Singh *et al.*, 1975; Singh and Harvey, 1975). This suggests that the characteristic distribution of the chromosome numbers reported here depends on different rates of survival of the aneuploid cells in the culture system used. Singh *et al.* (1972) showed that in suspension cultures from mature seeds of *Visia*

FIG. 2. Chromosome number in cells of *C. japonica*. A, The root-tip cell of an intact plant with 18 chromosomes; B and C, suspension culture cells of ABA containing 34 and 36 chromosomes; D, cloned cells of line 2 with 31 chromosomes; E, cell of line 3 with 34 chromosomes; F, cell of line 5 with 66 chromosomes (bar, 10 μm).

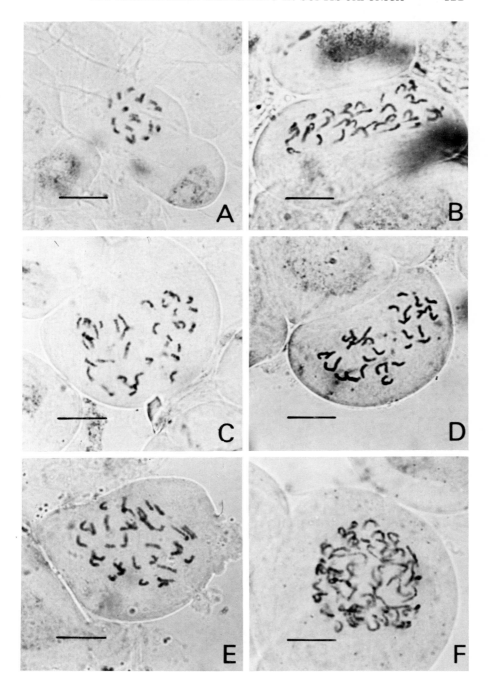

hajastana nonrandom selection of a particular chromosome took place in aneuploid cells. In suspension culture, our ABA cells grew at a constant rate for each subculture generation. After all the genetically variant cells divided evenly during culture, the distribution patterns of chromosome variation in the cell population became constant in subsequent generations, as first reported by Orton (1980). Thus, the steady-state distribution of chromosome number in our ABA cells was due to long-term subculturing with constant proliferation of each cell type.

Although we did not monitor the detailed structure of the chromosomes in our ABA cells, a number of abnormalities was observed; some cells showed chromosome stickiness during metaphase and anaphase (Fig. 3A and B). The frequencies of aberrant cells, however, were low; less than 3% of the metaphase cells showed this stickiness, and only about 5% of the cells counted had chromosome bridges. Because no fragment was observed in cells with bridges, we concluded that a pseudobridge was formed due to chromosome stickiness. Neither colchicine treatment nor heating (to remove starch granules) caused these aberrations, since abnormal chromosomes were not formed in root-tip cells when these were treated in the same manner.

The protoplasts isolated from the ABA cells eventually were cultured in agar media. The 27 calluses that were formed from these single protoplasts were transferred separately to liquid media and cultured for about 6 months (to the tenth generation). Six clones were used to count chromosome numbers. Of these, the chromosome numbers of five clones (lines 1, 2, 3, 9, and 22) ranged from triploid to tetraploid, and the chromosome number of line 5 was scattered over the hexaploid and octoploid range (Figs. 2D–F and 4). Low frequencies of stickiness and pseudobridges also were found for each clone (less than 2%). One cell from line 3 had a long dicentric chromosome (Fig. 3C and D). Thus, our observations show that chromosome variation increases in cells cultured from a single protoplast.

The distribution pattern of chromosome numbers in line 2 was very similar to that in ABA. The chromosome number of line 1 basically resembled that of ABA, but the $4x - 1$ fraction as well as the tetraploid fraction were larger (Fig. 4). The frequency of the $4x - 1$ fraction in line 3 also was high, whereas it was low in lines 2, 9, and 22. Why the frequency of the $4x - 1$ fraction differs among these clones is not clear. Singh and Harvey (1975) reported that for a diploid line of cultured *Haplopappus gracilis* cells, cells from which chromosome was deleted were more viable than cells without chromosome 1 in the particular culture system they used. Therefore, the genetic factors present in

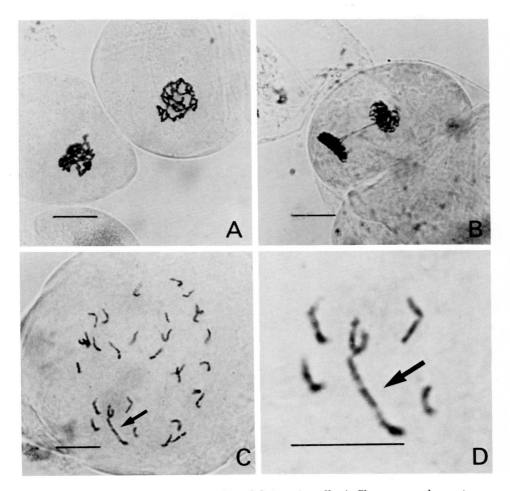

FIG. 3. Karyotypic variability in cultured *C. japonica* cells. A, Chromosome clump at metaphase in an ABA cell; B, chromosome bridge at anaphase in an ABA cell; C and D, a long dicentric chromosome (arrows) in a cell of line 3 with 31 chromosomes (bar, 10 μm).

each chromosome apparently affect differently cell viability in culture. Thus, it is likely that the $4x - 1$ fractions of lines 1 and 3 had a specific genetic composition that made for higher adaptability than the other $4x - 1$ fractions. Changes in the chromosome complement during culture may have induced a generation with this specific genetic composition.

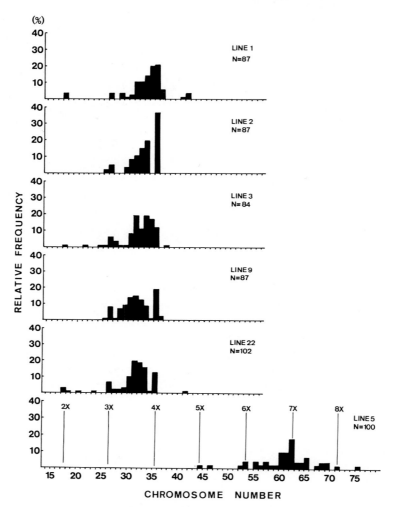

Fɪɢ. 4. Composite distributions of chromosome number in six cloned cell lines derived from single protoplasts from an ABA culture. N is the number of cells observed.

Chromosome variations in cultured cells can be explained by three different responses during culture (D'Amato, 1978): (1) nuclear variation in the explant, (2) nuclear response to the induction of callus, and (3) nuclear response during an established culture. In our ABA culture each of these factors might have generated the chromosome variations found, because this line had been derived from the induced callus of rootlet explants. For clones, responses 2 and 3 would be the usual

TABLE I

Type of Nucleus in Protoplasts
Prepared from an ABA Line
of Cultured *C. japonica* Cells[a]

Relative frequency of nuclear type (%)			
Monoploid	Diploid	Triploid	Tetraploid
96.38	3.15	0.37	0.09

[a] Protoplasts were fixed on slide glasses by heating them, after which they were stained with carbol fuchsin. A total of 1,079 protoplasts were counted.

catalysts of change. Isoploid (tetra-, hexa-, and octoploid) cells should be produced from other lower isoploid (diploid and tetraploid) and anisoploid (triploid) cells by polyploidization, i.e., due to spindle failure, chromosome lag, and nuclear fusion in dinucleate cells.

The first two causes were not discernible in our study, but the dinucleate protoplasts from our ABA cultures amounted to about 3% (Table I). Production of an aneuploid chromosome number depended on chromosome lag, nondisjunction, and the presence of multipolar spindles and dicentric chromosomes. Multipolar spindles (less than 1%) and dicentric chromosomes (Fig. 3C and D) were both present in our cultured cells. These factors have been reported to be the origins of variations in chromosome number (Partanen, 1965; D'Amato, 1975; Sunderland, 1977).

III. Instability of Berberine Production

Table II shows the berberine content in some of *C. japonica* cell lines derived from single protoplasts of ABA cells. The cell line numbers in Fig. 4 are to be referred to the numbers in Table II for comparison of chromosome number variation with the content of berberine in each cell line. Cell lines 1, 2, 3, and 9 produce much less berberine than the ABA parent cell line, whereas cell lines 5 and 22 produce amounts similar to the value found for ABA cells. Only a few cell lines derived from single ABA protoplasts, however, produce amounts of berberine similar to the value for the ABA parent cell line.

As yet, we do not know why almost all the cell lines derived from single protoplasts produce less berberine than the parent cell line. The cells in lines derived from single protoplasts are heterogeneous, both white and yellow cells being present. Probably, during cell division from a single, berberine-producing protoplast (yellow), heterogeneous

TABLE II

BERBERINE CONTENT (% DW) IN *C. japonica* CELL LINES DERIVED FROM
SINGLE PROTOPLASTS AND THEIR PARENTAL LINE ABA

Line	First	Second	Fourth	Fifth	Sixth	Seventh	Tenth	Average
1	2.54	2.84	3.09	2.79	2.66	2.30	2.30	2.32
2	1.20	1.35	2.02	1.92	1.99	1.44	1.94	1.42
3	4.04	3.19	3.34	3.23	3.97	3.26	4.37	3.00
5	NE	7.82	8.10	7.80	8.10	5.77	5.19	7.13
9	3.75	2.81	3.54	4.00	3.22	2.16	3.30	3.25
22	6.78	5.85	5.41	6.82	7.54	5.73	5.73	6.27
ABA	NE	NE	8.56	6.90	7.57	7.19	4.80	7.00

a NE, Not examined.

cells (white) that produce no berberine are formed along with berberine-producing cells.

IV. Concluding Remarks

In the multicellular higher plant, heterogeneous constituents are
natural during development; single cells also tend to become heterogeneous constituents of cell aggregates during cell division. The instability of berberine production in cloned protoplasts apparently is
due to the "singleness" of cells isolated from the homogeneous berberine-producing cell aggregates, in which cells interact.

The results reported here have led us to conclude that in protoplast
cultures of *C. japonica,* chromosome variation and the instability of
berberine production increase during callus formation and subculture
similarly to their increase in ordinary callus cultures from intact explants. Although critical studies of the relation between berberine
content and chromosome variation have yet to be finished, we have
evidence that indicates there is little correlation between the two. Specifically, five of six clones (lines 1, 2, 3, 9, and 22) showed different
berberine productivities though the variations in their chromosome
numbers were similar (Fig. 4). Furthermore, the content of berberine
in line 5, which showed higher numbered polyploid chromosomes than
the others, is similar to that for line 22 (Table II). This is evidence that
change in chromosome number in cultured *C. japonica* cells is not as
important for the production of berberine as such other factors as the
genetic composition of the genome and/or gene expression for the functional differentiation of secondary metabolites.

ACKNOWLEDGMENT

This work has been done in cooperation with Mr. Hiroyuki Morimoto and Miss Yukie Tanaka in our laboratory. This study was in part supported by Grant-in-Aid for Special Project Research, Multicellular Organization from the Ministry of Education, Science and Culture.

REFERENCES

Bayliss, M. W. (1975). *Chromosoma* **51,** 401.
Bright, S. W. J., Neleson, R. S., Karp, A., Jarrett, V. A., Greissen, G. P., Ooms, G., Miflin, B. J., and Thomas, E. (1982). *Proc. Int. Congr. Plant Tissue Cell Cult., 5th* p. 413.
D'Amato, F. (1975). *In* "Crop Resources for Today and Tomorrow" (O. Frankel and J. G. Hawks, eds.), p. 333. Cambridge Univ. Press, London and New York.
D'Amato, F. (1978). *In* "Frontiers of Plant Tissue Culture" (T. A. Thorpe, ed.), p. 287. Calgary Univ. Press, Calgary.
Deus, B., and Zenk, M. (1982). *Biotechnol. Bioeng.* **24,** 956.
Evans, D. A., and Gamborg, O. L. (1982). *Plant Cell Rep.* **1,** 104.
Inomata, N. (1982). *Jpn. J. Genet.* **57,** 59.
Langlet, O. F. J. (1932). *In* "Chromosome Atlas of Glowering Plants" (C. D. Darlington and A. P. Wylie, eds.), p. 26. Allen & Unwin, London.
Ogihara, Y. (1982). *Jpn. J. Genet.* **57,** 499.
Ogura, H. (1982). *Proc. Int. Congr. Plant Tissue Cell Cult., 5th* p. 433.
Orton, T. J. (1980). *Theor. Appl. Genet.* **56,** 101.
Partanen, C. R. (1965). *Proc. Int. Congr. Plant Tissue Cult.* p. 463.
Sato, F., and Yamda, Y. (1984). *Phytochemistry* **23,** 281.
Shepard, J. F., Bindney, D., and Shahin, E. (1980). *Science* **208,** 18.
Singh, B. D., and Harvey, B. L. (1975). *Cytologia* **40,** 347.
Singh, B. D., Harvey, B. L., Kao, K. N., and Miller, R. A. (1972). *Can. J. Genet. Cytol.* **14,** 65.
Singh, B. D., Harvey, B. L., Kao, K. N., and Miller, R. A. (1975). *Can. J. Genet. Cytol.* **17,** 109.
Sunderland, N. (1977). *In* "Plant Cell and Tissue Culture" (H. E. Street, ed.), Vol. 2, p. 177. Univ. of California Press, Berkeley.
Yamamoto, Y., Mizuguchi, R., and Yamada, Y. (1982). *Theor. Appl. Genet.* **61,** 113.

INDEX

A

Abscisic acid
cell line culture in, *Coptis japonica,*
410–416
pollen embryogenesis induction, tobacco, 401
Abscission zone, dwarf bean
ethylene induction of
dictyosome conformation, auxin effect, 391–392
β-1,4-glucanhydrolase, 389, 391–392
inhibition by auxin, 390–392
specific antigen, 390
secondary in cortical parenchyma,
induction by auxin, 393–394
Acetylcholine
in neural crest cells on cell-free medium, 187
in trunk and cephalic crest cultures,
179–181
Acid phosphatase 2, isozyme in prestalk
cells, *Dictyostelium discoideum,* 245
Actinomycin D, pollen embryogenesis induction, 401–402
Adenohypophysis, δ-crystallin, chicken
embryo, 138, 147–148
Adrenal chromaffin cells, conversion into
neuronal cells
in culture from neonatal rat
catecholamine redistribution, 102
cholinergic synaptic transmission,
105–107
delayed outgrowth induction, 100–
102
inhibition by dexamethasone, 108–
109
dendritic network, 102–103
ultrastructure, 102, 104–105

in situ, induction by
autologous transplantation in adult
guinea pig, rat, 107–108
nerve growth factor injection to neonatal rat, 107
γ-Aminobutyric acid, uptake by RT4-D
line
induction by dibutyryl cAMP with testololactone, 218–219
Anaerobiosis, pollen embryogenesis induction, 399–400
Antibodies
monoclonal, *see* Monoclonal antibodies
production by B cells, bursectomy effect, avian embryo, 302–303
Antigens
abscission zone-specific, ethylene-induced, dwarf bean, 390
B6, spore-specific, in *Dictyostelium discoideum,* 253–254
C1, stalk-specific, in *Dictyostelium discoideum* during development, 249,
252–254
glucose effect on synthesis, 249–
251
flowering-specific, long day-induced,
mustard, 387
in hydra epitheliomuscular cells, differentiation changes and, 263,
265–267
Ia on thymic cells in cortex and
medulla, avian embryo, 306–
309
MB1, on hemangioblasts, avian embryo, 296–297
MP26, lens cell membrane, chicken
embryo, 14
appearance in gliocytes in monolayer
culture, 14–16